Advances in Photochemistry

Volume 10

Advances in Photochemistry

Volume 10

Editors

JAMES N. PITTS, JR., *Department of Chemistry and Statewide Air Pollution Research Center, University of California, Riverside, California*

GEORGE S. HAMMOND, *Merrill College, University of California, Santa Cruz, California*

KLAUS GOLLNICK, *Institut für Organische Chemie, Universitat Munchen, Munchen, West Germany*

An Interscience® Publication

John Wiley & Sons, New York · London · Sydney · Toronto

An Interscience® Publication

Library of Congress Catalog Card Number 63-13592
ISBN 0-471-02145-8

Printed in the United States of America

10 9 8 7 6 5 4 3 2 1

Foreword

Volume 1 of this series, Advances in Photochemistry, appeared in 1963. The stated purpose of this series was to explore the frontiers of photochemistry through the medium of chapters written by pioneers who are experts. Frontiers in photochemistry are at times full of confusion and for this reason they are fascinating!

In the first nine volumes the editors have not acted as censors. We have followed the principle set forth in the original introduction to this series, that is, "the authors are free to make any statement they wish which cannot be proven wrong." We have solicited articles from experts who have strong personal points of view concerning fundamental aspects of photochemistry and spectroscopy. We have asked these authors to present their points of view through critical discussion and evaluation of existing data. In no sense have the articles been simply literature surveys although in some cases they have fulfilled both purposes.

We trust that, during the past few years when photochemistry has truly come of age, this series has proven to be stimulating and has served our original purpose of bringing together points of views of photochemists and spectroscopists who only a short five to ten years ago did not even use the same vocabulary.

JAMES N. PITTS, JR., Editor
Statewide Air Pollution Research Center and Department of Chemistry
University of California
Riverside, CA 92521
Telephone: (714)787-4584

GEORGE S. HAMMOND, Editor
Merrill College
University of California
Santa Cruz, CA 95060
Telephone: (408)429-2004

KLAUS GOLLNICK, Editor
Institut für Organische Chemie
Chemisches Laboratorium der
Universität München
Karlstrasse 23
D-8000 München, Germany
Telephone: 0811-590-2228

The editors are deeply indebted to Mrs. Glenna Paschal for her painstaking effort in final typing this volume of the Series.

Contents

Advances in Photochemistry

Volume 10

Vapor Phase Photochemistry of the Neutral Oxides and Sulfides of Carbon

STEPHEN V. FILSETH
Department of Chemistry
York University
Downsview, Ontario, Canada

CONTENTS

I. INTRODUCTION

The purpose of this article is the presentation of a summary of the contributions that have been made to our understanding of the gas phase photochemistry of the simple oxides and sulfides of carbon. An effort has been made to cover the published accounts of research in the English language literature through December 1974.

An understanding of the photochemistry of selected members of this group of molecules is essential to an appreciation of their abundance in both planetary atmospheres and the interstellar medium. The current state of our knowledge is the result of contributions by many groups employing a variety of traditional and some not-so-familiar techniques. We may expect that the near future will bring us very much more detailed spectroscopic and photochemical information than we presently have. It seems likely that the development of new light sources will be the single most important factor in bringing this about. The use of commercially available tunable dye lasers in the ultraviolet and visible, both for fluorescence excitation and for photochemistry, is becoming widespread. In the vacuum ultraviolet, the increasing availability of the synchrotron radiation produced by electron storage-rings has been recognized and it is clear that these intense pulsed sources of plane-polarized radiation will make many new experiments possible.

Thus it seemed likely that others would find useful a summary of our present understanding of the photochemistry of some molecules of particular interest to the author. In the following sections the several oxides and sulfides of carbon are considered by wavelength interval. In each case a low-resolution absorption spectrum is given together with a table of wavelength thresholds for selected photochemical primary processes. A discussion of the relevant experimental information then completes each section.

II. C_3O_2

Carbon suboxide is a linear symmetric molecule in its ground state and is of considerable interest from a variety of viewpoints. Its electronic structure, for example, has been subjected to considerable theoretical analysis because of the extensive delocalized π network. Because its constituent atoms are doubly bonded C and O, it is the source of intensely chemiluminescent flames with atomic oxygen which have been shown to result in inverted distributions in product CO. The molecule has also been proposed as (i) a product of photochemical reactions in the primitive earth atmosphere, (ii) a constituent of the Venus atmosphere, and (iii) a possible explanation for the variety of colors associated with the Mars surface.

A. Spectroscopy

The absorption spectrum of C_3O_2 has not been investigated extensively; however, the principal features have been established from the near to far ultraviolet.

The existence of a weak absorption maximum with considerable vibrational structure near 270 nm was noted by Thompson and Healey (2) and confirmed by subsequent investigations. A low-resolution spectrum recorded by Bayes (3) is given in Fig. 1. The absorption in this interval appears to consist of bands beginning at 330 nm and extending to lower wavelengths and a continuum between 280 and 240 nm. Kim and Roebber (4) extended the measured spectrum to 110 nm. They observed several strong bands with maxima at 158, 137, 127, and 123 nm which could be analyzed in a Rydberg series converging to an ionization limit at 117 nm (10.6 eV). The most prominent feature of the spectrum is the band at 158 nm which has an oscillator strength of 1.5. The photoelectric record obtained by Kim and Roebber is given in Fig. 2. The small feature located at 265 nm has subsequently (5) been identified as due to contaminant SO_2. The spectrum has been studied photographically by Bell et al. (6) who found substantial agreement with the earlier vacuum ultraviolet work.

Roebber et al. (5) have restudied the spectrum photoelectrically below 200 nm extending the observations below the ionization limit to 60 nm. This work clearly reveals the

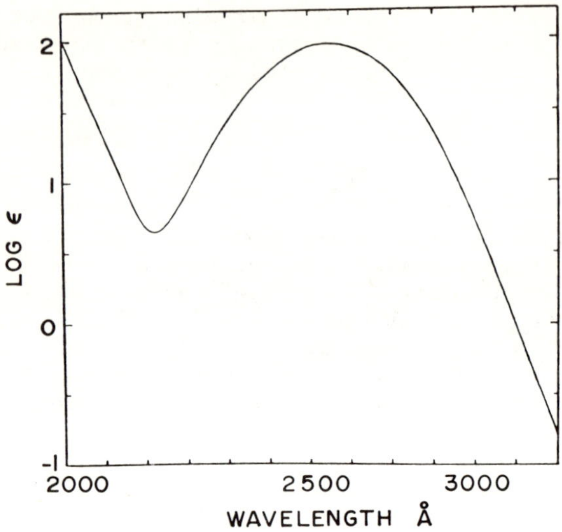

FIGURE 1. The common logarithm of the molar extinction coefficient, ε(liter $mole^{-1}cm^{-1}$) for C_3O_2 vapor at 300^oK obtained photoelectrically with a Cary model 14 spectrometer. Reprinted with permission from K.D. Bayes, J. Am. Chem. Soc., 84, 4077 (1962). Copyright by the American Chemical Society.

vibrational structure that overlies the strong continuum in the bands between 180 and 60 nm.

B. Photochemistry

Two primary processes may be considered in the C_3O_2 photochemistry. These are breaking of a carbon-carbon bond (eq. 1) and elimination of a carbon atom (eq. 2):

$$C_3O_2 + h\nu \rightarrow C_2O + CO \quad (1)$$

$$\rightarrow C + 2\ CO \quad (2)$$

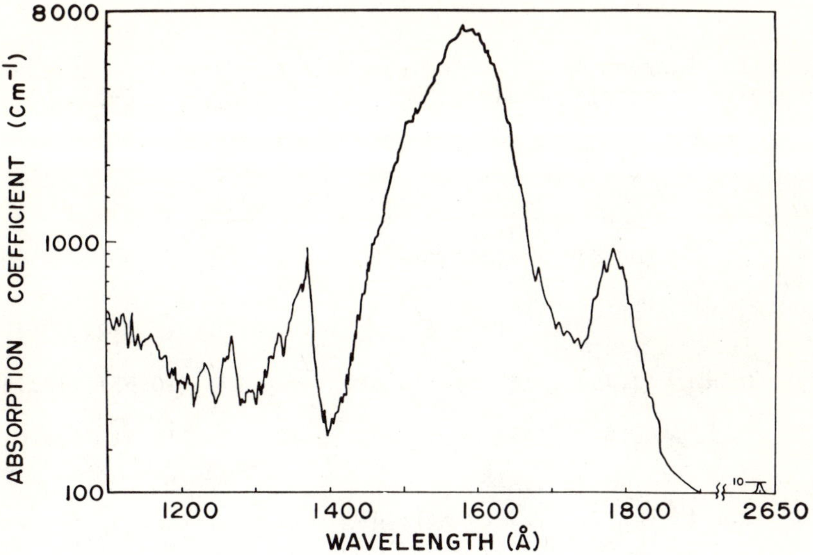

FIGURE 2. Absorption coefficients of C_3O_2 by photoelectric detection with an instrumental band width of about 0.1 nm. Curve is a smoothed line drawn through 300 data points. Absorption coefficient is to the base e for 1 atm of gas at 273°K (from Kim and Roebber[4] with permission).

The energetics and threshold wavelengths for these processes are given in Tables 1 and 2, respectively. Breaking of a carbon-oxygen bond would produce the species C_3O which has not been detected in any system. While the species C_2 and C_3 have been observed, it is unlikely that they result from a primary photochemical step. Evidence for processes 1 and 2 is reviewed below for the different regions of absorption.

1. Wavelengths longer than 220 nm. The absence (2) of fluorescence or phosphorescence in carbon suboxide excited in the ultraviolet would suggest that it is efficiently photodecomposed in this region of the spectrum. From its structural similarity to ketene, it is expected that this will involve production of C_2O in step 1. The work of Bayes has in fact shown that this is the most probable process in the absorption region with a maximum at 265 nm. Bayes (11) found that photolysis of C_3O_2/C_2H_4 mixtures led to production of allene rather than the acetylene, which is known (12) to be formed by insertion of carbon atoms into C_2H_4.

TABLE 1. Energetics of photodissociative primary processes for C_3O_2 with threshold wavelengths > 100 nm.[a] $C_3O_2 \rightarrow C_2O + CO$

$C_2O \downarrow$	$CO \rightarrow$	$X^1\Sigma^+$	$a^3\Pi$	$a'^3\Sigma^+$	$d^3\Delta$	$e^3\Sigma^-$	$A^1\Pi$
		(0.00)	(6.01)	(6.86)	(7.52)	(7.90)	(8.03)
$\tilde{X}^3\Sigma$	(0.00)	(2.75)	(8.76)	(9.61)	(10.27)	(10.65)	(10.78)
		450.84	141.53	129.01	120.72	116.41	115.01
$\tilde{A}^3\Pi$	(1.44)[b]	(4.19)	(10.20)	(11.05)	(11.71)	(12.09)	(12.22)
		295.89	121.55	112.20	105.88	102.55	101.46

[a] Energies as listed within parentheses in electron volts; threshold wavelengths appear directly underneath in nanometers. Assumes a value of 2.75 ± 0.67 eV[1] for the bond dissociation enthalpy of C_3O_2 at 0°K, ΔH_0^o; energies of states of CO are taken from ref. 7.

[b] Energy of the $\tilde{A}^3\Pi$ state taken from ref. 8; energies of the $\tilde{a}^1\Delta$ and $\tilde{b}^1\Sigma$ states are estimated as approximately 0.5 and 0.8 eV, respectively, by Bayes (9) on the basis of a semi-empirical MO calculation.

TABLE 2. Energetics of photodissociative primary processes for C_3O_2 with threshold wavelengths > 100 nm.[a]
$C_3O_2 \rightarrow C + CO + CO$

	C $\rightarrow$	$^3P_0$	$^1D_2$	$^1S_0$	$^5S^o$
CO		(0.00)	(1.26)	(2.68)	(4.18)
$\downarrow$					
$X^1\Sigma^+$	(0.00)	(6.00)	(7.26)	(8.68)	(10.18)
		206.63	170.77	142.83	121.79
$a^3\Pi$	(6.01)	(12.01)	-	-	-
		103.23			

[a]Energies are listed within parentheses in electron volts; threshold wavelengths appear directly underneath in nanometers. Assumes a value of 6.00 ± 0.04 eV[1] for the bond dissociation enthalpy of C_3O_2 at 0^oK, ΔH_0^o; energies of states of CO are taken from ref. 7; energies of states of C are taken from ref. 10.

However, it is not energetically possible to produce carbon atoms in a single primary photochemical process with radiation of wavelength longer than 207 nm. While carbon atoms could be formed in the secondary photolysis of C_2O, Bayes (3) found that allene was produced without delay in comparison to CO, which would require an absorption coefficient 1000 $atm^{-1}cm^{-1}$ for C_2O, an unreasonably large value. He concluded that reaction of C_2O with C_2H_4 was a more likely source of allene.

In subsequent work, Bayes (9) found that formation of allene was suppressed by NO and O_2, well-known triplet scavengers, at wavelengths above 260 nm. No suppression was observed below 260 nm. From this observation he concluded that C_2O in a low-lying singlet state was the photodecomposition product below 260 nm, but that triplet C_2O was formed above this wavelength. This behavior with O_2 and NO is similar to that found with singlet and triplet methylene. The relative importance of the two channels of reaction 1 which lead to singlet or triplet C_2O is in some doubt near 260 nm, however, since Cundall et al. (13) found that butene-2 suppressed the major part of processes which yield CO at 253.7 nm while Bayes found that O_2 and NO had little effect at 254 nm on allene formation. The photolysis of C_3O_2 at 253.7 nm in the presence of H_2 has been studied by Forchioni and Willis (14) who propose that singlet C_2O formed exclusively reacts with H_2 in a spin-allowed process to produce singlet methylene.

The flash photolysis of C_3O_2 has been studied by Morrow and McGrath (15) in the wavelength interval from 230 to 290 nm. The absorption spectrum of the C_3 radical was identified at short times following the flash and is attributed to secondary reactions of the primary product, singlet C_2O. In addition, diffuse transient spectra were observed whose assignment is uncertain but may include singlet C_2O or its reaction products.

2. Wavelengths shorter than 220 nm. The photochemistry of C_3O_2 in the vacuum ultraviolet was studied initially by Stief and De Carlo (16). In mixtures with CH_4 at 147 nm they found that both acetylene and ethylene were produced while confirming that only ethylene was formed at 253.7 nm. These authors concluded that carbon atoms were formed at 147 nm, although reactions of electronically excited C_2O radicals were not excluded.

Jacox et al. (17) have observed the three vibrational fundamentals of C_2O in absorption in matrix-isolated C_3O_2 at 4.2°K which had been irradiated with 147 nm light. The same spectra were produced by photolysis of matrix-isolated cyanogen azide/CO mixtures with wavelengths shorter than 280 nm. An absorption continuum was observed near 500 nm which they attributed to C_2O. Reactions of C_2O under matrix-isolation conditions seem to be of little importance relative to the reactions of carbon atoms (18) also produced.

Braun et al. (19) studied the vacuum ultraviolet photochemistry of C_3O_2 by kinetic absorption spectroscopy and identified $C(^3P)$, $C(^1D)$ and in lesser yield, $C(^1S)$ as products. Under their conditions more than 80% of the decomposition was effected by wavelengths between 170 and 140 nm. Absorption by ground state C_2O was only observed after addition of CO to the photolysis mixture. This is presumably due to formation of C_2O in a three-body recombination process. Through spectroscopic comparison of the yield of CO and the consumption of C_3O_2 at times shorter than collision times, Braun et al. found that 2 molecules of CO were formed for each C_3O_2 removed. This indicates that primary process 2 dominates process 1 in this wavelength interval. An upper limit of 25% is placed on the yield of C_2O. The dependence of carbon atom yields on flash intensity suggest that $C(^3P)$ and $C(^1D)$ are formed in a primary photolytic process, while $C(^1S)$ is the result of some secondary reaction. The relative yields of $C(^3P)$, $C(^1D)$ and $C(^1S)$ were determined to be 1.00, 0.25, <0.025, respectively.

In further studies of the 147 nm photolysis of C_3O_2/CH_4 mixtures first investigated by Stief and De Carlo, Tschuikow-Roux et al. (20) found that C_2H_4 yields were suppressed by addition of H_2 which had a lesser effect on C_2H_2 yields, whereas addition of CO eliminated C_2H_2 production with little effect on the C_2H_4 yield. From these effects, the authors concluded that C_2H_2 results from $C(^1D)$ insertion into CH_4 followed by H_2 elimination and that C_2O reaction with CH_4 is the source of C_2H_4.

Forchioni and Willis (14) have reported the only study of the C_3O_2 photochemistry at wavelengths as short as 123.6 nm. Their observation that the product distribution in mixtures with excess H_2 was the same at 123.6, 147.0, and 253.7 nm provides a qualitative indication that reactions of C_2O may be important at all three wavelengths. However, other evidence summarized above indicates persuasively that carbon atoms are the dominant species involved at 147 nm, while singlet C_2O is most important at 253.7 nm.

III. CO

A. Spectroscopy

Carbon monoxide is a constituent of stellar atmospheres and the solar chromosphere as well as being a product of the combustion of hydrocarbons and other organic compounds. Its spectroscopy has, therefore, been extensively investigated. The absorption spectrum above 100 nm is dominated by the intense Fourth Positive system, $A^1\Pi \leftarrow X^1\Sigma^+$. The rotational structure of the bands of this system have been studied by Tanaka et al. (21) and by Simmons et al. (22). A recent photoelectric recording by Myer and Samson (23) of the absorption spectrum at 0.025 nm resolution above 105 nm is given in Fig. 3. An atlas

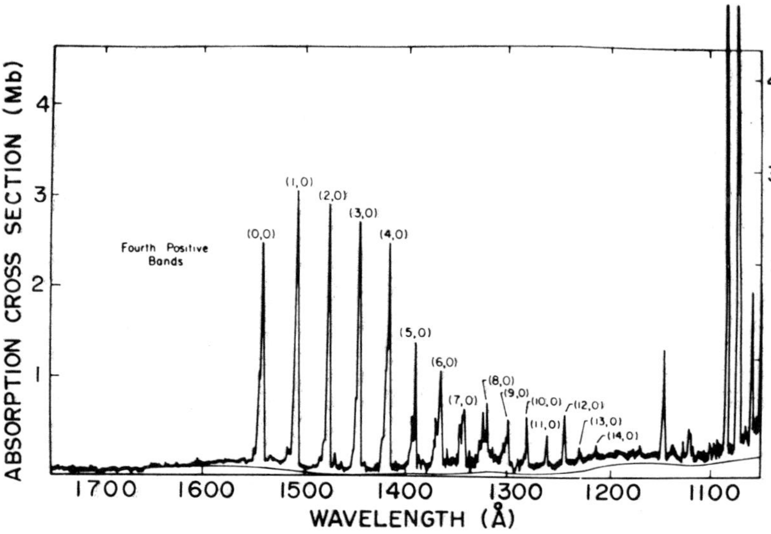

FIGURE 3. Absorption cross sections for CO by photoelectric detection with an instrumental band pass of 0.025 nm. Absorption cross section is to the base e; 1 Mb = 10^{-18}cm^2 (from Myer and Samson[23] with permission).

of the observed absorption spectrum between 106.0 and 190.0 nm has been compiled by Tilford and Simmons (24).

The absorption spectrum has also been studied below 100 nm. A photoelectric absorption spectrum obtained by Huffman et al. (25) between 100 and 60 nm at 0.05 nm resolution is given in Fig. 4. In the wavelength region above 96 nm the absorption spectrum is discrete. Below this wavelength, discrete bands overlie a continuum which becomes stronger toward shorter wavelengths. An additional continuum which may be attributed to ionization makes a contribution for wavelengths below 88.47 nm, the ionization limit. Below this wavelength the spectrum consists of numerous strong bands above strong ionization continua.

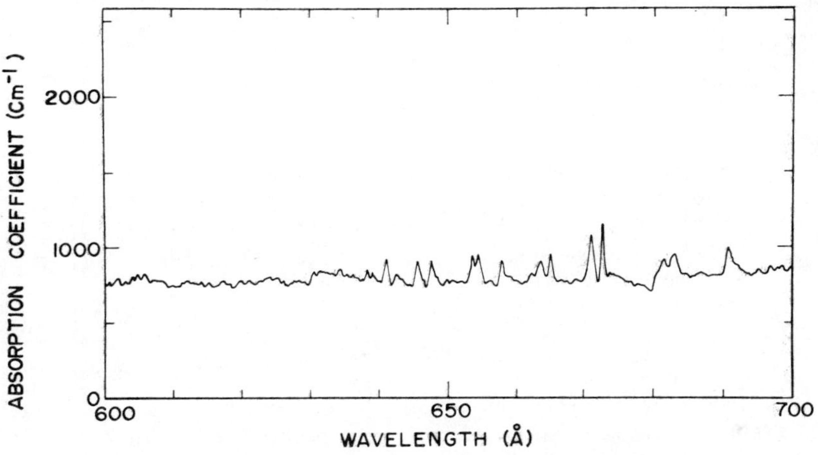

FIGURE 4a. Absorption coefficients of CO by photoelectric detection with an instrumental band width of 0.05 nm. Absorption coefficient is to the base e for 1 atm of gas at 273°K (from Huffman et al.[25] with permission).

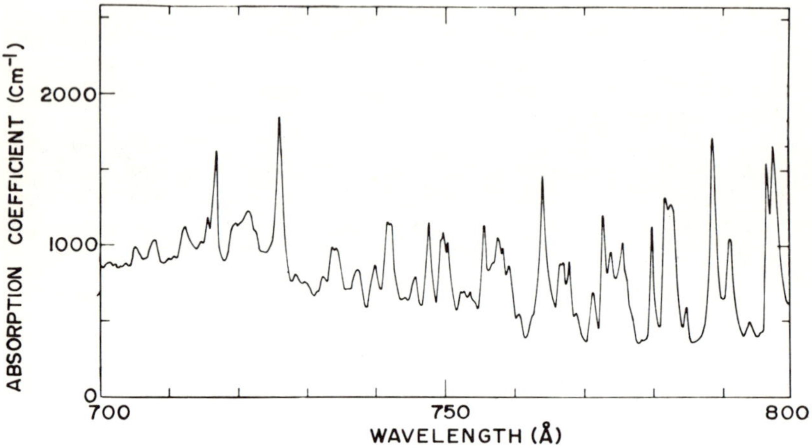

FIGURE 4b. Absorption coefficients of CO by photoelectric detection with an instrumental band width of 0.05 nm. Absorption coefficient is to the base e for 1 atm of gas at 273°K (from Huffman et al.[25] with permission).

B. Photochemistry

As the most tightly bound of diatomic molecules, CO will not be subject to photodissociation except in the short vacuum ultraviolet. The energetics of the various possible primary photodissociative processes are given in Table 3 together with the corresponding threshold wavelengths. Photodissociation is not energetically possible at wavelengths longer than 111.8 nm and in fact would not seem to occur to an appreciable extent before the onset of the first continuum near 96 nm. Photodissociation presumably occurs between 96 nm and the onset of the first ionization continuum near 89 nm and possibly at shorter wavelengths. Within this wavelength interval the absorption continuum is fairly strong (25), indicating a spin-allowed process,

$$CO(X^1\Sigma^+) + h\nu \rightarrow C(^3P) + O(^3P), \qquad (3)$$

rather than one involving production of one of the ^1D atomic states.

TABLE 3. Energetics of photodissociative primary processes for CO with threshold wavelength > 60.0 nm.[a] $CO \rightarrow C + O$

O → / C ↓		$^3P_2$	$^1D_2$	$^1S_0$	$^5S^o$
		(0.00)	(1.97)	(4.19)	(9.14)
$^3P_0$	(0.00)	(11.09)	(13.06)	(15.28)	(20.23)
		111.79	94.93	81.14	61.29
$^1D_2$	(1.26)	(12.35)	(14.32)	(16.54)	-
		100.39	86.58	74.96	
$^1S_0$	(2.68)	(13.77)	(15.74)	(17.96)	-
		90.04	78.77	69.03	
$^5S^o$	(4.18)	(15.27)	(17.24)	(19.46)	-
		81.19	71.91	63.71	
$^3P^o$	(7.48)	(18.57)	(20.54)	-	-
		66.76	60.36		
$^1P^o$	(7.69)	(18.78)	-	-	-
		66.02			

[a]Energies are listed within parentheses in electron volts; threshold wavelengths appear directly underneath in nanometers. Assumes a value of 11.09 ± 0.02 eV[1] for the bond dissociation enthalpy of CO at 0^oK, ΔH_0^o; energies of states of C and O are taken from ref. 10.

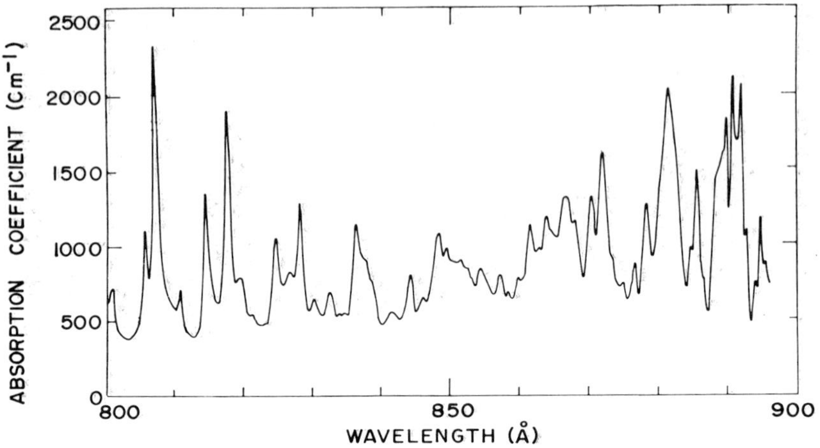

FIGURE 4c. Absorption coefficients of CO by photoelectric detection with an instrumental band width of 0.05 nm. Absorption coefficient is to the base e for 1 atm of gas at 273°K (from Huffman et al.[25] with permission).

Recently Simmons and Tilford (126) have presented spectroscopic evidence for an accidental predissociation of CO at 94,872 cm^{-1}. This energy is below the 99,650 cm^{-1} threshold energy for production of $O(^{1}D)$ and just above that for process 3. They observe that the R(30) doublet in the 0,0 band of the E-X system is enhanced in absorption and missing in emission and attribute the predissociation to a perturbing state which correlates with ground state atoms.

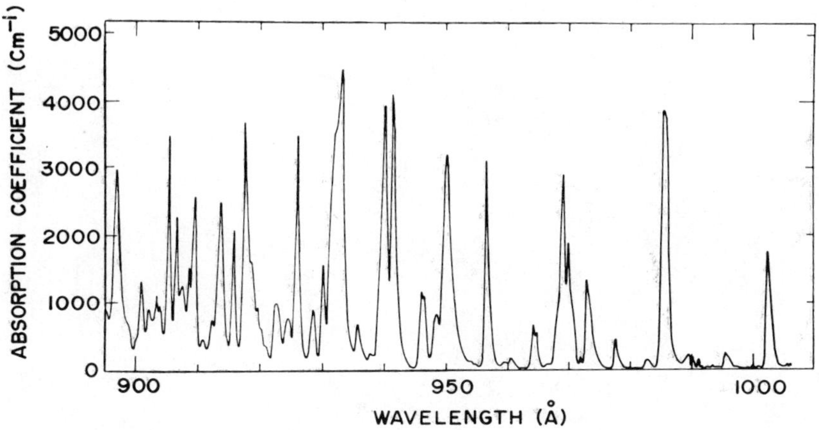

FIGURE 4d. Absorption coefficients of CO by photoelectric detection with an instrumental band width of 0.05 nm. Absorption coefficient is to the base e for 1 atm of gas at $273^{\circ}K$ (from Huffman et al.[25] with permission).

C. Excited States

At wavelengths longer than 100 nm, near the LiF transmission limit, CO appears to be completely stable with respect to dissociation in a primary step. The reactivity of optically excited CO in this region is thus due to the reactions of its excited states. Tilford and Simmons (24) have presented a set of RKR potential energy curves for CO calculated by R.H. Howard and this is given in Fig. 5 for the excited states.

The lowest-lying triplet state is $a^3\Pi$ which is the upper state of the well-known Cameron system $a^3\Pi \rightarrow X^1\Sigma^+$. Despite the low oscillator strength of this transition, Harteck et al. (26) found that absorption by CO of the 206.2 nm iodine line led to the production of carbon suboxide. Liuti, Dondes, and Harteck (27) observed that this C_3O_2 was isotopically enriched in ^{13}C at the center carbon and suggested as a mechanism,

$$CO^* + CO \rightarrow CO_2 + C \qquad (4)$$

$$C + CO + M \rightarrow C_2O + M \qquad (5)$$

$$C_2O + CO + M \rightarrow C_3O_2 + M \qquad (6)$$

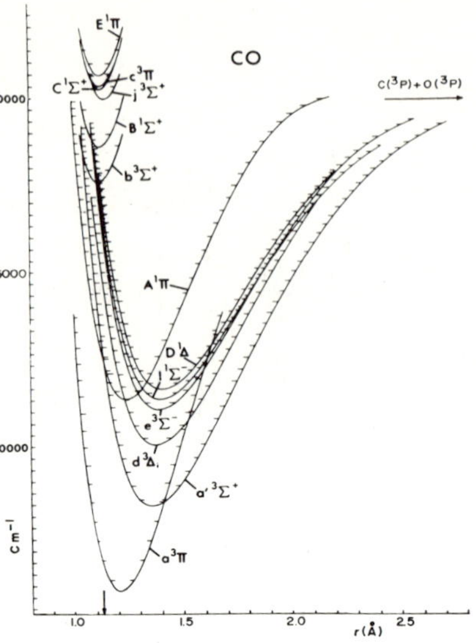

FIGURE 5. Potential energy curves for the excited states of CO below 11.77 eV (from Tilford and Simmons[24] with permission).

At wavelengths shorter than 155 nm and extending to about 120 nm, absorption by CO in the Fourth Positive system leads to production of the $A^1\Pi$ state. This is followed by fluorescence with approximately 10 ns lifetime or collision-induced vibrational relaxation and quenching with near gas-kinetic cross-sections (28,29). Absorption by CO of radiation from a CO discharge has been shown to produce other emissions as well as the Fourth Positive. Slanger and Black (30) have observed emissions in the Herman, Triplet, and Asundi systems corresponding to the transitions, $e^3\Sigma^- \rightarrow a^3\Pi$, $d^3\Delta \rightarrow a^3\Pi$, and $a'^3\Sigma^+ \rightarrow a^3\Pi$, respectively. This is attributed to direct production of the three upper states by absorption of $e \rightarrow x$, $d \rightarrow x$, and $a' \rightarrow x$ radiation from the discharge. Absorption of radiation from rare gas line sources has been shown to lead to CO fluorescence also. Becker and Welge (31) observed the fluorescence of CO excited by radiation from a Xe lamp (129.5, 147.0 nm). No fluorescence was observed when radiation from a Kr lamp (116.5, 123.6 nm) was used. The Xe lamp resulted in strong Fourth Positive fluorescence and weaker emission in the three triplet systems mentioned above. Slanger (32) and Slanger and Black (28) have shown that absorption of the Xe 147 nm line by CO results in production of CO $d^3\Delta$, v'=7 via the spin-forbidden transition, $d^3\Delta \leftarrow X^1\Sigma^+$. It appears that the Fourth Positive emission observed by Becker and Welge was due to resonance fluorescence of this system caused by its presence in the Xe light source.

Population of the CO $d^3\Delta$, v'=5 level by absorption of Fourth Positive radiation from a CO discharge has been attributed (30) to rotational perturbations between the $d^3\Delta$ and $A^1\Pi$ states. The existence of these perturbations has been demonstrated by Slanger and Black (33) who investigated the emissions from the three sublevels of the $d^3\Delta$ state to the $a^3\Pi$ state following absorption by CO of $d^3\Delta \rightarrow X^1\Sigma^+$ radiation from a CO discharge. This work has been extended by these authors (34) to perturbations between $A^1\Pi$ and $e^3\Sigma^-$. Elastic cross-relaxation was demonstrated between $A^1\Pi$ and $d^3\Delta$ and $e^3\Sigma^-$. Perturbations coupling the $a'^3\Sigma^+$ state to v' = 1 of $A^1\Pi$ have been observed by Burnham et al. (35) in a level-crossing experiment.

Flash photolysis in this wavelength interval has been studied by Donovan and Husain (36) who followed the decay of CO ($X^1\Sigma^+$, v" = 1) by absorption. Vibrationally excited CO($X^1\Sigma$) was proposed to result from CO quenching of CO($a^3\Pi$) which was produced directly in the flash by absorption in the Cameron system.

Overall quantum yields for decomposition have been measured at the wavelengths of Xe and Kr resonance lamps. Faltings et al. (37) found a quantum yield for CO removal of near unity at 129.5 nm and near zero at 147 nm. On the other hand, Becker and Welge (31) find a yield between 0.001 and 0.01 for both Xe wavelengths and a similar result at the combined Kr wavelengths. The mechanism of this decomposition has not been established but is presumably given by (4) through (6) above.

It is possible to state in summary that there is no direct evidence to indicate that CO is photolytically decomposed at any wavelength in its absorption spectrum, except near 108 nm as discussed by Simmons and Tilford (126).

IV. CO_2

Much of the current interest in the photochemistry of gaseous CO_2 is due to its presence as the principal constituent of the atmospheres of Mars and Venus. While laboratory studies continue to confirm its efficient photodecomposition and inefficient recombination, neither ground-based or planetary-mission spectroscopy reveal appreciable concentrations of either CO or O_2. Thus it is toward a resolution of these apparently incompatible results that much recent work has been directed.

A. Spectroscopy

While the absorption coefficients for CO_2 are appreciable only at wavelengths below 200 nm, long wavelength absorption is important as well to a consideration of planetary atmospheres. The absorption spectrum of CO_2 has been studied frequently in the past and this work has been summarized recently by Rabalais et al. (38). The absorption spectrum they have reported for CO_2 is given in Fig. 6. Nakata et al. (39) have

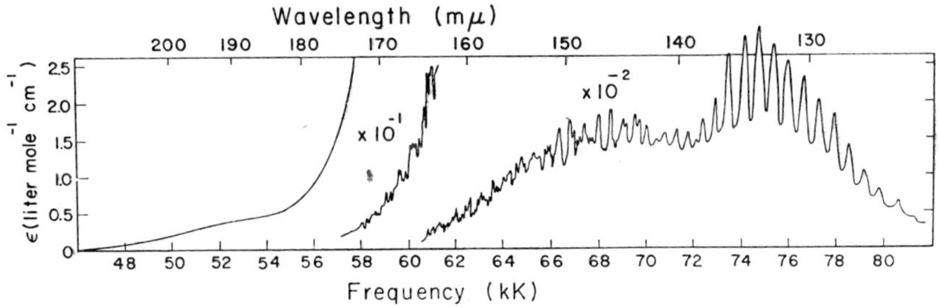

FIGURE 6. Molar absorptivity of CO_2 by photoelectric detection, defined as $(bc)^{-1} \log_{10} I_0/I$ where b is the path length in cm and c is the concentration in moles/liter. Reprinted with permission from J.W. Rabalais et al., Chem. Rev., 71, 73 (1971). Copyright by the American Chemical Society.

measured the absorption spectrum between 58 and 167 nm with a resolution of 0.02 nm. Absorption coefficients at longer wavelength have been measured by Heimerl (40), Ogawa (41), and Shemansky (42). The results of Shemansky are reproduced in Fig. 7. The extinction cross section is dominated by Rayleigh scattering at wavelengths longer than about 204 nm.

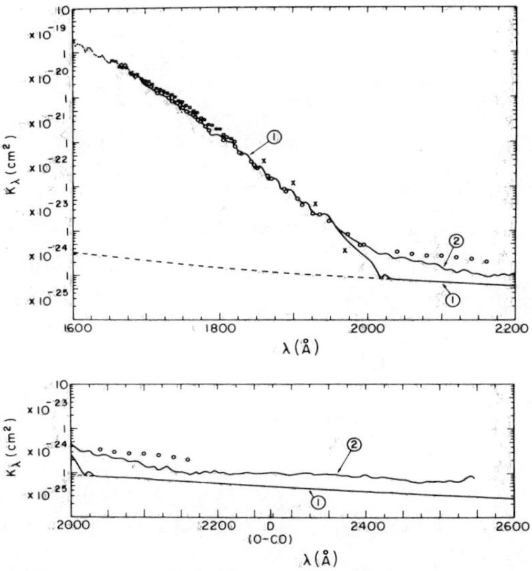

FIGURE 7. Extinction coefficient, k_λ, for CO_2 by photoelectric detection with an effective instrumental resolution of about 0.4 nm. Extinction coefficient determined through the relation $I_0/I_\lambda = \exp(k_\lambda n\ell)$. Curve (1): estimated extinction coefficient, curve (2): measured values, see reference 42 (from Shemansky[42] with permission).

The absorption spectrum is characterized by several continua with superimposed vibrational structure that have been only incompletely analyzed. A summary of the assignments (based on $\pi_g^3\ 2\pi_u$ and $\pi_g^3\ 6\sigma_g$ configurations identified by Walsh (43) and Mulliken (44)) that have been made for these continua has been given (38). The spectrum can be divided on the basis of appearance into several different regions whose onsets coincide approximately with the threshold wavelengths for different photochemical processes. These regions are:

1. above 170 nm, $f = 10^{-5}$, ${}^3\Sigma_u^+({}^3B_2) \leftarrow {}^1\Sigma_g^+({}^1A_1)$, >200 nm;

   ${}^3\Delta_u \leftarrow {}^1\Sigma_g^+({}^1A_1)$, 180-210 nm; ${}^1\Sigma_u^-({}^1A_2) \leftarrow {}^1\Sigma_g^+({}^1A_1)$, 160-190nm.

2. 140 to 170 nm, f = 6.2×10^{-3}, ${}^1\Delta_u({}^1B_2) \leftarrow {}^1\Sigma_g^+ ({}^1A_1)$.

3. 120 to 140 nm, f = 7.5×10^{-3}, ${}^1\Pi_g({}^1A_2) \leftarrow {}^1\Sigma_g^+ ({}^1A_1)$.

4. 110 to 120 nm, f = 1.2×10^{-1}, ${}^1\Sigma_u^+ ({}^1B_2) \leftarrow {}^1\Sigma_g^+ ({}^1A_1)$.

5. 90 to 110 nm, unassigned overlapping bands.

6. below 90 nm.

The energetics of the various primary photodissociative processes leading to CO and O and the corresponding threshold wavelengths are given in Table 4. There is no evidence for a photochemical primary process that produces O_2 and C that would have a wavelength threshold near 108 nm. For wavelengths longer than 108 nm, corresponding approximately to the transmission limit of LiF, the only energetically allowable processes are:

$$CO_2(X^1\Sigma_g^+) + h\nu \rightarrow CO(X^1\Sigma^+) + O(^3P,\ ^1D,\ ^1S). \qquad (7\ a,b,c)$$

Thus the considerable majority of laboratory work reported in the literature has dealt with production of atomic oxygen in the ground state or in the first two metastable excited states. The conclusions which can be drawn on the basis of this work are summarized below for the different spectral regions identified above.

B. Photochemistry

In general, the information bearing on the nature of the CO_2 primary photodissociative steps is of two kinds: (i) indirect inference based upon analysis of final products following the completion of secondary reactions, and (ii) direct spectroscopic observation of primary and intermediate products. Both approaches have made major contributions to the present level of understanding of the CO_2 photochemistry which has been summarized recently by Welge (45), and Preston and Cvetanovic (46) for wavelengths longer than 100 nm.

TABLE 4. Energetics of Photodissociative Primary Processes for CO_2 with Threshold Wavelength > 60.0 nm[a]. $CO_2 \rightarrow CO + O$

CO ↓ \ O →		$^3P_2$	$^1D_2$	$^1S_0$	$^5S^o$	$^3S^o$
		(0.00)	(1.97)	(4.19)	(9.14)	(9.52)
$X^1\Sigma^+$	(0.00)	(5.45) 227.49	(7.42) 167.09	(9.64) 128.61	(14.59) 84.98	(14.97) 82.82
$a^3\Pi$	(6.01)	(11.46) 108.18	(13.43) 92.32	(15.65) 79.22	(20.60) 60.18	-
$a'^3\Sigma^+$	(6.86)	(12.31) 100.71	(14.28) 86.82	(16.50) 75.14	-	-
$d^3\Delta$	(7.52)	(12.97) 95.59	(14.94) 82.99	(17.16) 72.25	-	-
$e^3\Sigma^-$	(7.90)	(13.35) 92.87	(15.32) 80.93	(17.54) 70.68	-	-
$A^1\Pi$	(8.03)	(13.48) 91.97	(15.45) 80.25	(17.67) 70.16	-	-
$I^1\Sigma^-$	(8.14)	(13.59) 91.23	(15.56) 79.68	(17.78) 69.73	-	-
$^1\Pi(^1\Delta)$	(8.21)	(13.66) 90.76	(15.63) 79.32	(17.85) 69.46	-	-
$b^3\Sigma^+$	(10.39)	(15.84) 78.27	(17.81) 69.61	(20.03) 61.90	-	-
$B^1\Sigma^+$	(10.77)	(16.22) 76.44	(18.19) 68.16	(20.41) 60.74	-	-
$C^1\Sigma^+$	(11.39)	(16.84) 73.62	(18.81) 65.91	-	-	-

[a]Energies are listed within parentheses in electron volts; threshold wavelengths appear directly underneath in nanometers. Assumes a value of 5.453 ± 0.003 eV[1] for the bond dissociation enthalpy of CO_2 at 0^oK, ΔH_0^o; energies of states of CO are taken from ref. 7; energies of states of O are taken from ref. 10.

1. Wavelengths longer than 170 nm. In the Mars atmosphere, photodissociation of CO_2 in this wavelength region is the dominant source of atomic oxygen below 40 km (47). Although the absorption cross section in this wavelength region is very low, atomic oxygen production rates exceed those from solar radiation at shorter wavelengths at higher altitude.

For wavelengths longer than 170 nm, only ground state products are energetically possible in a spin-forbidden process. Photochemical investigations of CO_2 in this wavelength region have not been conducted until recently. Photolysis of CO_2 at 184.9 nm has been studied at room temperature in the pressure ranges 200 to 800 torr by Krezenski et al. (48) and 65 to 800 psi by DeMore and Mosesman (49). In addition, photochemical decomposition has been observed at room temperature with a continuous source in the interval 174 to 210 nm by Inn and Heimerl (50) and at the longest wavelength studied, 213.9 nm, by Krezenski et al. (48).

In the high pressure photolysis (49), a quantum yield of 1.0 is obtained for the production of oxygen atoms and, therefore, for the process,

$$CO_2(X^1\Sigma_g^+) + 184.9\text{ nm} \rightarrow CO(X^1\Sigma) + O(^3P) \qquad (7a)$$

The measurement was made in mixtures containing small amounts of O_2 which served to recover atomic oxygen as O_3. The rate of production of O_3 was measured by absorption and compared, at equal optical depths for 184.9 nm radiation, to that in the O_2 photolysis, for which a quantum yield of 2.0 was assumed.

At lower pressures (48), the decomposition at 184.9 nm was studied by measuring the yield of O_2 and CO chromatographically after fixed periods of irradiation. Using a value of 1.41 for the quantum yield of N_2 from N_2O at 184.9 nm as an actinometer, the quantum yield for CO, and thus for process 7a, was found to be 1.08 ± 0.12. In common with the results of many other studies of the CO_2 photochemistry, a variable deficiency of O_2 is observed compared to the stoichiometric ratio of 0.5 for $(O_2)/(CO)$.

The primary quantum yield has been found to be unity for a wavelength integrated measurement in the interval 174 to 210 nm (50). With a high-pressure Xe continuum source filtered at the short wavelength end by an atmosphere of air and at the long-wavelength end by the declining absorption coefficient of CO_2 itself, a value of 1.0 ± 0.25 was obtained for CO production. Resonance fluorescence of the CO Fourth Positive system was employed to measure the yield of CO, and absolute light intensity measurements were made against a National Bureau of Standards lamp.

Thus the results of two studies at 184.9 nm yield a photodissociation quantum yield of unity and a third investigation suggests this conclusion and indicates that it may be true as well for wavelengths as short as 175 nm. On the other hand, in a single study at 213.9 nm (48), a CO quantum yield of 0.16 was found.

2. **Wavelengths between 140 and 170 nm.** In this wavelength interval, production of the aeronomically important metastable species $O(2^1D_2)$ becomes energetically possible in a spin-allowed process and abundant evidence exists for its formation,

$$CO_2(X^1\Sigma_g^+) + h\nu \rightarrow CO(X^1\Sigma^+) + O(^1D). \qquad (7b)$$

The excited state produced in absorption in this region is presumably $^1\Delta_u(^1B_2)$, which correlates with $O(^1D)$ and $CO(X^1\Sigma^+)$. Experimental evidence for this process and its quantum yield is entirely of the product analysis variety since $O(^1D)$ has not yet been detected directly in the photolysis of CO_2. This seems correct despite two studies in which observation of $O(^1D)$ was reported. Noxon (51) reported the observation of 630.0 nm radiation accompanying radiative decay of the ^1D state to the ground state. This observation, however, has since been found by Clark and Noxon (52) to be due to scattered radiation from the photolysis lamp. In a flash-photolysis study of CO_2 decomposition, Clerc and Barat (53) detected absorption at 115.2 nm and attributed it to the transition, $O(3s^1\ ^1D^o \leftarrow 2p^4\ ^1D_2)$. The experimental description was incomplete and the observation has not been confirmed by a subsequent report. In view of the efficiency with which $O(^1D)$ is now known to be collisionally deactivated by CO_2, it seems at least as likely that this absorption feature was due to CO, not $O(^1D)$. The band head of the 0,0 band in the $B^1\Sigma^+ \leftarrow X^1\Sigma^+$ system of CO lies very near this wavelength.

The experimental work prior to 1963 has been reviewed previously by McNesby and Okabe (54) and a subsequent review by Clark (55) comments on work up to 1970. Studies of CO_2 photochemistry in the 140-170 nm wavelength interval have been made by measurement of the final products O_2 and CO by gas chromatography, mass spectrometry, and resonance fluorescence (CO) and by measurement of atomic oxygen by resonance fluorescence and chemiluminescence. Evidence for quantum yields of unity for CO at 147 nm has been provided by several workers. Mahan (56) measured the CO quantum yield and found it to be 1.1 ± 0.1 in an experiment in which CO and O_2 were analyzed by mass spectrometry and the photolysis of O_2 was employed as an actinometer. Failure to observe the NO_2 chemiluminescence when NO was added was interpreted as evidence against the formation of $O(^3P)$ in a primary process. It is now known (57,58), however, that $O(^1D)$ is rapidly deactivated to $O(^3P)$ and that therefore chemiluminescence would be expected. In subsequent work, Young et al. (57) detected the NO_2 chemiluminescence and measured a quantum yield for $O(^3P)$ of 0.97 ± 0.05.

Warneck (59) obtained a CO quantum yield of 1.17 at 147 nm and 1.0 with a hydrogen discharge (near 160 nm) relative to an assumed quantum yield of 2.0 for production of O_3 in the O_2 photolysis. Ung and Schiff (60), however, found a value of only 0.25 at 147 nm using the O_2/CO actinometer employed by Mahan. It has been suggested that their low results were due to the presence of water vapor (61) or long-wavelength contaminant emission (62). The importance of water vapor has been demonstrated by Loucks and Cvetanovic (63), who showed that a satisfactory material balance could be obtained in CO_2 photolysis at 163.3 nm if O_3 production was considered. The product distribution was changed substantially in the presence of water vapor.

Yang and Servedio (61) have measured a value for the CO quantum yield of 1.0 at 147 nm relative to a quantum yield of 1.4 for nitrogen production in the 147 nm N_2O photolysis. Similarly, Thompson et al. (64) found a quantum yield of unity for CO_2 removal and hence for the primary photodissociation. Against these several measurements of unity or near-unity may be set the more recent determination by Inn (65) of a CO quantum yield of 0.75 in experiments employing Fourth Positive resonance fluorescence and an NBS calibrated vacuum photodiode.

In addition to the value of 1.0 for the CO quantum yield obtained by Warneck near 160 nm, Quick and Cvetanovic (66) identified $O(^1D)$ as a primary product of the photolysis at the emission wavelengths of a bromine lamp, 157.5 and 163 nm. It was previously (67) established that insertion of $O(^1D)$ into neopentane leads to a 65 percent production of neopentanol. Thus the yield of neopentanol in combination with the effect of various additive gases serves as a qualitative test for the presence of $O(^1D)$. The quantum yield of neopentanol can then be used to measure the quantum yield for $O(^1D)$. These authors find the $O(^1D)$ quantum yield to be near unity at these two wavelengths. This result is in conflict with that of Inn and Heimerl (50), who measured a quantum yield of 0.5 for CO production in the interval from 150 to 167 nm. These latter measurements were made by Fourth Positive fluorescence with a hydrogen continuum serving as the photolytic source. They have recently been repeated by Inn (65) who obtained a value of 0.53 in agreement with the previous work.

Of considerable importance to a resolution of the planetary CO_2 problem is the possibility that primary process 7 occurs via an excited CO_2 molecule which may either radiate or at higher pressures be collisionally dissociated,

$$CO_2 + h\nu \rightarrow CO_2^* ,$$

$$CO_2^* \rightarrow CO_2 + h\nu ,$$

$$CO_2^* + M \rightarrow CO + O + M .$$

In an effort to assess the importance of this process, Clark and Noxon (68) searched for fluorescent emissions in the wavelength interval 100 to 800 nm upon excitation of CO_2 with radiation in the 150 to 170 nm range. Their failure to detect any emissions led them to place a limit of 0.1% upon the quantum efficiency for fluorescence at a CO_2 pressure of 0.1 torr. Thus it appears that photodissociation in this wavelength interval occurs as a single-step mechanism with near unit efficiency. This conclusion was extended to the CO_2 photolysis at 147 nm by Felder et al. (69) who found the quantum yield for CO production to be approximately the same for CO_2 pressures between 0.002 and 20 torr.

In recent work, Slanger et al. (70) found values for the CO quantum yield relative to that at 147 nm as unity of 0.77 ± 0.07 at the combined atomic nitrogen lines, 149.3, 147.5 nm. Assuming that the value 0.75 obtained by Inn (65) at 147 nm is correct leads to a CO quantum yield of 0.58 ± 0.05 at these combined wavelengths, in agreement with the work of Inn (65). These measurements were made by comparing the rate of production of CO from CO_2 by photolysis at selected wavelengths under conditions of approximately constant optical depth. Light intensities were compared by means of a calibrated vacuum photodiode and CO production rates were measured by Fourth Positive fluorescence.

Thus the results of many investigations of the CO_2 photochemistry in the 140 to 170 nm wavelength range contradict one another with respect to the value for the primary quantum yield for photodecomposition. The most recent work suggests that while photodecomposition is efficient it is not complete, and that under appropriate conditions, CO_2 fluorescence may be expected.

3. Wavelengths between 120 and 140 nm. For wavelengths shorter than 128 nm, production of $O(^1S)$ in a spin-allowed process has been confirmed by direct observation of the 557.7 nm emission corresponding to the transition $O(^1S_0 \rightarrow {}^1D_2)$,

$$CO_2(X^1\Sigma_g^+) + h\nu \rightarrow CO(X^1\Sigma^+) + O(^1S) . \qquad (7c)$$

Evidence for the identity of the primary products is largely spectroscopic while information on the quantum yield comes primarily from measurement of the yield of stable products.

Observation of $O(^1S)$ in the CO_2 flash photolysis has been reported between 105 and 125 nm by Stuhl and Welge (71) and Lawrence (72). Filseth et al. (73) and Young et al. (74) have employed the photolysis of CO_2 as a source of $O(^1S)$ for the purpose of making direct measurements of its removal coefficient in the presence of various gases.

Lawrence (72) has measured the photolytic production efficiency for $O(^1S)$ from 81.2 to 121.6 nm by observation of the 557.7 nm emission line. He finds a value of approximately 0.15 at 121.6 nm. A similar study has been reported by Welge (75) between the wavelength limits of 108 and 121.6 nm, and it is in essential agreement with that of Lawrence. Ridley et al. (76) have measured the $O(^1S)$ yield at 121.6 nm relative to that from the photolysis of N_2O. They obtained $\phi(^1S,CO_2)/\ \phi(^1S,N_2O) = 0.31 \pm 0.06$. Welge and Gilpin (77) have utilized the Auger emission of electrons from Cs surfaces by metastable species to detect the production of metastable atomic oxygen $O(^1D)$ and/or $O(^1S)$, in the pulsed photolysis of CO_2 between 105 and 116 nm in a time-of-flight experiment. From an analysis of the distribution of arrival times it was determined that a large fraction (> 50%) of the dissociation produces $CO(X^1\Sigma)$ with considerable internal energy. Observation of $O(^1S)$ during the photolysis of CO_2 with the 116.5/123.6 nm radiation from a krypton discharge has been attributed by Slanger (78) to absorption of radiation at 116.5 nm rather than at 123.6 nm on the basis of the effect of CO_2 pressure on 557.7 nm intensity.

For wavelengths longer than 128 nm, $O(^1S)$ production is not energetically possible and the primary product is presumably $O(^1D)$ via process 7b as at wavelengths of 147 nm and longer. No experimental information is available in this interval which establishes $O(^1D)$ as a primary product. However, if the assignment of the predominant continuum in this region to a ${}^1A_2 \leftarrow {}^1\Sigma_g^+(^1A_1)$ transition is correct, ^1D is the reasonable product, since $^1A_2$ does not correlate with an S atomic state if the molecular product is Σ^+. Since production of $O(^3P)$ would be spin-forbidden, it is possible that absorption in the principal continuum in this region is accompanied by formation of $O(^1D)$. This conclusion is consistent with the low efficiency for production of $O(^1S)$ at 121.6 nm reported by Lawrence (72) and Ridley et al. (76), which may be entirely due to absorption in the overlapping long-wavelength portion of the continuum whose maximum lies at 112 nm.

The overall quantum yield of dissociation in the 120 to 140 nm region has been measured most frequently at 123.6 nm. Mahan (56) found a value of 1.2 ± 0.1 by mass spectrometric analysis for CO with NO photoionization being used to monitor lamp intensity. Warneck (59) found a value of 1.0 with mass spectrometric analysis for CO and lamp intensity measurements by O_2 actinometry with an assumed value of 2.0 for the quantum efficiency of O_3 formation. Recently Slanger et al. (70) have measured the rate of CO production in the CO_2 photolysis at different wavelengths and expressed the results relative to the rate at 147 nm. Their results are: 0.76 ± 0.11 at 121.6 nm, 1.13 ± 0.13 at 123.6 nm, 0.28 ± 0.08 at the 130.2 atomic oxygen triplet, and 0.61 ± 0.05 at the 136 to 140 nm atomic chlorine quartet. Assuming that Inn's (65) value of 0.75 at 147 nm is correct leads to the following quantum efficiencies at these wavelengths, respectively, 0.57 ± 0.09, 0.85 ± 0.10, 0.21 ± 0.07, and 0.46 ± 0.04. These results are illustrated in Fig. 8 from Slanger et al. The comparison of the photodissociation coefficient with the photoabsorption coefficient suggests that the absorption spectrum consists of a dissociative continuum with superimposed bands corresponding to bound CO_2 states. Thus the most recent work leaves no basis for the assumption of unit quantum efficiency for dissociation in this wavelength interval.

4. Wavelengths between 110 and 120 nm. This wavelength interval includes the region of strongest absorption by CO_2 and corresponds at least in part to the production of $O(^1S)$, in process 7c. The wavelength resolved measurements of the efficiency of production of $O(^1S)$ between 120.0 and 81.2 nm by Lawrence (72) and between 116.0 and 108.0 nm by Welge (75) suggest a quantum efficiency close to unity in the interval 116 to 110 nm dropping off abruptly on either side of this interval. The quantum efficiencies obtained by Lawrence (72) between 80 and 120 nm are given in Fig. 9.

5. Wavelengths between 90 and 110 nm. At wavelengths shorter than 110 nm several new primary processes are energetically possible and some have been detected. Of particular interest is the production of $CO(a^3\Pi)$ via the spin-allowed process,

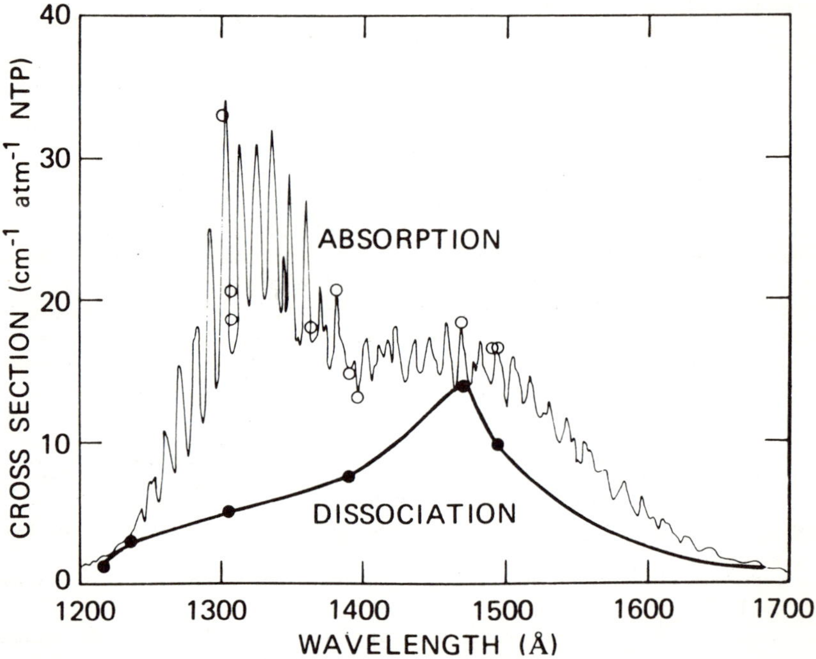

FIGURE 8. CO_2 photoabsorption and photodissociation coefficients; CO_2 pressure range 0.38 - 4.0 torr. Open circles - measured absorption coefficients. Filled circles - measured dissociation coefficients. Light solid line - CO_2 absorption coefficients.[39] Heavy solid line has no experimental basis (from Slanger et al.[70] with permission).

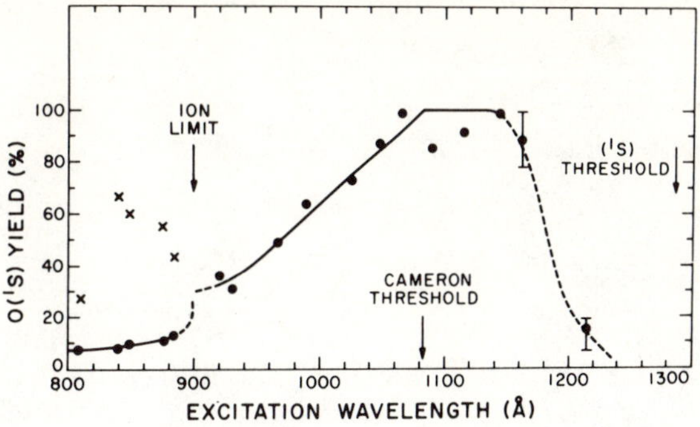

FIGURE 9. Percentage yield of O(^{1}S) from photodissociation of CO_2. Dots are yields with respect to absorption; crosses are yields with respect to dissociation. Threshold wavelengths for photodissociation with production of O(^{1}S) or CO($a^3\Pi$) are indicated (from Lawrence[72] with permission).

$$CO_2(X^1\Sigma_g^+) + h\nu \rightarrow CO(a^3\Pi) + O(^3P) \; . \qquad (8)$$

Appreciable emission has been observed in the ultraviolet dayglow of Mars in the Cameron system CO($a^3\Pi \rightarrow X^1\Sigma^+$) (79). Production of CO($a^3\Pi$) in a spin-allowed process is energetically possible at wavelengths shorter than 108.2 nm and has been detected by Lawrence (80) through observation of the subsequent emission of the Cameron system. He finds that the fractional yield of CO($a^3\Pi$) increases linearly from zero at the threshold at 108 nm to a value of approximately 0.5 near the ionization limit of 90.0 nm. The wavelength dependence is illustrated in Fig. 10.

In the wavelength region near 105 nm Lawrence finds the yield of CO($a^3\Pi$) to be near 15%. Other photolytic processes occur in this region as was demonstrated by the observation of O(^{1}S) emission by Young et al. (74) at the 104.8/106.7 nm argon resonance lines. This emission is presumably a consequence of primary process 7c. The quantum efficiency with which O(^{1}S) is produced has been measured in this wavelength interval by Lawrence (72). A prominent feature of the results of Lawrence (72,80), (Figs. 9 and 10) is the wavelength dependence of the yields of O(^{1}S) and CO($a^3\Pi$), the former decreasing smoothly between 108 nm and the ionization limit as the latter rises. The quantum yield of CO has been measured by Warneck (81) at 104.8/106.7 nm, and he finds a value of 1.06 ± 0.10. Product

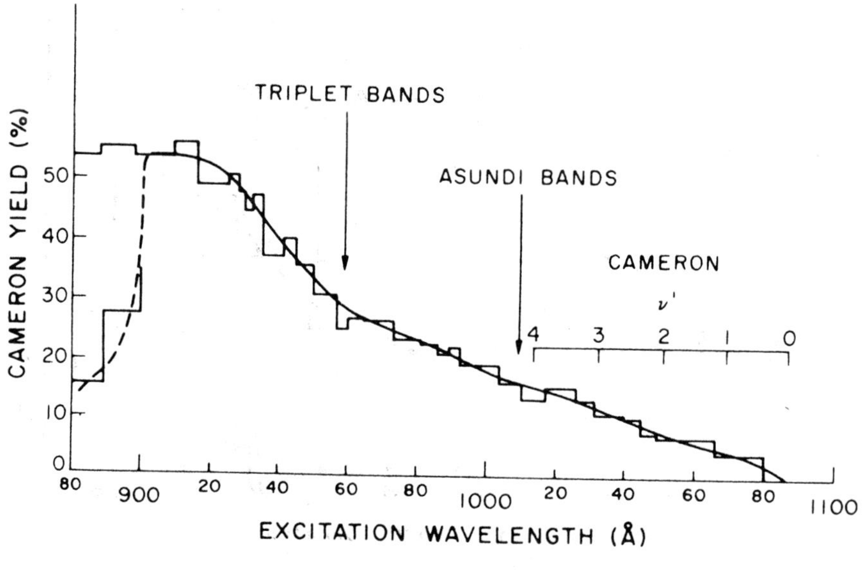

FIGURE 10. Fractional yield of CO Cameron, ($a^3\Pi - X^1\Sigma^+$), bands from the photodissociation of CO_2. Upper curve is with respect to dissociation only; dotted curve is with respect to total absorption. Thresholds of possible cascading states are shown (from Lawrence[80] with permission).

analysis was by mass spectrometry and actinometry was with reference to the photoionization yields of NO at Lyman α and O_2 at Lyman β.

These observations imply that at least near its threshold, 107 nm, there is very little contribution from the molecular elimination primary step,

$$CO_2(X^1\Sigma_g^+) + h\nu \rightarrow C(^3P) + O_2(X^3\Sigma_g^-) \qquad (9)$$

There is in fact no evidence for the participation of this primary process in the CO_2 photochemistry at any wavelength.

In the wavelength interval between 100 nm and the ionization threshold at 90 nm, several other excited states of CO, $a'^3\Sigma^+$, $d^3\Delta_i$, $e^3\Sigma^-$, and $A^1\Pi$, may be produced in spin-allowed processes. The emission from the triplet states to the lower-lying $a^3\Pi$ state, the Asundi, Triplet, and Herman bands, respectively, have been employed by Judge and Lee (82) and Lee and Judge (83) to measure cross sections for production of these excited states at selected wavelengths from 76.4 to 92.3 nm. Adopting the data of Cook et al. (84) for the continuum photodissociation cross section between 70 and 90 nm, Judge and Lee (82) conclude that production of the triplet states together account for greater than 85 percent of the photodissociative processes in the wavelength interval from 76.4 to 92.3 nm. This value is about a factor of two higher than the result of Lawrence (80) which measures the overall cross section (primary + cascade) for production of CO $(a^3\Pi)$.

At wavelengths of 90.1 and 92.3 nm Judge and Lee find that the highest vibrational levels energetically possible in $CO(a'^3\Sigma^+)$, $v' = 11$ and 8, respectively, are produced together with the corresponding highest vibrational levels in $CO(d^3\Delta)$, $v' = 6$ and 3, respectively. These authors find that the vibrational distribution in these two electronic states of CO can be reproduced at each of the two excitation wavelengths by a Poisson distribution in the excess energy. A first-order time-dependent perturbation theory calculation by Berry (85) has yielded vibrational distributions for the CO triplets in good agreement with the results of Lawrence (80) and Lee and Judge (82,83).

6. Wavelengths shorter than 90 nm. Absorption by CO_2 at wavelengths shorter than 90 nm is dominated by photoionizing transitions involving production of CO_2^+. The fluorescence that results has been studied by several authors. Both electron- and photon-excited fluorescence in the A-X and B-X systems of CO_2^+

is important in the Mars and Venus airglows. The discussion below is restricted to photodissociative processes producing neutral fragments.

At wavelengths below 92 nm, production of $CO(A^1\Pi)$ becomes energetically possible in a spin-forbidden process,

$$CO_2(X^1\Sigma_g^+) + h\nu \rightarrow CO(A^1\Pi) + O(^3P) . \quad (10a)$$

While the wavelength interval extending below 92 nm includes that studied by Judge and Lee (82,83), they did not study emission in the region below 200 nm in which CO Fourth Positive, $A^1\Pi \rightarrow X^1\Sigma_g^+$, fluorescence would be expected. This emission has, however, been detected previously. Beyer and Welge (86) have observed fluorescence attributed to the CO Fourth Positive system when CO_2 was photodissociated with discrete excitation wavelengths between 90 and 65 nm.

In recent work, Gentieu and Mentall (87) have studied the vacuum ultraviolet fluorescence produced by wavelength resolved (0.05 nm resolution) photodissociation of CO_2 with a pulsed Hopfield continuum. Cross sections for wavelength integrated Fourth Positive emission have been measured from threshold at 92.5 nm to 63.6 nm. At wavelengths longer than 80 nm, process 10a is the only energetically possible process producing $CO(A^1\Pi)$ although it is not spin allowed. At wavelengths less than 80 nm, $O(^1D)$, and at wavelengths less than 70 nm, $O(^1S)$ production are energetically possible in a photolytic primary process also involving $CO(A^1\Pi)$ formation,

$$CO_2(X^1\Sigma_g^+) + h\nu \rightarrow CO(A^1\Pi) + O(^1D, {}^1S) . \quad (10b,c)$$

While Gentieu and Mentall have not observed either $O(^1D)$ or $O(^1S)$ directly, their presence was inferred from an apparent vibrational cooling in the CO Fourth Positive emission near the $O(^1D)$ and $O(^1S)$ wavelength thresholds of 80 and 70 nm, respectively. A quantitative evaluation of this effect suggests that dissociation via 10b and 10c occurs at the expense of 10a and accounts for at least half the $CO(A^1\Pi)$ formed below these thresholds.

Below 85 nm it is energetically possible for photodissociation to leave atomic oxygen in the metastable excited state, $^5S^o$,

$$CO_2(X^1\Sigma_g^+) + h\nu \rightarrow CO(X^1\Sigma_g^+) + O(^5S^o) . \quad (11)$$

This process has been suggested to account for a portion of the 135.6 nm radiation, $O(^5S^o \rightarrow {}^3P)$, observed in the Mars and Venus atmospheres (88). Gentieu and Mentall do not report the observation of this emission; however, the approximately 200 μs radiative lifetime (89) of this state would have made its observation unlikely.

At wavelengths shorter than 83 nm, production of $O(^3S^o)$ is energetically possible in a spin and parity forbidden process,

$$CO_2(X^1\Sigma_g^+) + h\nu \rightarrow CO(X^1\Sigma^+) + O(^3S^o). \quad (12)$$

While Gentieu and Mentall have observed the emission of the 130.2 nm triplet that should follow production of $O(^3S^o)$, they have not measured the wavelength threshold, and it may be produced by cascade following primary production of $O(2p^33p\ ^3P)$, which has a wavelength threshold near 75 nm. They measure cross sections of 1.0×10^{-20} and $< 0.1 \times 10^{-20} cm^2$ at 74.4 and 58.4 nm, respectively, for production of the 130.2 nm triplet. An upper limit for this cross section from threshold to the 63.6 nm limit of their measurements was calculated as about 6 percent of the $CO(A^1\Pi)$ cross section. Thus it does not seem that photoproduction will be an important source of $O(^3S^o)$ in the Mars upper atmosphere.

C. Photodissociation in Planetary Atmospheres

The considerable amount of work that has been devoted to the laboratory study of CO_2 photochemistry has established that CO_2 is efficiently, although not necessarily completely, photodissociated in the wavelength interval from less than 90 to greater than 160 nm. At the same time, the results of several highly successful planetary missions to Mars and Venus indicate a surprisingly low abundance of the ultimate products of this photodissociation, CO and O_2. In the case of Mars for example, mixing ratio limits of 7×10^{-4} for O_2 (90) and 8×10^{-4} for CO (91) have been obtained spectroscopically. From electron density profiles derived from occulation experiments, it can be concluded that there is little atomic oxygen below 200 km. A number density of the order of 10^8 cm^{-3} at 125 km has been deduced from the Mariner 6 and 7 dayglow measurements (92).

The problem of the stability of the CO_2 atmospheres of Mars and Venus has been considered by several authors. Early mechanisms proposed to account for low abundance of CO and O_2 in these atmospheres have been discarded or modified as laboratory measurements have shown that the rates of one or more of the elementary steps were too low to provide an adequate recombination rate.

Clark (55) reviewed the laboratory data and concluded that there was no homogeneous gas phase process capable of accounting for the CO and O_2 mixing ratio limits at any altitude. More recently, McElroy and McConnell (93) have concluded that vertical transport with high eddy diffusion coefficients may explain the mixing ratios in the Mars upper atmosphere. The same mechanism cannot, however, account for the lower atmosphere. The situation was revewed by Donahue (94) who concluded, by elimination of homogeneous gas phase processes, that recombination of O, O_2, and CO must occur heterogeneously.

A recombination mechanism considered previously by Donahue and by McElroy and Hunten has been reconsidered in modified form by Parkinson and Hunten (90). The elementary step in this sequence is the removal of CO catalytically by hydroxyl radicals which are regenerated in subsequent reactions involving H_2O, H_2O_2, and odd hydrogen. This sequence appears to be capable of accounting for the low CO mixing ratio. Maintenance of low O_2 mixing ratios is attributed to long wavelength photodissociation of the minor constituents, HO_2 and H_2O_2. This mechanism can be summarized as,

$$H + O_2 + CO_2 \rightarrow HO_2 + CO_2 \tag{13}$$

and

$$HO_2 + O \rightarrow OH + O_2 \tag{14a}$$

or

$$2HO_2 \rightarrow H_2O_2 + O_2 \tag{14b}$$

followed by

$$H_2O_2 + h\nu \rightarrow 2OH \tag{15}$$

and finally

$$CO + OH \rightarrow CO_2 + H \ .$$

The CO_2/O_2 balance in the Mars atmosphere has also been considered by McElroy and Donahue (95) and the low CO and O_2 mixing ratios have been attributed to the same mechanism. In their model, CO and O_2 are taken as given and the model then predicts H_2O and O_3 abundances that agree with observation. It appears that this model, featuring CO and O transport to lower altitudes where recombination occurs, provides a satisfactory explanation for the stability of the Martian atmosphere.

Considerable differences exist between the Venus and Mars atmospheres. The predominance of CO_2 is the most apparent similarity between them. While the CO mixing ratio is similar to that on Mars, the O_2 abundance seems much lower. The stability of the Venus CO_2 atmosphere has been consdered by Prinn (96) who suggests a sequence of reactions in which OCl replaces OH as the catalytic species,

$$C\ell + O_2 + CO_2 \rightarrow C\ell O_2 + CO_2,$$

$$C\ell O_2 + O \rightarrow C\ell O + O_2,$$

$$CO + C\ell O \rightarrow CO_2 + C\ell.$$

The rates of the elementary steps of this mechanism have not been measured, but Prinn concludes that $C\ell O_2$ is sufficiently stable thermodynamically to permit it to play an important role in CO oxidation. In addition, McElroy et al. (97) have found that photodissociation of $HC\ell$ provides an adequate source of atomic hydrogen to permit the mechanism proposed for Mars, (eqs. 13 through 15), to serve for Venus as well.

V. OCS

Carbonyl sulfide is an important gas kinetic source of ground state and electronically excited sulfur atoms, and its photodissociation has been employed by several groups to study the rates and products of reactions of $S(3^3P_2)$, $S(3^1S_0)$, and $S(3^1D_2)$. It is, in addition, one of the interstellar molecules, has been proposed as a minor constituent of the Venus atmosphere, and leads to laser emission in CO following photodissociation.

A. Spectroscopy

The absorption spectrum of carbonyl sulfide consists of three distinct transitions between the ultraviolet and 140 nm followed by a series of Rydberg transitions extending toward shorter wavelengths. A photoelectric recording obtained with a McPherson 1-M scanning monochromator is given in Fig. 11 (38). This work also refers to earlier studies of OCS. The absorption spectrum between 107 and 170 nm has been recorded photoelectrically by Matsunaga and Watanabe (98) at a resolution of 0.05 nm. The assignments that have been made (38) for the longer wavelength transitions are:

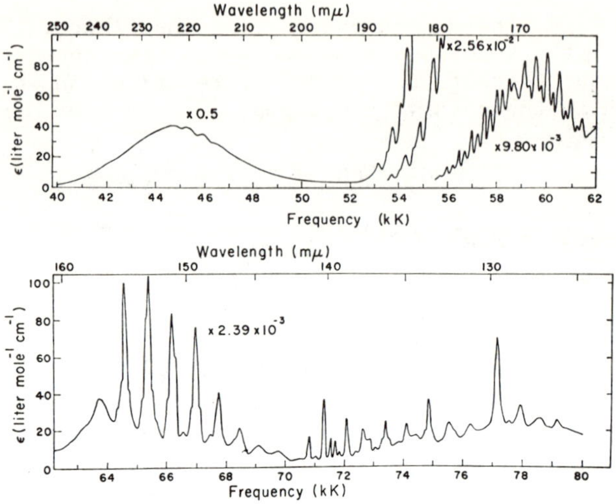

FIGURE 11. Molar absorptivity of OCS by photoelectric detection, defined as $(bc)^{-1} \log_{10} I_0/I$, where b is the path length in cm and c is the concentration in moles/liter. Reprinted with permission from J.W. Rabalais et al., Chem. Rev., 71, 73 (1971). Copyright by the American Chemical Society.

1. above 200 nm, $f = 1.8 \times 10^{-3}$, $^1\Delta \leftarrow {}^1\Sigma^+$

2. 190 to 160 nm, $f = 0.13$, $^1\Pi \leftarrow {}^1\Sigma^+$

3. 140 to 160 nm, $f = 0.38$, $^1\Sigma^+ \leftarrow {}^1\Sigma^+$.

At wavelengths shorter than 142 nm, an extensive set of Rydberg transitions is observed extending to the ionization threshold which lies near 110 nm.

The energetics of the primary photodissociative processes that lead to CO and S are presented in Table 5. There is very little evidence for a corresponding primary process to produce CS and O. A set of wavelength thresholds for this process (17a, b,c) is given in Table 6.

The experimental evidence which bears on the participation of primary process 16,

$$OCS(X^1\Sigma^+) \rightarrow CO(X^1\Sigma^+) + S(^3P, {}^1D, {}^1S), \qquad (16a,b,c)$$

is given below for the different spectral regions above and below 200 nm.

TABLE 5. Energetics of photodissociative primary processes for OCS with threshold wavelength > 100 nm.[a] $OCS \rightarrow CO + S$

CO ↓ \ S →		$^3P_2$	$^1D_2$	$^1S_0$	$^5S^o$	$^3S^o$	^5P
		(0.00)	(1.15)	(2.75)	(6.52)	(6.86)	(7.87)
$X^1\Sigma^+$	(0.00)	(3.12)	(4.27)	(5.87)	(9.64)	(9.98)	(10.99)
		397.37	290.35	211.21	128.61	124.23	112.81
$a^3\Pi$	(6.01)	(9.13)	(10.28)	(11.88)	-	-	-
		135.79	120.60	104.36			
$a'^3\Sigma^+$	(6.86)	(9.98)	(11.13)	-	-	-	-
		124.23	111.39				
$d^3\Delta$	(7.52)	(10.64)	(11.79)	-	-	-	-
		116.52	105.16				
$e^3\Sigma^-$	(7.90)	(11.02)	(12.17)	-	-	-	-
		112.50	101.87				
$A^1\Pi$	(8.03)	(11.15)	(12.30)	-	-	-	-
		111.19	100.80				

[a]Energies are listed within parentheses in electron volts; threshold wavelengths appear directly underneath in nanometers. Assumes a value of 3.12 ± 0.03 eV[1] for the bond dissociation enthalpy of OCS at 0^oK, ΔH_0^o; energies of states of CO are taken from ref. 7; energies of states of S are taken from ref. 99.

TABLE 6. Energetics of photodissociative primary processes for OCS with threshold wavelengths > 100 nm.[a] $OCS \rightarrow CS + O$

	O →	$^3P_2$	$^1D_2$	$^1S_0$
CS		(0.00)	(1.97)	(4.19)
↓				
$X^1\Sigma^+$	(0.00)	(6.34)	(8.31)	(10.53)
		195.55	149.19	117.74
$a^3\Pi$	(3.42)	(9.76)	(11.73)	-
		127.03	105.69	
$A^1\Pi$	(4.81)	(11.15)	-	-
		111.19		

[a]Energies are listed within parentheses in eV; threshold wavelengths appear directly underneath in nm. Assumes a value of 6.34 ± 0.23 eV[1] for the bond dissociation enthalpy of OCS at 0°K, ΔH_0^o; energies of states of O are taken from ref. 10; energies of $CS(a^3\Pi)$ and $CS(A^1\Pi)$ are taken from ref. 100. The energies of $CS(a'^3\Sigma^+)$ and $CS(e^3\Sigma^-)$ have been determined to be ≤ 4.82 and ≤ 4.85 eV respectively[101].

B. Photochemistry

1. Wavelengths longer than 200 nm. The photochemistry of OCS has been investigated extensively through analysis of final products. Gunning and Strausz have given a review of the production of S atoms in this wavelength region in Volume 4 of this series. While the energetic limit on the OCS decomposition corresponds to a wavelength of 397 nm, absorption is not observed at wavelengths longer than about 260 nm. The energy of radiation in the broad absorption feature whose maximum lies near 224 nm will permit production of $S(^1D)$ atoms in a primary process, and there is abundant evidence that this process is an important one,

$$OCS(X^1\Sigma^+) + h\nu \rightarrow CO(X^1\Sigma^+) + S(^1D) . \qquad (16b)$$

Much of the evidence has been provided by Gunning and co-workers. In experiments (102) in which OCS is photodissociated in the presence of increasing amounts of olefin, they demonstrated that quantum yields of CO production are reduced to a limiting value equal to half that observed in the absence of olefin. It is expected, therefore, that the quantum yield for the production of CO should be 2 in the absence of olefin unless the quantum efficiency of the primary process is less than unity. Sidhu et al. (103) found that the CO quantum yield at 253.7 nm was pressure independent between 80 and 680 torr with a value of 1.81. This leads to a primary quantum efficiency of 0.90 for 16a + 16b. The same value was obtained at a wavelength of 228.8 nm.

Photolysis of OCS in the presence of olefins leads to production of both episulfides analogous to the addition reactions observed with atomic oxygen, $O(^3P)$, and to mercaptans analogous to the products of atomic oxygen, $O(^1D)$, reactions. Thus, the episulfides and mercaptans are identified as products of $S(^3P)$ addition and $S(^1D)$ insertion, respectively. A comparison of the yields of these products then indicates the fractional yield of these two atomic states in the primary photodissociation. In the photolysis of 200 torr of OCS with up to 1800 torr of propane, Knight et al. (102) showed that at least 74 percent of the sulfur atoms were $S(^1D)$ since they could be removed by insertion reactions. This leads, (103), to the following quantum yields for the two energetically allowable primary processes in this wavelength region: $\phi_{16a} \leq 0.24$ and $\phi_{16b} \geq 0.67$. Breckinridge and Taube (104) have studied the photochemistry of OCS in the presence of N_2O and conclude that the quantum efficiency for the production of $S(^1D)$ at 253.7 nm in the gas phase is 0.74 ± 0.04.

An investigation of the reaction products from the photodissociation of OCS in mixtures with olefins in both liquid and solid solution (102) indicates that 16b continues to participate as a primary process but with reduced quantum efficiency in comparison to the gas phase. It is suggested (102) that intersystem crossing between the singlet excited state of OCS and a repulsive triplet state is more important in condensed than in gas phases. On the other hand, Gollnick and Leppin (105) have studied the photolysis of OCS at 253.7 nm in solution with a variety of solvents, find the quantum efficiency of CO formation to be 0.90 ± 0.05 in all solvents, and suggest that process 16b is the only primary process.

There have been several contributions that have focused on the nature and energetics of the electronically excited states of OCS. An LCAO-MO-SCF calculation by Clementi (106) showed that the first excited states above the $^1\Sigma^+$ ground state arise from a $3\pi^3 4\pi^1$ (Π-Π^*) configuration from which six singlet and triplet sigma and delta states can be derived, the lowest being $^3\Sigma^+$. The very low f-value for the near ultraviolet transition suggests that the upper bound state of this transition is $^1\Delta$.

A semi-empirical MO calculation by Sidhu et al. (103) with the inclusion of d-orbitals of sulfur is in agreement that the near uv transition corresponds to absorption to a Π-Π^* state. If the upper state is $^1\Delta$, Leppin and Gollnick (107) suggest that $S(^1D)$ results from a predissociation via curve crossing to a $^1\Pi$ state correlating with $CO(^1\Sigma) + S(^1D)$. This appears to be the most important process in the gas phase at all pressures, while crossing to a repulsive $^3\Sigma$ state which correlates with $CO(^1\Sigma) + S(^3P)$ may participate in condensed phases. The negative results of a search for fluorescence or phosphorescence in low-temperature EPA glasses (107) suggests that the difference between a quantum efficiency of 0.9 and unity is due to radiationless processes.

Production of CS + O in primary process 17a is not energetically possible above 200 nm and no absorption from CS was found following the flash photolysis of OCS in this wavelength region (108). Thus 16a, b are believed to account for at least 90 percent of the primary photochemical processes above 200 nm.

2. Wavelengths shorter than 200 nm. Production of the second excited state of sulfur, $S(^1S)$, is energetically possible in process 16c at wavelengths shorter than 211 nm. Donovan (109) has observed this state in absorption at the shortest time delays possible following the flash photolysis of mixtures of Ar and OCS with wavelengths transmitted through LiF windows and

thus extending down to or somewhat below 110 nm. The production of $S(^1S)$ at wavelengths as long as 147 nm is indicated by Dunn et al. (110) who employed OCS photolysis at this wavelength as a $S(^1S)$ source for the purpose of kinetic measurements of its gas-phase reactivity. Recently, Black et al. (128) have employed a pulsed H_2 continuum and a vacuum monochromator to study the relative yield of $S(^1S)$ in the 110 to 170 nm wavelength range. By making time-resolved measurements of the forbidden transition $S(3^1S_0 \rightarrow 3^3P_1)$ at 458.9 nm and comparing these signal intensities to the $O(^1S)$ yield from N_2O at 129 nm, they find a yield near unity for $S(^1S)$ in the wavelength interval 142 to 160 nm. Their results are given in Fig. 12.

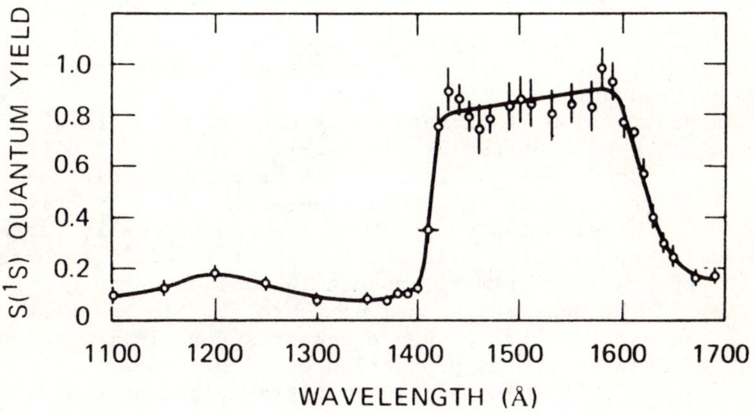

FIGURE 12. Quantum yield of $S(^1S)$ from the photodissociation of OCS with a wavelength resolution of approximately 2.5 nm (from Black et al.[128] with permission).

At wavelengths shorter than 196 nm, production of $CS(X^1\Sigma^+)$ is energetically possible and its absorption has been observed by Donovan (109) at long delays following the flash photolysis of OCS with wavelengths transmitted through LiF windows. The absorption was weak and Donovan estimates that production of CS in process 17a and/or 17b accounts for no more than 10 percent of the primary processes in the 110 to 200 nm wavelength interval employed. This estimate is presumably based on a sensitivity estimate for the absorption measurement coupled with failure to detect CS immediately after the light pulse. Very recent work by Klemm et al. (127) places a substantially lower upper limit upon the participation of 17a and/or 17b as a primary process. By the technique of flash photolysis and resonance fluorescence detection of both $O(^3P)$ and $S(^3P)$, an upper limit of 2 percent relative to the yield of $S(^3P)$ was obtained.

In addition to the $S(^1S)$ signal, Donovan (109) detected absorption due to vibrationally excited (v" = 2) CO and $S(^3P_2$, $^3P_1$ and $^3P_0)$, the latter in a non-Boltzmann distribution which does not appear as quickly following the flash lamp discharge as does the $S(^1S)$ absorption. This delayed appearance of the $S(^3P_J)$ atoms suggests that they are, at least in part, produced in reactions subsequent to the photolysis. This is confirmed by Donovan et al. (111) who studied the flash photolysis with added He as the buffer gas rather than with added Ar. The strong spectrum of $S(^3P_J)$ observed in the presence of Ar at short delay was virtually absent in the presence of He. Thus it seems that process 16a is not an important process at the wavelengths employed by Donovan.

The observation of vibrationally excited CO up to v" = 13 has recently been reported by Lin (129) in the vacuum ultraviolet (λ > 165 nm) photolysis of OCS. It is proposed by Lin that process 16b is responsible for production of the vibrationally excited CO. Wavelengths near 170 nm are energetically capable of producing CO up to v" = 13 and no higher levels were observed.

It appears that primary process 16 dominates the OCS photochemistry in the near and vacuum ultraviolet. In the near ultraviolet, the experimental evidence supports certain limits on the quantum efficiency of the production of $S(^1D)$, namely, $0.74 \leq \phi_{16b} \leq 0.90$. In the vacuum ultraviolet the quantum yield for production of $S(^1S)$ in process 16c rises from about 0.8 at 142 nm to 0.9 at 161 nm and is much lower, $\leq$ 0.2, at wavelengths outside this interval.

VI. CS_2

The photochemistry of carbon disulfide has been studied in only moderate detail. It has long been interesting to spectroscopists because of the complexity of its near ultraviolet absorption spectrum and the fact that its radiative lifetime is substantially longer than that expected on the basis of integrated oscillator strengths. Its fluorescence has been observed whereas that of the two other analogous 16 valence-electron molecules, CO_2 and OCS, has not.

A. Spectroscopy

The absorption spectrum of CS_2 has been studied frequently. References to earlier work are included in the review of Rabalais et al. (38). There are at least seven distinctive absorption regions which can be distinguished between the long wavelength limit of absorption in the visible and the ionization threshold near 123 nm. A photoelectric record obtained by Rabalais et al. (38) is reproduced in Fig. 13 and one obtained by Okabe (112) and extending to shorter wavelengths is given in Fig. 14.

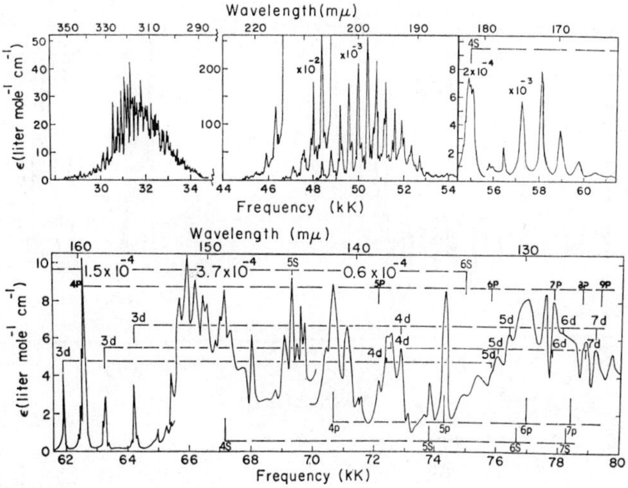

FIGURE 13. Molar absorptivity of CS_2 of photoelectric detection, defined as $(bc)^{-1} \log_{10} I_0/I$, where b is the path length in cm and c is the concentration in moles/liter. Reprinted with permission from J. W. Rabalais et al., Chem. Rev., 71, 73 (1971). Copyright by the American Chemical Society.

The assignments which have been made (38) for the non-Rydberg transitions are,

1. Above 350 nm, $f \cong 8 \times 10^{-5}$, ${}^1\Sigma_u^- \leftarrow {}^1\Sigma_g^+$
2. 290 to 350 nm, $f = 2.7 \times 10^{-4}$, ${}^1\Delta_u \leftarrow {}^1\Sigma_g^+$
3. 180 to 220 nm, $f = 1.1$, ${}^1\Sigma_u^+ \leftarrow {}^1\Sigma_g^+$
4. 160 to 180 nm, $f = 2.9 \times 10^{-2}$, ${}^1\Pi_g \leftarrow {}^1\Sigma_g^+$
5. 150 to 154 nm, $f = 4.4 \times 10^{-2}$, ${}^1\Pi_u(?) \leftarrow {}^1\Sigma_g^+$.

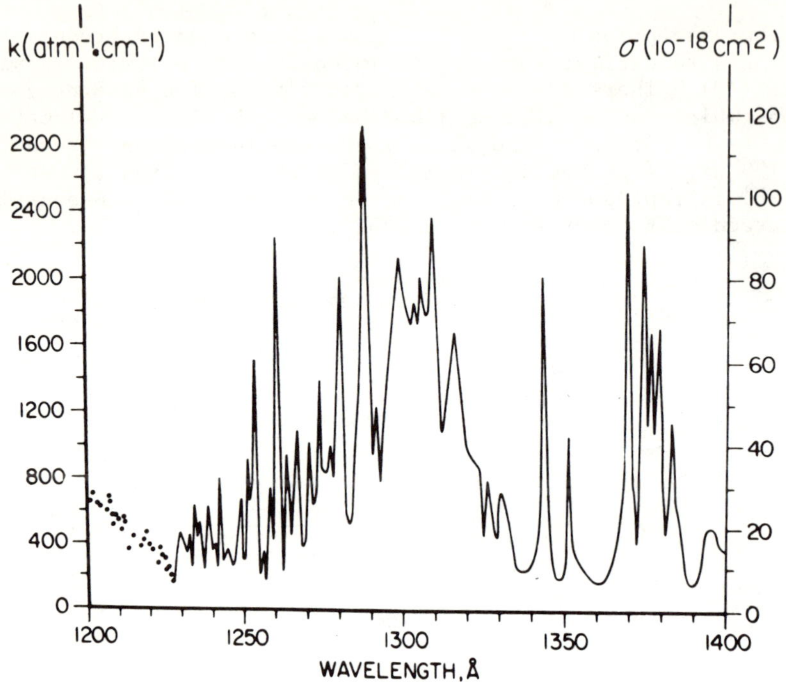

FIGURE 14. Absorption coefficients of CS_2 by photoelectric detection with an instrumental resolution of about 0.1 nm. Data taken at 0.1-nm intervals and expressed to the base e for 1 atm of gas (from Okabe[112] with permission).

Seven Rydberg series have been observed with the first member at 182 nm.

The energetics of the primary photodissociative processes which produce CS and S are presented in Table 7. Experimental evidence suggesting the participation of these processes is given below.

TABLE 7. Energetics of photodissociative primary processes for CS_2 with threshold wavelength > 100 nm.[a] $CS_2 \rightarrow CS + S$

CS ↓ \ S →		$^3P_2$	$^1D_2$	$^1S_0$	$^5S^o$	$^3S^o$	^5P
CS		(0.00)	(1.15)	(2.75)	(6.52)	(6.86)	(7.87)
$X^1\Sigma^+$	(0.00)	(4.01) 309.18	(5.16) 240.27	(6.76) 183.40	(10.53) 117.74	(10.87) 114.06	(11.88) 104.36
$a^3\Pi$	(3.42)	(7.43) 166.86	(8.58) 144.50	(10.18) 121.79	-	-	-
$A^1\Pi$	(4.81)	(8.82) 140.57	(9.97) 124.35	(11.57) 107.16	-	-	-

[a]Energies are listed within parentheses in electron volts; threshold wavelengths appear directly underneath in nanometers. Assumes a value of 4.01 ± 0.25 eV[1] for the bond dissociation enthalpy of CS_2 at 0°K, ΔH_0^o; energies of states of S are taken from ref. 99, energies of $CS(a^3\Pi)$ and $CS(A^1\Pi)$ are taken from ref. 100, and energies of $CS(a'^3\Sigma^+)$ and $CS(e^3\Sigma^-)$ have been determined to be ≤ 4.82 and ≤ 4.85 eV, respectively (101).

B. Photochemistry

1. Wavelengths longer than 290 nm. Photodissociation of CS_2 is not energetically possible for wavelengths longer than 309 nm and does not appear to occur at wavelengths longer than 290 nm. The absorption spectrum at wavelengths longer than 290 nm has been studied at high resolution by Klemen (113). It appears to consist of two systems representing transitions from the ground state to two or more excited states. One system lies in the region from 380 to 330 nm and the other in the region from 330 to 290 nm. It is only in the latter that there is any direct evidence of the fate of the excited molecule produced in absorption.

The fluorescence of CS_2 excited at eight wavelengths between 288 and 346 nm was studied by Heicklen (114). He observed that the emission efficiency was wavelength dependent and concluded that collisional deactivation of electronically excited CS_2

occurred on every collision. From measurements of the integrated absorption coefficient a lifetime of 3 μs was calculated. This can be compared with the measured value of 15 μsec obtained by Douglas (115). The difference is attributed by Douglas to interelectronic level mixing which leads to an anomalously long experimental lifetime in comparison to that expected on the basis of the absorption spectrum. This problem has been discussed by Bixon and Jortner (116) who attribute the level mixing to intramolecular vibronic interactions arising from favorable Franck-Condon factors despite the low density of vibronic states in CS_2.

2. Wavelengths between 190 and 290 nm. There have been many studies of the photochemistry of CS_2 by flash photolysis in the wavelength region above the quartz transmission limit near 200 nm. Porter (117) reported the observation of vibrationally excited CS in early work in 1950. Callear and Norrish (118) observed CS with up to four vibrational quanta and studied its relaxation. Callear (119) found that three-fourths of the CS produced in the primary process is produced in vibrational levels higher than v" = 2.

While both processes 18a and 18b are energetically possible at these wavelengths,

$$CS_2(X^1\Sigma_g^+) + h\nu \rightarrow CS(X^1\Sigma^+) + S(^3P,\ ^1D), \qquad (18a,b)$$

Callear observed only absorption due to $S(^3P_2)$. This negative evidence for process 18b cannot be considered definitive however in view of the rapidity with which $S(^1D)$ would be deactivated by CS_2. The photolysis of CS_2 in mixtures with various other gases was studied by deSorgo et al. (120) in two different wavelength intervals: 195 to 225 nm, and greater than 230 nm. The long wavelength (>230 nm) continuous photolysis of CS_2 alone or with added ethylene or propane yielded only a solid polymeric material. This would appear to exclude the participation of process 18a in the wavelength interval covering the 320 nm bands and is in agreement with the earlier work which reported only fluorescence in this wavelength interval. The shorter wavelength photolysis (195-225 nm) led to the production of ethylene episulfide when ethylene was present, however no volatile product was produced if propane was the added gas. This is entirely consistent with the participation of only process 18a at these wavelengths since the expected mercaptan products of $S(^1D)$ insertion reactions are absent.

3. Wavelengths shorter than 190 nm. The single study of the CS_2 photochemistry at wavelengths below 190 nm is that of Okabe (112) who observed the production of $CS(A^1\Pi)$ for wavelengths shorter than 133.7 nm,

$$CS_2(X^1\Sigma_g^+) + h\nu \rightarrow CS(A^1\Pi) + S(^3P). \quad (19)$$

This wavelength corresponds closely to the threshold energy for the process, which is 140 nm. Since the efficiency for production of $CS(A^1\Pi)$ parallels the Rydberg structure in the CS_2 absorption spectrum in the wavelength interval below 133.7 nm and above the ionization limit, it is suggested that the dissociation comes from these states rather than from a repulsive state and proceeds with high quantum efficiency.

VII. C_3S_2, CS AND C_2O

A. C_3S_2

Carbon subsulfide is a red liquid of low volatility at room temperature. The ground electronic state is $^1\Sigma$ as is that of C_3O_2. Its absorption spectrum has not been extensively investigated above 200 nm and is not known at all below 200 nm. The photoelectric absorption spectrum obtained by Robertson and Merritt (121) in the gas phase and in solution is reproduced in Figs. 15 and 16. No information is available regarding the photochemistry of C_3S_2.

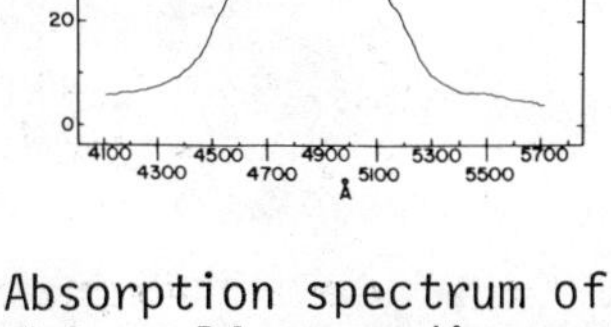

FIGURE 15. Absorption spectrum of C_3S_2 solution obtained with a Cary 14 recording spectrophotometer (from Robertson and Merritt[121] with permission).

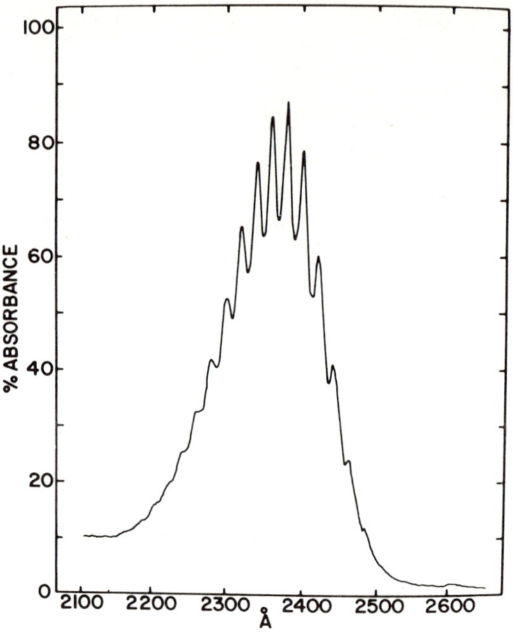

FIGURE 16. Near-ultraviolet absorption spectrum of C_3S_2 vapor obtained with a Cary 14 recording spectrophotometer (from Robertson and Merritt[121] with permission).

B. CS

Carbon monosulfide is one of the interstellar molecules, and its electronic spectroscopy has been studied in some detail. The $A^1\Pi \rightarrow X^1\Sigma^+$ system, which is analogous to the CO Fourth Positive system has been observed in emission (122) and the lifetime of $v' = 0$ measured as 176 ± 14 ns (123).

The emission spectrum of $CS(a^3\Pi)$, the system analogous to the CO Cameron system, has been observed by Tewarson and Palmer (100) in a low-pressure flame. There is no direct evidence to suggest that CS is photodissociated at any wavelength in its absorption spectrum. The ease with which it may be produced, however, and the fact that major absorption systems lie in the relatively long wavelength ultraviolet are not likely to permit this situation to prevail for long.

C. C_2O

Optical absorption due to C_2O has been described by Jacox et al. (17) in condensed phase and by Devillers (124) in the gas phase. The spectrum of the $A^3\Pi_i \rightarrow X^3\Sigma$ system has been analyzed by Devillers and Ramsay (8).

C_2O is known to be a primary product of the photolysis of C_3O_2 as described above. In similar fashion it is to be expected that C_2S will be a primary product of the C_3S_2 photolysis. From the dependence of $C(^1S)$ concentration on flash intensity, Braun et al. have suggested that metastable atomic carbon is formed in a secondary process rather than as a primary photoproduct in the photolysis of C_3O_2. They have suggested photodissociation of C_2O as a reasonable process. Photodissociation of C_2O to produce $CO(X^1\Sigma)$ and $C(^3P_2)$, $C(^1D)$, and $C(^1S)$ would have wavelength thresholds of approximately 380, 270, and 210 nm respectively.

In a recent study of the kinetics of some reactions of $C(^1S)$, Husain and Kirsch (125) have observed that as CO is added during the flash photolysis of C_3O_2, the $C(^1S)$ yield first decreases then increases. This latter increase is attributed to the three-body recombination of $C(^3P)$ and CO followed by photodecomposition of C_2O with the production of $C(^1S)$.

It seems likely that the photochemistry of C_2O will be investigated more thoroughly in the near future as will that of C_2S on which there is presently no information.

ACKNOWLEDGMENT

The author would like to thank Dr. T. Carrington and Dr. L. J. Stief for their helpful criticism.

REFERENCES

(1) D. R. Stull, H. Prophet, Ed., "JANAF Thermochemical Tables," 2nd ed., NSRDS-NBS 37, U.S. Government Printing Office, 1971.

(2) H. W. Thompson, N. Healey, Proc. Roy. Soc. (London), A157, 331 (1936).

(3) K. D. Bayes, J. Am. Chem. Soc., 84, 4077 (1962).

(4) H. H. Kim and J. L. Roebber, J. Chem. Phys., 44, 1709 (1966).

(5) J. L. Roebber, J. C. Larrabee, and R. E. Huffman, J. Chem. Phys., 46, 4594 (1967).

(6) S. Bell, T. S. Varadarajan, A. D. Walsh, P. A. Warsop, J. Lee, and L. Sutcliffe, J. Mol. Spectrosc., 21, 42 (1966).

(7) P. H. Krupenie, "The Band Spectrum of Carbon Monoxide", NSRDS-NBS 5, U. S. Government Printing Office, 1966.

(8) C. Devillers and D. A. Ramsay, Can. J. Phys., 49, 2839 (1971).

(9) K. D. Bayes, J. Am. Chem. Soc., 85, 1730 (1963).

(10) W. L. Wiese, M. W. Smith, and B. M. Glennon, "Atomic Transition Probabilities", Vol. I, Hydrogen Through Neon, NSRDS-NBS 4, U. S. Government Printing Office, 1966.

(11) K. D. Bayes, J. Am. Chem. Soc., 83, 3712 (1961).

(12) C. MacKay, P. Polack, H. E. Rosenberg, and R. Wolfgang, J. Am. Chem. Soc., 84, 308 (1962).

(13) R. B. Cundall, A. S. Davies, and T. F. Palmer, J. Phys. Chem., 70, 2503 (1966).

(14) A. Forchioni and C. Willis, J. Phys. Chem., 72, 3105 (1968).

(15) T. Morrow and W. D. McGrath, Trans. Faraday Soc., 62, 3142 (1966).

(16) L. J. Stief and V. J. DeCarlo, J. Chem. Phys., 43, 2552 (1965); J. Am. Chem. Soc., 91, 839 (1969).

(17) M. E. Jacox, D. E. Milligan, N. G. Moll, and W. E. Thompson, J. Chem. Phys., 43, 3734 (1965).

(18) N. G. Moll and W. E. Thompson, J. Chem. Phys., 44, 2684 (1966).

(19) W. Braun, A. M. Bass, D. D. Davis, and J. D. Simmons, Proc. Roy. Soc., A312, 417 (1969).

(20) E. Tschuikow-Roux, Y. Inel, S. Kodama, and A. W. Kirk, J. Chem. Phys., 56, 3238 (1972).

(92) D. J. Strickland and T. M. Donahue, Planet. Space Sci., 18, 661 (1970).

(93) M. B. McElroy and J. C. McConnell, J. Atmos. Sci., 28, 879 (1971).

(94) T. M. Donahue, J. Atmos. Sci., 28, 895 (1971).

(95) M. B. McElroy and T. M. Donahue, Science, 177, 986 (1972).

(96) R. G. Prinn, J. Atmos. Sci., 29, 1004 (1972).

(97) M. B. McElroy, N. D. Sze, and Y. L. Yung, J. Atmos. Sci., 30, 1437 (1973).

(98) F. M. Matsunaga and K. Watanabe, J. Chem. Phys., 46, 4457 (1967).

(99) W. L. Wiese, M. W. Smith and B. M. Miles, "Atomic Transition Probabilities", Vol. II, Sodium Through Calcium, NSRDS-NBS 22, U. S. Govt. Printing Office, 1969.

(100) A. Tewarson and H. B. Palmer, J. Mol. Spectrosc., 27, 246 (1968).

(101) R. F. Barrow, R. N. Dixon, A. Lagerqvist, and C. V. Wright, Ark. Fys., 18, 543 (1960).

(102) A. R. Knight, O. P. Strausz, S. M. Malm, and H. E. Gunning, J. Am. Chem. Soc., 86, 4243 (1964).

(103) K. S. Sidhu, I. G. Csizmadia, O. P. Strausz, and H. E. Gunning, J. Am. Chem. Soc., 88, 2412 (1966).

(104) W. H. Breckinridge and H. Taube, J. Chem. Phys., 53, 1750 (1970).

(105) K. Gollnick and E. Leppin, J. Am. Chem. Soc., 92, 2217, 2221 (1970).

(106) E. Clementi, J. Chem. Phys., 36, 750 (1962).

(107) E. Leppin and K. Gollnick, Mol. Photochem., 2, 177 (1970).

(108) W. D. McGrath, J. J. McGarvey, and D. N. Dempster, Can. J. Chem., 45, 2454 (1967).

(109) R. J. Donovan, Trans. Faraday Soc., 65, 1419 (1969).

(110) O. J. Dunn, S. V. Filseth and R. A. Young, J. Chem. Phys., 59, 2892 (1973).

(111) R. J. Donovan, L. J. Kirsch, and D. Husain, Nature, 222, 1164 (1969).

(112) H. Okabe, J. Chem. Phys., 56, 4381 (1972).

(113) B. Klemen, Can. J. Phys., 41, 2034 (1963).

(114) J. Heicklen, J. Am. Chem. Soc., 85, 3562 (1963).

(115) A. E. Douglas, J. Chem. Phys., 45, 1007 (1966).

(116) M. Bixon and J. Jortner, J. Chem. Phys., 50, 3284 (1969).

(117) G. Porter, Proc. Roy. Soc. (London), A200, 284 (1950).

(118) A. B. Callear and R. G. W. Norrish, Nature, 188, 53 (1960).

(119) A. B. Callear, Proc. Roy. Soc. (London), A276, 401 (1963).

(120) M. deSorgo, A. J. Yarwood, O. P. Strausz, and H. E. Gunning, Can. J. Chem., 43, 1886 (1965).

(121) L. C. Robertson and J. A. Merritt, J. Chem. Phys., 56, 5428 (1972).

(122) S. Bell, T. L. Ng, and C. Suggitt, J. Mol. Spectrosc., 44, 267 (1972).

(123) S. J. Silvers and C. Chiu, J. Chem. Phys., 56, 5663 (1972).

(124) C. Devillers, C. R. Acad. Sci. Paris, 262c, 1485 (1966).

(125) D. Husain and L. J. Kirsch, J. Photochem., 2, 297 (1973/74).

(126) J. D. Simmons and S. G. Tilford, J. Mol. Spectrosc., 49, 167 (1974).

(127) R. B. Klemm, S. Glicker, and L. J. Stief, Chem. Phys. Lett., 33, 512 (1975).

(128) G. Black, R. L. Sharpless, T. G. Slanger, and D. C. Lorents, J. Chem. Phys., 62, 4274 (1975).

(129) M. C. Lin, Chem. Phys., 1, 433 (1975).

Photolysis of Saturated Alcohols, Ethers, and Amines

CLEMENS VON SONNTAG AND HEINZ-PETER SCHUCHMANN
Institut für Strahlenchemie, Max-Planck-Institut
für Kohlenforschung
D-4330 Mülheim a.d. Ruhr
Stiftstraße 34-36
West Germany

CONTENTS

I. INTRODUCTION

The scope of this article is the photolysis of saturated alcohols, ethers, acetals, and amines. These compounds are transparent in the near ultraviolet, and except for the amines they noticeably begin to absorb only near 200 nm. Although this wavelength region does not present serious experimental difficulties, as evidenced by the wealth of material available on aliphatic hydrocarbons, the knowledge of the photolytic behavior of this class of saturated compounds bearing heteroatoms is comparatively limited. Hence, it seems worthwhile to give an account on its present state in order to stimulate research in this field.

In the context of a review on the direct photolysis dealing with alcohols, ethers, acetals, and amines, it could be argued, carbohydrates should be included also. It seems to us, however, that the material available on carbohydrate photolysis (1-20) indicates the involvement of the open-chain carbonyl form and/or carbonyl function-bearing impurities, and it is, therefore, left out of consideration.

Furthermore, photosensitized reactions, except those sensitized by mercury and rare gases, will not be dealt with here. In contrast to other photosensitized systems, many of the secondary processes occurring in the Hg-photosensitized decompositions are identical to those in the direct photolyses, and the conclusions drawn from the former are often applicable to the latter.

Molecular oxygen forms charge transfer complexes with the compounds in question, giving rise to strong absorptions at longer wavelengths than do the oxygen-free compounds. The photochemistry of these CT-complexes is also a subject of this review.

II. HISTORICAL

A. Photolysis of Alcohols

The liquid phase photolysis of several alcohols was first systematically studied in the years between 1910 and 1913 by Berthelot and Gaudechon using a medium pressure mercury arc (21). They recognized that the main photolysis products of primary and secondary alcohols were H_2 and an equivalent amount of aldehyde and ketone, respectively (21a,b). They also found (21d) that the part of the ultraviolet spectrum active on alcohols must lie below 250 nm, while carbonyl compounds are readily decomposed by light around 250 nm and above, yielding much CO. It was concluded (21d,22) that CO was a secondary product of the alcohol photolyses and due to the photolysis of the primarily produced carbonyl compound. In aqueous solutions of the alcohols the rate of decomposition was lower (21d) but led to the same products as those found in the neat photolyses. Similar to the liquid phase results Bates and Taylor (1927) found H_2 as the major product in the methanol and ethanol gas phase photolysis (23).

In 1929 Paneth and Hofeditz (24) proved the existence of free radicals, and Rice and co-workers (1934) realized that such species play an important role in many thermal decomposition reactions (25). Hence the concept that free radicals might be equally important in photochemical reactions gained ground. The mechanism of a photochemical reaction would thus consist of a number of steps rather than a single step (Bonhoeffer (26), 1934). In particular, the study of alcohol photolysis was taken up again in this new light, and efforts were made to determine whether hydrogen was formed in a molecular, or in a free radical process. Patat (1933-1935) investigated the methanol and ethanol gas phase photolysis and attempted unsuccessfully to establish the participation of free H atoms in these systems using the para-hydrogen method (27,28). The molecular hydrogen produced in the ethanol photolysis was found (29) to be an equilibrium mixture of para- and ortho-H_2, as one would have expected. Terenin and Neujmin (1935) spectroscopically detected the OH radical in the methanol and ethanol photolysis (30,31). Farkas and co-workers (1937, 1939) showed that liquid and solid phase alcohol photolysis yielded the same primary products, H_2 and the aldehyde (32,33); the formation of dehydrodimers was later (1958) reported by Pfordte and Leuschner (34-36). Employing D_2O as a solvent Farkas and Hirschberg found that the hydrogen evolved had a 50 percent deuterium content, and they concluded that a C-H and the O-D bond were broken concurrently (32).

The photolysis of deaerated aqueous solutions of various polyols was initiated after 1910 by Berthelot and Gaudechon (5e), and by Henri and co-workers (4a,37,38,10). Polyols such as glycerol and mannitol formed products reducing Fehling's solution, that is, glyceraldehyde (37) and, in general, aldoses (4a,10,11, 39). Their photolysis yielded H_2 and CO in the ratio 1:1. This result (5a) is in good accord with what one would expect if the primary products are sugars which in turn undergo photolysis from their carbonyl forms. α-Diols were found also to produce water; for example, ethylene glycol formed water and acetaldehyde (40).

B. Photolysis of Ethers

The photolysis of dimethyl ether (gas phase) and diethyl ether (liquid phase) was first undertaken by Berthelot and Gaudechon (21c). The main products detected were hydrocarbons (methane from dimethyl ether, ethane from diethyl ether) with some CO and H_2. Ethers were later shown also to form dehydrodimers (41).

C. Photolysis of Amines

Amine photolysis, just as that of alcohols and ethers, seems to have been studied for the first time by Berthelot and Gaudechon (1911) who discovered that piperidine produced H_2 on ultraviolet irradiation (42). Bates and Taylor (1927) established that in the gas phase the main product of the direct photolysis of ethylamine was H_2 (23). Further studies on the gas phase photolysis of methylamine and ethylamine showed that products were H_2, methane, ammonia, N_2, ethane, and a nonvolatile liquid (43,44). The fact that in the presence of O_2 the H_2 yield decreased by one order of magnitude suggested the participation of free H atoms.

D. Hg-Sensitized Photolysis

The possibility of mercury-photosensitized reactions was first predicted in 1922 by Franck (45) and experimentally verified in the sensitized decomposition of H_2 by Cario and Franck (46). They found that free H atoms were produced when a mixture of Hg vapor and H_2 was irradiated with the 254 nm Hg resonance line at room temperature. Bates and Taylor (23) studied the Hg-sensitized decomposition of methanol, ethanol, and ethylamine, and showed that with these compounds the rate of the sensitized reaction was faster than the direct photolysis by about two orders of magnitude. Again hydrogen was the major product. Aldehydes were formed from the alcohols.

Wetmore and Taylor (47) investigated the methylamine photolysis and found the decomposition rate of the Hg-photosensitized reaction equal to that of the direct one where a plug of gold wire was interposed between the vacuum system and the photolysis cell to exclude Hg. It seems possible though that Hg was present in both cases. Photolyses carried out with the light of low-pressure Hg arc lamps were, at least in part, Hg-sensitized if the contamination of the photolysis vessels by the Hg vapor present in the vacuum system was not rigorously excluded. This holds true especially for the early work.

E. Photolysis of O_2-CT Complexes

The phenomenon of the concurrent action of light and air on organic compounds, leading under certain conditions to the formation of substances that are more powerful oxidants than oxygen itself, has interested chemists for a long time. Peroxidic compounds were obtained from alcohols (48-56) and ethers (57-64) through the action of ultraviolet and/or visible light in the presence of O_2. It is certain that for the most part these reactions did not reflect the photochemistry of O_2-CT complexes but were sensitized by carbonyl compounds and other impurities. This must be especially true for aqueous and neat alcohols since it was practically impossible at the time to ensure the absence of such impurities.

In a thorough study of the diethyl ether autoxidation carried out in 1924, Clover (59) demonstrated that all experimental facts could be adequately explained on the assumption that the ether hydroperoxide, $CH_3CH(OOH)-OC_2H_5$, is formed. Under the influence of light or moisture the hydroperoxide gives rise to alcetaldehyde which then further catalyses the autoxidation, even in the dark. Besides carbonyl compounds, traces of Fe^{3+} can be a powerful photooxidation catalyst (62). Diethyl ether carefully purified of Fe^{3+} showed only negligible formation of peroxides in the daylight.

Studies on acetals and amines (65) are much less numerous. Formaldehyde diethyl acetal was found to form a peroxide of the composition $C_5H_{12}O_4$ (60). Paraldehyde gave rise to peroxides faster than did alcohols or ethers in the presence of oxygen and ultraviolet light (66). A theoretical approach to the problem of photooxidation was attempted in 1929 by Milas (67). Not only atoms, but also molecules could add oxygen. In particular, a molecule carrying lone pair electrons might under suitable conditions form a dative bond to an oxygen molecule. Bond formation would be eased if the donor molecule was energized in its lone pair:

$$D: \xrightarrow{h\nu} D^*:$$

$$D^*: + \ddot{\underset{..}{O}}:\ddot{\underset{..}{O}}: \longrightarrow D:\ddot{\underset{..}{O}}:\ddot{\underset{..}{O}}:$$

A rearrangement of this dative peroxide could then lead to ordinary peroxide. A thought along similar lines had already been expressed in 1897 by Bach (68).

III. UV ABSORPTION SPECTRA

Saturated alcohols begin to absorb near 200 nm, with the maximum of this first, unstructured band near 185 nm. This band has usually been ascribed to an $n \rightarrow \sigma^*$ transition (69,70) but is assigned by some to a Rydberg transition $n \rightarrow 3s$ (71). Gas phase values of ε_{max} (near 185 nm) are listed in Table 1. The absorption coefficient at shorter wavelengths, increasingly determined by the alkyl part of the molecule, tends to be the larger, the longer and the more highly branched the alkyl chain (73,79), paralleling the order of the ionization potentials (80).

It has been claimed that the molecular extinction coefficient of methanol vapor is increased in the presence of noble gases (75). Solvents often strongly influence the extinction coefficient of alcohols. In aqueous solutions of methanol, ethanol, and isopropanol ε_{185} rises with increasing dilution, methanol and ethanol showing slight maxima at molar fractions x_{ROH} near 0.01 (77). The extinction coefficient at 185 nm of t-butanol in aqueous solution, however, remains constant ($\varepsilon_{185} \simeq 90$) over the entire concentration range (78). Mixtures of t-butanol and isopropanol also exhibit ideal behavior at 185 nm: $\varepsilon_{mix} = x_{i\text{-}PrOH}\, \varepsilon_{i\text{-}PrOH} + x_{t\text{-}BuOH}\, \varepsilon_{t\text{-}BuOH}$ (78). In mixtures with hydrocarbons the rise of ε with increasing dilution can be considerable (81-83). These phenomena reflect the change in the state of association due to H-bonding.

TABLE 1. Values of ε ($M^{-1}cm^{-1}$) near 185 nm for some saturated alcohols.[a]

	Vapor (ε_{max})	Neat liquid (ε_{185})
MeOH	150 (72), 155 (73), 171 (74), 356 (70), 610 (75)	6 (76), 8 (77)
EtOH	300 (73), 320 (72), 326 (74), 670 (70)	
n-PrOH	240 (72,73), 248 (74)	
i-PrOH	560 (73), 600 (74), 620 (72)	32 (78)
n-BuOH	330 (73)	
i-BuOH	450 (73)	
t-BuOH	1150 (73)	90 (78)

[a] References in parentheses.

The first ultraviolet band of saturated ethers begins around 200 nm (e.g., cyclic ethers except ethylene oxide), or slightly below (e.g., open-chain ethers and ethylene oxide). It often shows structure and tends to be more intense than the first band of the alcohols, with which it shares assignment to an $n \rightarrow \sigma^*$ (70) or an $n \rightarrow 3s$ transition (84). Extinction coefficients (ε_{185}) of some open-chain and cyclic ethers are shown in Table 2. Further analyses of spectra have been presented: dimethyl ether (93,94) diethyl ether (95,96), ethylene oxide (97), oxetane (98), tetrahydrofuran (98,99), tetrahydropyran (98,100), and p-dioxane (98). Ethers absorb so strongly near 185 nm that the extinction coefficient of the neat

TABLE 2. Values of ε_{185} of some saturated ethers and acetals (vapor).[a]

	$\varepsilon_{185}(M^{-1}cm^{-1})$		$\varepsilon_{185}(M^{-1}cm^{-1})$
Me_2O	2000 (73)	ethylene oxide	<100 (90)
	2200 (85)		
Et_2O	1200 (73)	oxetane	1200 (90)
	1500 (86)		1500 (91)
	1800 (85)		
	2600 (70)		
n-Pr_2O	1300 (73)	tetrahydrofuran	600 (84)
			700 (92)
i-Pr_2O	500 (73)	tetrahydropyran	850 (92)
n-Bu_2O	1000 (73)	2,2-dimethyloxetane	900 (91)
t-Bu_2O	2200 (87)	p-dioxane	4500 (92)
MeO-n-Pr	1600 (88)	1,3-dioxane	<100 (101)
MeO-n-Bu	4000 (89)	$CH_2(OMe)_2$	50 (103a)
MeO-t-Bu	200 (86)	$CH_3CH(OMe)_2$	50 (103b)
		$C(CH_3)_3CH(OMe)_2$	400 (104)

[a] References in parentheses.

liquid cannot be measured at those wavelengths except from very thin films. However, open-chain ethers apart from showing more structure in the gas phase seem to behave quite similarly in both the liquid and gas phase, judging from the fact that both give ε-values identical within experimental error in those wavelength regions where both have been measured (86-89). With 1,4-dioxane in fluorocarbon solution, a slight blue shift of the absorption spectrum relative to the vapor has been observed (183).

Only very few examples of acetal ultraviolet absorption spectra are known. With 1,3-dioxane (101), 1,3-dioxolane (102), formaldehyde dimethyl acetal (103a), acetaldehyde dimethyl acetal (103b), and pivalaldehyde dimethyl acetal (104), it is remarkable that absorption sets in below 200 nm, that is, at shorter wavelengths than ethers, and ε_{185} is lower (Table 2). This could be a consequence of the fact that the ionization potentials of saturated acetals are quite high (1,3-dioxane, 10.15 eV; 1,4-dioxane, 9.47 eV; (101)).

Saturated alkylamines start to absorb at considerably longer wavelengths than do alcohols and ethers. The limit is shifted particularly far towards the red in tertiary amines, as shown by a few examples (Table 3). Absorption sets in at remarkably short wavelength (225 nm) in the case of ethyleneimine (97). This behavior parallels that of ethylene oxide. Spectra of azacyclobutane (97), pyrrolidine (84), and some six-membered cyclic amines (106) have been described. The absorption bands of bicyclic tertiary amines (107-110) show pronounced structure, in contrast to tertiary amines such as trimethylamine, triethylamine, and N,N,N',N'-tetramethylethylenediamine. This is attributed to the rigid, cagelike configuration of the bicyclic compounds. The first absorption band of aliphatic amines is assigned to an $n \rightarrow \sigma^*$ state (70), to an $n \rightarrow 3s$ state (105), or a combination of both (111,112). The first excited state configuration of methylamine (113) and t-butylamine (114) is planar.

TABLE 3. Methyl and ethylamines in the gas phase. Wavelength (nm) where $\varepsilon = 10\ M^{-1}cm^{-1}$ is reached (105).

$MeNH_2$	Me_2NH	Me_3N	$EtNH_2$	Et_2NH	Et_3N
245	247	264	242	252	265

There are solvent effects on the optical properties of amines. The first absorption band undergoes a marked blue shift when amines are mixed with H donor liquids (115) (Table 4). With alcohols, 1:1 amine/alcohol complexes seemed to be formed.

Starting in the early fifties, the concept of the CT complex was developed largely by Mulliken and his co-workers (116).

TABLE 4. Solvent effects on the first absorption band of ethylamines; relative extinction coefficients (115)

Solvent	$EtNH_2$(λ 238nm)	Et_2NH(λ 248nm)	Et_3N(λ 250nm)
Gas phase	1	1	1
Isoöctane	0.89	0.44	0.51
Diethyl ether	1	0.55	0.69
t-Butanol	0.071	0.105	0.30
Methanol	0.023	0.024	0.039
Water	0.014	0.024	0.021

Spectral evidence that O_2 also forms CT complexes was subsequently obtained for a variety of aromatic compounds, and solvents such as cyclohexane (117). The ground state strength of CT complexes with O_2 is very low even with very strong electron donors. They are a type of contact CT complex (118).

CT absorption bands are found with alcohols (119-121), ethers (120,122), and amines (120,123). Their onset is well to the red when compared with the oxygen free compounds, and with tertiary amines reaches just into the visible near 400 nm (123,124).

IV. PHOTOLYSIS OF ALCOHOLS

A. Primary and Secondary Alcohols in the Condensed Phase

When GC had become available as a powerful tool for the analysis of trace components, photolysis experiments with saturated alcohols were resumed by Yang and his group (125,126). They investigated methanol, ethanol, isopropanol (125), and n-pentanol (126). Photolysis became negligible when the 185 nm line of the Hg low-pressure arc was filtered off by means of a vycor quartz tube. This is as expected from the characteristics of alcohol spectra, and is in contrast to the findings of Sullivan and Koski who observed ESR signals from samples of methanol (289a) and ethanol (289b) at 77 K on irradiation at λ 254. In the earlier work conversion of the starting material was still too high and considerable photolysis of the products occurred, thus preventing firmer conclusions regarding the primary processes to be reached. It was shown by von Sonntag (127-129) that at conversions below 0.1 percent the yield/dose curves become linear. Product quantum yields have been determined for the 185 nm irradiation. It has been recognized that with methanol (127), ethanol (128), and isopropanol (129) hydrogen is the most prominent gaseous product (Table 5). The corresponding products are dehydrodimers and carbonyl compounds. Their relative yields vary with the nature of the compound studied (Table 5). Products which arise from C-O and C-C bond fragmentation are of minor importance. A number of O-deuterated n-alcohols and six differently deuterated isopropanols have been photolyzed, and the $H_2/HD/D_2$, CH_4/CH_3D, or CD_3H/CD_4 ratios determined (127,128,130-132). The results are compiled in Table 6. In the series of O-deuterated alcohols the HD content always exceeded 80 percent of the total hydrogen. Virtually no D_2 was formed. This implies that the major primary processes always involve the hydrogen bound to oxygen. This may either occur through a molecular process forming HD and the

TABLE 5. Quantum yields of key products in the UV photolysis (λ 185 nm) of liquid primary and secondary alcohols.

	MeOH (127)	EtOH (128)	i-PrOH (129)
$\phi(H_2)$	0.83	0.84_5	0.75
ϕ(dehydrodimer)	0.78	0.30_5	0.30_6
ϕ(carbonyl)[a]	0.05_8	0.54	0.72
ϕ(methane)	0.05	0.01	0.04_6
ϕ(ethane + ethylene)	0.002	0.014	0.002
ϕ(propane + propene)	-	-	0.03

[a]Aldehyde or ketone of the same carbon number as the starting alcohol.

Table 6. Relative yields of $H_2/HD/D_2$, CH_4/CH_3D, and CD_3H/CD_4 in the photolysis (λ 185 nm) of deuterated alcohols in the liquid phase. Deuteration $\geq$ 96% (127,128,130-132)

Alcohol	H_2	HD	D_2	CH_4	CH_3D	CD_3H	CD_4
CH_3OD	14.3	84.7	1.0	98.5	1.5		
CH_3-CH_2-OD	10.5	88.8	0.7	47	53		
$CH_3-CH_2-CH_2-OD$	16.3	82.9	0.8				
$CH_3-CH_2-CH_2-CH_2-OD$	11.4	87.2	1.4				
$CH_3-CH_2-CH_2-CH_2-CH_2-OD$	18.7	80.6	0.7				
$(CH_3)_2CHOD$	14.1	85.0	0.9	54	46		
$(CH_3)_2CDOH$	24.4	75.0	1.6	94	6		
$(CH_3)_2CDOD$	2.3	25.2	72.5	46	54		
$(CD_3)_2CHOH$	91.0	9.0	<0.1			81.5	18.5
$(CD_3)_2CDOH$	8.5	88.6	2.9			76	24

$$R_1R_2CH\text{-}OD \xrightarrow{h\nu} HD + R_1R_2C{=}O \qquad (1)$$

$$R_1R_2CH\text{-}OD \xrightarrow{h\nu} R_1R_2CH\text{-}O^{\bullet} + D^{\bullet} \qquad (2)$$

$$D^{\bullet} + R_1R_2CH\text{-}OD \longrightarrow HD + R_1R_2\dot{C}\text{-}OD \qquad (3)$$

corresponding carbonyl compound (reaction 1), or by homolytic cleavage of the O-D bond (reaction 2) followed by abstraction of a carbon bound hydrogen, most probably that in α-position to the hydroxyl group (reaction 3). The infrared spectrum of $^{\bullet}CH_2OH$ was observed in the 147 nm photolysis of methanol at $14^{o}K$ (290).

In the photolysis of O-deuterated alcohols some (about 15 - 20%) H_2 is also formed. This could be caused by elimination of carbon bound hydrogen atoms (e.g., reaction 4), or the formation of molecular hydrogen and a carbene. Carbenes have been shown to be intermediates in the photolysis of hydrocarbons

$$R_1R_2CHOD \xrightarrow{h\nu} H^{\bullet} + R_1R_2\dot{C}OD \qquad (4)$$

(e.g., 133,134). However, judging from the small yield of D_2 in the photolysis of hexadeuteroisopropanol (Table 6) this process appears to be unimportant in the alcohols. With ethanol and isopropanol, elimination of molecular hydrogen and formation of an enol is another possible route (e.g., reaction 5). Because of

$$CH_3\text{-}CH_2OH \xrightarrow{h\nu} H_2 + CH_2{=}CHOD \qquad (5)$$

intermolecular elimination of molecular hydrogen encountered in the photolysis of t-butanol (135,136,83) (see below), another process leading to hydrogen and eventually to the corresponding carbonyl compound as depicted in reactions 6 and 7 must be envisaged. The hemiacetal [1] is expected to decompose into the

$$R_1R_2CHOH^{*} + R_1R_2CHOH \longrightarrow H_2 + HC(R_1)(R_2)\text{-}O\text{-}C(R_1)(R_2)\text{-}OH \qquad (6)$$

[1]

$$[1] \longrightarrow R_1R_2CHOH + R_1R_2C{=}O \qquad (7)$$

alcohol and the carbonyl compound (reaction 7). Therefore, reaction 6 cannot be distinguished from reaction 1.

Another route of hydrogen elimination occurs with the formation of epoxides. It has been shown (137) that only those alcohols give epoxides in which the hydrogen bound to the oxygen is suitably positioned relative to the hydrogen on the β-carbon atom. This is the case, for example, with isopropanol but not with n-propanol where the hydroxyl group can orient itself such as to yield to steric repulsion by the alkyl chain (c.f. the Newman projections [2] and [3]).

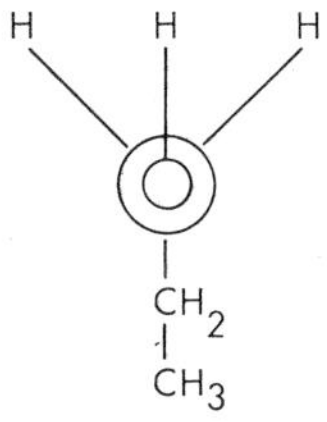

n-propanol
[2]

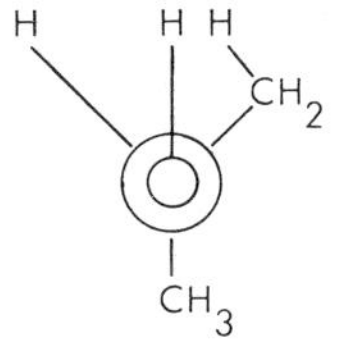

i-propanol
[3]

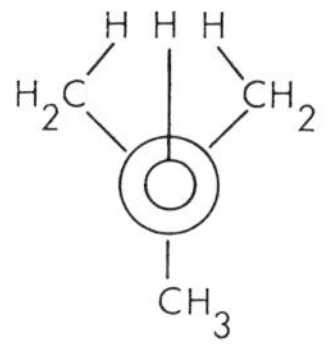

t-butanol
[4]

The most prominent example is t-butanol (see below). In this molecule [4] only β-hydrogens but no α-hydrogens are available for an intramolecular hydrogen elimination process.

In reaction 2 an alkoxyl radical is formed. The alkoxyl radicals are rather reactive and react with the alcohols by hydrogen abstraction to give the alcohol and an α-hydroxyalkyl radical (reaction 8). For example, the rate constant for the reaction of the methoxyl radical with methanol (138,139) is $5 \times 10^{5}\ M^{-1}s^{-1}$ (139). Hence, under the experimental conditions of the 185 nm liquid phase photolysis the alkoxyl radicals practically do not react with other radicals.

$$R_1R_2CH\text{-}O^{\bullet} + R_1R_2CH\text{-}OH \longrightarrow R_1R_2CHOH + R_1R_2\dot{C}OH \qquad (8)$$

The yield of the carbonyl compound gives an upper limit for the molecular process 1. The disproportionation reaction of the α-hydroxyalkyl radicals (reaction 9) yields further aldehyde or ketone. Since in all cases studied other radicals play a minor role in the radical reactions the yield of this additional carbonyl is

$$2R_1R_2\dot{C}\text{-OH} \longrightarrow R_1R_2C{=}O + R_1R_2CHOH \qquad (9)$$

$$2R_1R_2\dot{C}\text{-OH} \longrightarrow R_1R_2C(OH)\text{-}R_1R_2C\text{-OH} \qquad (10)$$

determined by the ratio k_9/k_{10}. These values are not sufficiently well established. It appears that the hydroxymethyl (from methanol) mainly dimerizes, and that less than 10 percent of these radicals disproportionate (140). Methyl substituents favor the disproportionation reaction. A value of $k_9/k_{10}=0.43$ has been reported for the α-hydroxyethyl radical (141). The k_9/k_{10} value for the 2-hydroxypropyl-(2) radical has been assumed at around 16* (129) but other determinations suggest that it is more likely between 4 and 7 (143,144). More recently a value of 4.4 has been reported (145).

In the disproportionation reactions of the hydroxyalkyl radicals enols are also formed as intermediates. Apparently reactions 11 and 12 both occur (145-147).

$$2\ \begin{array}{c} CH_3 \\ \cdot| \\ C\text{-OH} \\ | \\ CH_3 \end{array} \longrightarrow \begin{array}{c} CH_3 \\ | \\ C{=}O \\ | \\ CH_3 \end{array} + \begin{array}{c} CH_3 \\ | \\ HC\text{-OH} \\ | \\ CH_3 \end{array} \qquad (11)$$

$$2\ \begin{array}{c} CH_3 \\ \cdot| \\ C\text{-OH} \\ | \\ CH_3 \end{array} \longrightarrow \begin{array}{c} CH_2 \\ \| \\ C\text{-OH} \\ | \\ CH_3 \end{array} + \begin{array}{c} CH_3 \\ | \\ HC\text{-OH} \\ | \\ CH_3 \end{array} \qquad (12)$$

The scission of a C-C bond is negligible in the case of ethanol (1.5%) and of minor importance in the case of isopropanol (5.5%). Studies with deuterated isopropanols dissolved in undeuterated isopropanol indicate that the C-C bond is largely broken by the molecular processes 13 and 14 and only to a minor extent by homolytic scission (reaction 15) (132).

---

*This value has been obtained from radiolysis experiments with N_2O-saturated aqueous solutions of isopropanol. More recent studies indicate that a radical chain reaction yielding acetone occurs in this system (142) and is expected to account, in part, for this very high value.

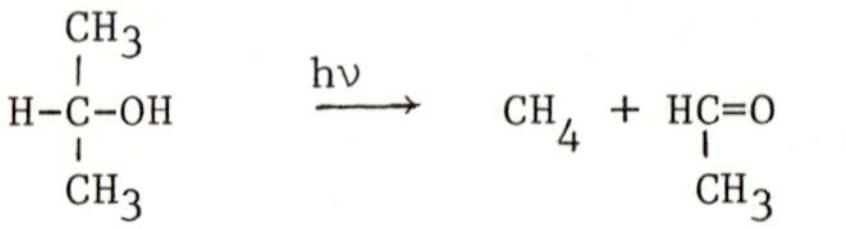

(13)

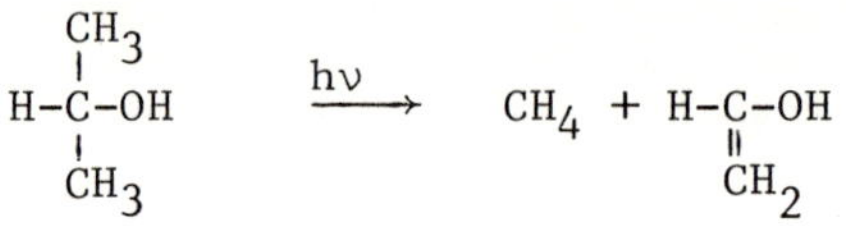

(14)

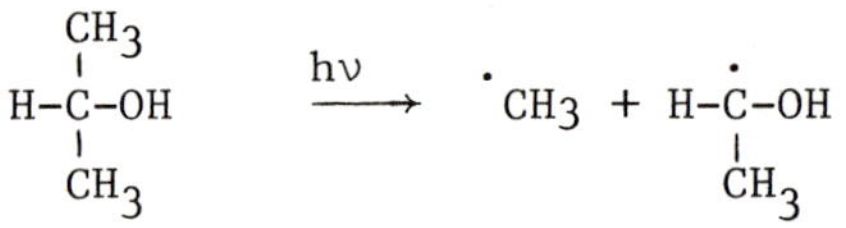

(15)

The scission of the C-O bond (reactions 16 and 17) does not contribute in these alcohols to more than about 6 percent.

$$R_1R_2CHOH \xrightarrow{h\nu} R_1R_2\dot{C}H + {}^{\bullet}OH \quad (16)$$

$$R_1R_2CHOH \xrightarrow{h\nu} \text{Olefin} + H_2O \quad (17)$$

In Schemes 1-3 the photolyses of liquid methanol, ethanol and isopropanol are summarized. They are derived using data on quantum yields, rate constants of radical reactions, and data from deuterium labeling experiments. The evaluation of the quantum yields of the molecular hydrogen-eliminating processes is somewhat uncertain in the case of ethanol (mainly through the uncertainty in k_9/k_{10}), and is problematic in the case of isopropanol, where the value k_9/k_{10} is even less certain, and where ϕ(dehydrodimer)=0.036 can only be obtained by extrapolation. It must be noted that even at low conversion (< 0.1%) acetone is

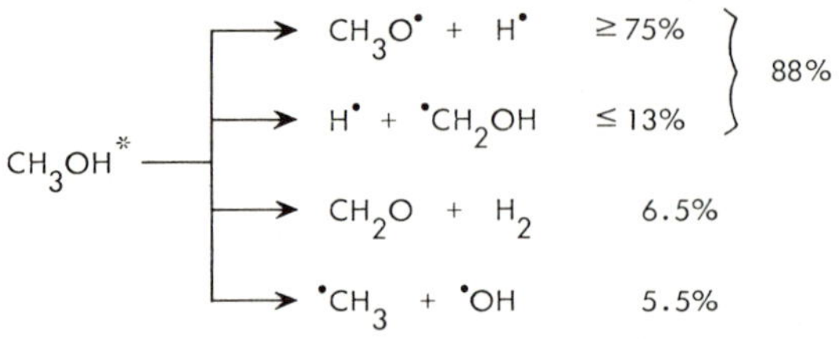

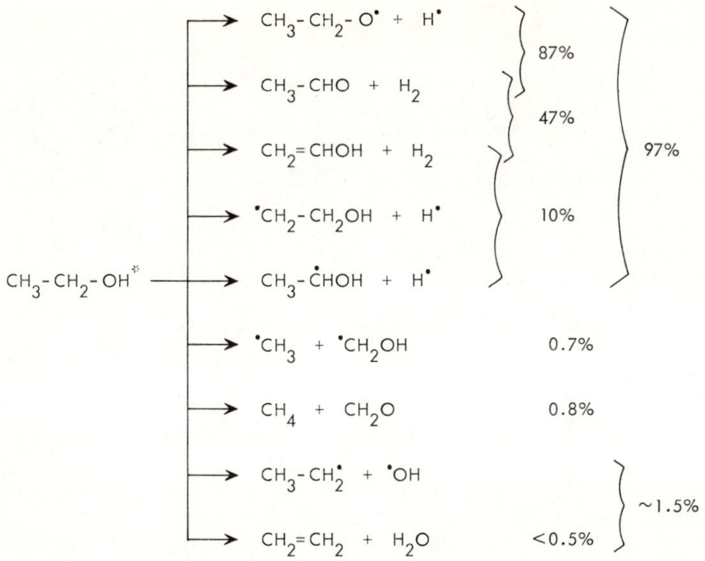

increasingly photolyzed causing an apparent increase in ϕ(dehydrodimer). However, accepting the lower values of k_9/k_{10} for the 2-hydroxy propyl-(2) radicals, the values given in Schemes 1-3 indicate that the elimination of molecular hydrogen increases from methanol (~ 6%) and ethanol (~ 47%) to isopropanol (~ 60%).

B. Tertiary Alcohols in the Liquid Phase

In the series of tertiary alcohols only t-butanol has been studied so far. That this compound differs quite drastically in its photolytic behavior from the primary and secondary alcohols was already suggested by the preliminary experiments of Yang and co-workers (125). They showed that C-C fragmentation is most prominent. A later study by Sänger and von Sonntag (136) confirmed these results and gave a more detailed analysis. The products and their quantum yields are given in Table 7. The photolysis of O-deuterated t-butanol yielded 95 percent HD and 5 percent H_2. The methane was about 80 percent CH_4 and 20 percent CH_3D. The decomposition pathways derived for the photolysis of t-butanol are shown in Scheme 4. The predominant process clearly is the scission of the C-C bond. Homolytic cleavage of this bond has been demonstrated but molecular processes also

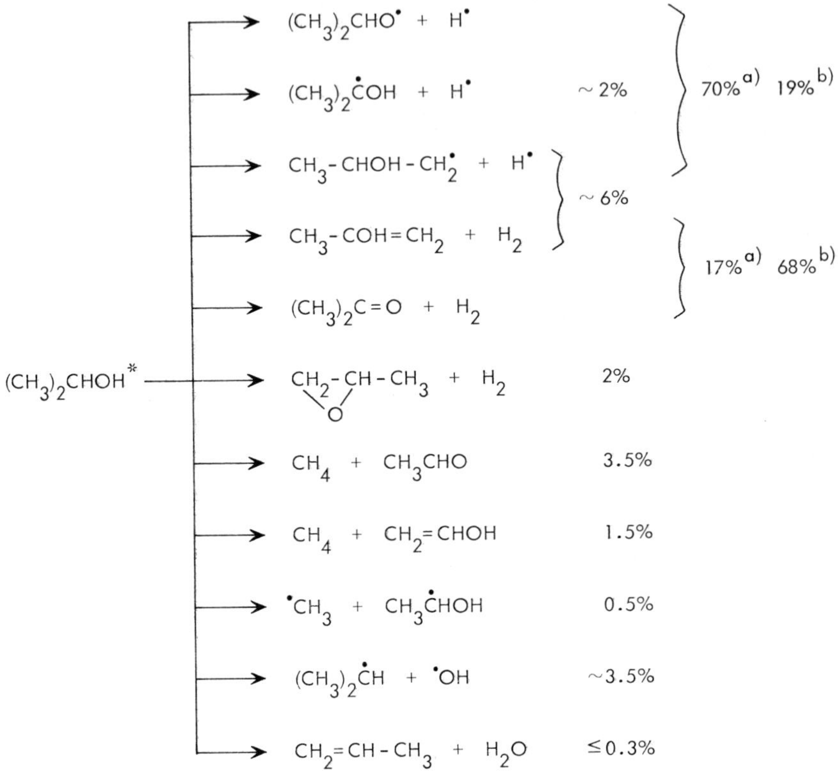

appear to play a major role. Some of the CH_3D in the photolysis of t-BuOD originates most likely in such a process. Molecular methane elimination processes have been well established in the isopropanol case (132). Material balance considerations and the experiments with t-BuOD suggest that free hydrogen atoms contribute to the hydrogen yield only to a marginal extent. Hydrogen is formed only intramolecularly (process 18), and intermolecularly (process 19) involving two neighboring molecules (135,136,83). In this process the ether [5] is formed. Free radicals do not appear to be intermediates in this process as

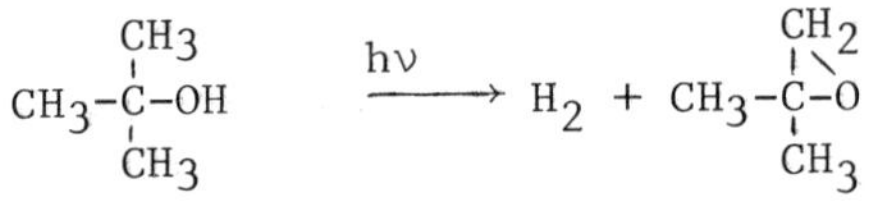

(18)

TABLE 7. Products and quantum yields of the 185 nm photolysis (25°) of deaerated liquid t-butanol (136).

Product	ϕ	Product	ϕ
H_2	0.11	$CH_3CHOHCH_3$	0.057
CH_4	0.27	$(t\text{-}Bu)_2O$	nil
C_2H_6	0.013	t-AmOH	0.05
$i\text{-}C_4H_8$	0.013	$(CH_3)_3C\text{-}C(CH_3)_2OH$	0.003
$i\text{-}C_4H_{10}$	0.015	$(CH_3)_3C\text{-}O\text{-}CH_2\text{-}C(CH_3)_2OH$	0.03
$C(CH_3)_4$	0.005	$(CH_3)_3C\text{-}CH_2\text{-}C(CH_3)_2OH$	0.012
t-Bu-O-Me	nil	$(CH_3)_2C(OH)\text{-}C(CH_3)_2OH$	0.015
Isobutene oxide	0.08	$[(CH_3)_2C(OH)]_2CH_2$	0.032
CH_3COCH_3	0.21	$[(CH_3)_2C(OH)CH_2]_2$	0.044

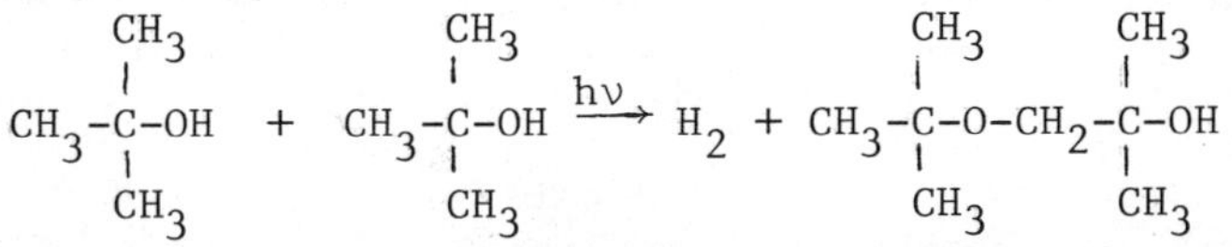

(19)

[5]

shown by dilution experiments with radical scavengers such as isopropanol and hexane (135). The intermolecular photoelimination of small molecules appears to be a more general phenomenon (148,149).

C. Solvent Effects

1. Alcohols in Water. The intermolecular H_2 elimination process observed in the 185 nm photolysis of liquid t-butanol (135,136) already suggests a strong interaction between the excited alcohol molecule and its surroundings. As early as 1913 Berthelot and Gaudechon (21d) reported that the hydrogen yield from pure ethanol was higher than that of its aqueous solutions. Similar phenomena

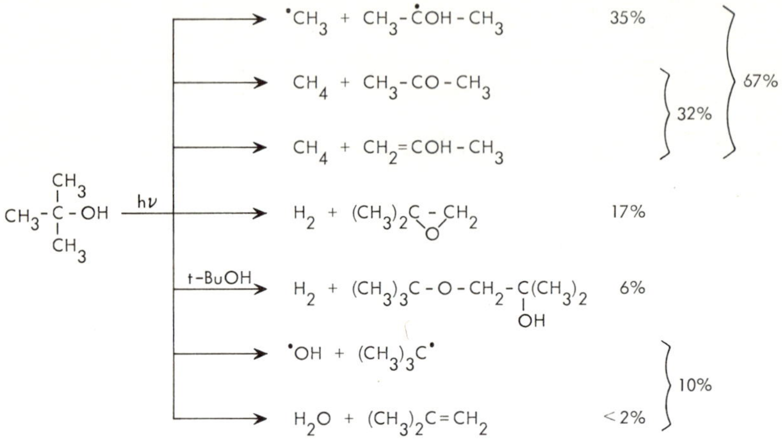

have been observed by Sokolov and Stein (150) with isopropanol. More extensive studies on the effect of water on the quantum yields of a larger series of products in the 185 nm photolysis of methanol (127), ethanol (128), isopropanol (129), and t-butanol (78,82) have revealed that the effect is not a mere quenching of the excited state. Thus, in the cases of methanol, ethanol, and isopropanol the hydrogen yield decreases linearly on dilution with water. With methanol, $\phi(H_2)=0.83$ (neat) is reduced to $\phi(H_2)=0.42$ (1M), with ethanol, $\phi(H_2)0.84$ (neat) to $\phi(H_2)=0.32$ (2M). The corresponding values for isopropanol are 0.75 and 0.26. The dehydrodimer and carbonyl yields follow, in sum, the hydrogen yields, but individually take a different course. Although a change in k_9/k_{10} with solvent composition is conceivable with these polar radicals it is thought not to be the only reason for this behavior, which may reflect, to some extent, changes in the contribution of free radical and molecular processes. Most striking is the effect on the scission of the C-O bond measured by the methane yield from methanol. $\phi(CH_4)$ drops from 0.05 (neat) to zero (1M), at first steeply, then more gradually. Similar results have been obtained for the ethane and ethylene yield in the case of ethanol (128). A strong effect of water as a solvent has also been found in the case of t-butanol. There, the solvent obviously affects the mode of methane formation (82).

The photolysis of ethylene glycol has been investigated in aqueous solution only, in order to avoid photolysis of the products facilitated by their slow diffusion out of the reaction layer in the viscous neat material. The quantum yields as reported by van der Linde and von Sonntag (151) are given in Table 8. The photolysis of 1,1,2,2-d_4-ethylene glycol yielded 91 percent HD, 8.5 percent H_2, and 0.5 percent D_2. On increasing the photolysis temperature from 22° to 70° the quantum yields of all products remained essentially the same except that of acetaldehyde which was increased by a factor of 6.5.

TABLE 8. Quantum yields (λ 185 nm) of products from deoxygenated 2M aqueous solutions of ethylene glycol at 22 and 70° (151).

Product	Quantum yield at 22°	Quantum yield at 70°
Hydrogen	0.20_4	0.19
Carbon monoxide	0.0[a]	Not detd.
Formaldehyde	0.19_4	0.21
Methanol	0.07_4	0.06_5
Glycolaldehyde	0.08	0.10
Acetaldehyde	0.02	0.13
Glycerol	0.06	0.05_3
Erythritol	0.03	0.03
3,4-Dihydroxybutanal	0.01	0.01_4
Succinaldehyde	0.001	0.003

[a] Value obtained by extrapolation to zero dose.

The free radical chemistry of ethylene glycol is somewhat complicated by the water elimination reaction from the 1,2-dihydroxyethyl radical (reaction 20) formed by hydrogen abstraction from the substrate. This reaction was first observed in 1966 in ESR experiments by Livingston and Zeldes (152),

$$\cdot CHOH\text{-}CH_2OH \longrightarrow CHO\text{-}\dot{C}H_2 + H_2O \qquad (20)$$

Voevodskii and co-workers (153), as well as Norman and his group (154). Later it was confirmed on the basis of product analytical studies by Seidler and von Sonntag (155) who suggested that the formylmethyl radical may initiate a chain reaction (reaction 21). A more detailed study of this system by von Sonntag and

$$CHO\text{-}\dot{C}H_2 + CH_2OH\text{-}CH_2OH \longrightarrow CH_3CHO + \cdot CHOH\text{-}CH_2OH \qquad (21)$$

Thoms (156), Burchill and Perron (157), as well as Venter, van der Linde, and Basson (158) confirmed the chain processes 20 and 21. Hence, the temperature effect on ϕ(acetaldehyde) is due to this chain reaction.

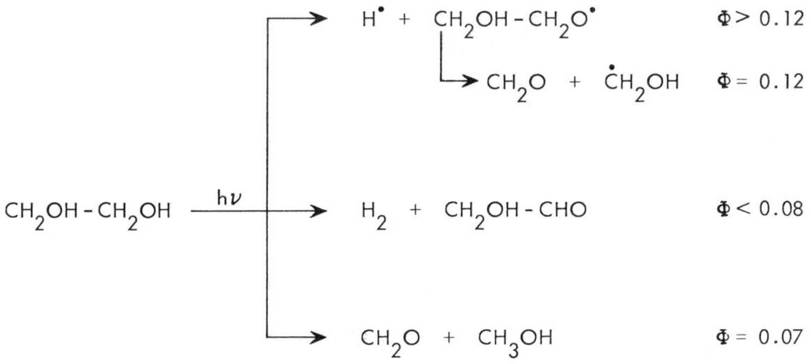

It has been shown that one of the major processes is the scission of an O-H bond. The resulting radical is unstable and largely fragments into formaldehyde and a hydroxymethyl radical (Scheme 5). The same reaction has been proposed to account for some products in the photolysis of 2-methoxyethanol (159). The formation of molecular hydrogen together with glycolaldehyde, as well as the fragmentation into formaldehyde and methanol (Scheme 5) were postulated as molecular processes in order to account for the products and their quantum yields.

2. The Farkas Actinometer (5M Ethanol in Water). In the Farkas actinometer the hydrogen evolved from 5M aqueous ethanol is measured. The Farkas actinometer appears to be very reliable for the 185 nm line emitted by the Hg low pressure arc.

The other major line at 254 nm contributes to the H_2 evolution only negligibly (128).

There is a long controversy regarding the correct value of its hydrogen formation quantum yield. Farkas and Hirshberg (32) give a value of 0.8 based on earlier work on chloroacetic acid (160) which itself was based on the decomposition of ammonia, $\phi(-NH_3)=0.25$ (161). The experimental results of Farkas and Hirshberg were reevaluated by Bernas et al. (162) who derived a quantum yield of 0.63. By an independent method Hagège and Vermeil (163) obtained a value of 0.6 ± 0.1. Getoff and Schenck (164) corrected the Farkas actinometer by substituting the old value of $\phi(-NH_3)=0.25$ for the ammonia decomposition by more recent ones ($\phi(-NH_3)=0.14$ to 0.18 (165)). Using the mean value of $\phi(-NH_3)=0.16$ they calculated for the Farkas actinometer a value of $\phi(H_2)=0.5$ but did not take into account the corrections which had been made by Bernas et al. (162). With these corrections a value of $\phi(H_2)=0.4$ is obtained. The same value of $\phi(H_2)=0.4$ results if the Farkas actinometer is based on $\phi(N_2)=1.0$ from the photolysis of N_2O in aqueous solution as has been proposed by Dainton and Fowles (166).

The value of $\phi(H_2)=0.4$ appears to be the most reliable. The higher values including 0.5 which are still currently used would give ϕ(sum of primary processes) > 1 for many systems such as N_2O (166), methanol (127), ethanol (128), isopropanol (129), di-t-butyl ether (87) and methyl-n-propyl ether (88), an unreasonable result. However, using $\phi(H_2)=0.4$, ϕ(sum of primary processes) approaches or equals but never exceeds unity. If normalized to the same actinometer quantum yield value, results from different laboratories, for example, the H_2 yield of ~1M methanol in water (cf. 127,166-168), agree within 10 percent.

It has been reported that $\phi(H_2)$ of the Farkas actinometer is strongly susceptible to temperature change (162). This must be an artefact. If the cell is thermostated separately and the Hg low-pressure arc is kept at constant temperature no change of the hydrogen yield should be observed in the range from 0 to 40° (128). Thus the Farkas actinomter deserves its position as a reliable actinometer as long as the ethanol used is free of impurities (162). The hydrogen evolved is most easily measured by a technique wherein the hydrogen formed is swept directly into the gas chromatograph by the carrier gas (169).

3. Alcohols in Hydrocarbons. There is not much material available on the photolysis of the primary and secondary alcohols in hydrocarbon solvents with the exception of a study on isopropanol (129), in which it was found that $\phi(H_2)$ remained unaltered or was slightly increased if isopropanol was diluted by n-hexane. The

HD content of the hydrogen formed (> 80% HD) from $(CH_3)_2CHOD$ remained constant over the whole range investigated, whereas that from $(CH_3)_2CDOH$ dropped from slightly below 80 to about 20 percent in the dilute system containing 15 mol % $(CH_3)_2CDOH$. These findings are in agreement with the above interpretation that the isopropanol molecule eliminates hydrogen by a molecular process (giving HD in both cases independent of solvent (H-donor) concentration, and/or splits the O-H (O-D) bond. The liberated H (D) atom then reacts with either the alcohol or the solvent, the importance of which reactions depending on the relative concentrations of the two components. However, as is more clearly shown in the case of t-butanol, one cannot rule out a drastic change in the photolytic behavior of the alcohol going from the neat alcohol to its hydrocarbon solutions. For this reason, it is not possible to assess the extent of molecular HD elimination in the neat alcohol (c.f. uncertainties discussed above, Scheme 3) from the value extrapolated to infinte dilution in the hydrocarbon system.

The t-butanol/hydrocarbon system has attracted more interest (135,82,83). As has been shown above neat t-butanol in its photolysis differs markedly from the primary and secondary alcohols in that the formation of hydrogen is not the major process. Rather, it is the scission of the C-C bond. On dilution with a hydrocarbon solvent, however, $\phi(H_2)$ increases and $\phi(CH_4)$ (which is an approximate measure for the C-C split) decreases strongly (82). More detailed studies have been undertaken in dilute solutions (0.04 to 1.16 mol/l) (83) in order to try and correlate the photolytic behavior of t-butanol, and the marked change in the absorption coefficient of t-butanol with dilution, and its state of aggregation in the hydrocarbon solvent. It is known that t-butanol forms various aggregates, but opinions as to their structure often differ; for example, trimer (170) or tetramer (171) species are considered to predominate if concentrations are not too low.

D. Alcohols in the Gas Phase

Some qualitative aspects of the gas phase photolysis of alcohols were recognized by Harrison and Lake (172). The gas phase photolysis of methanol has been studied in more detail by Porter and Noyes (173), by Hagège et al. (75,174,175), and more recently by Herasymowych and Knight (176a), who also studied isopropanol (176b).

In the case of methanol there is general agreement on the products and their relative yields. Hydrogen and ethylene glycol are the most important, followed by formaldehyde, methane, and water. Carbon monoxide has always been found but, rather than being a primary product, it probably stems from the photolysis of the product formaldehyde. The product distribution is moderately dependent on the photolyzing wavelength (λ 185,147, 124 nm) (75). With ethylene as a hydrogen atom scavenger the hydrogen yield is reduced to about the level of formaldehyde formation, that is, about 20 percent. The photolysis of CD_3OH (173,174) affords largely HD (∿71-76% HD, ∿20-25% H_2, and ~ 4% D_2). The formation of D_2 is fully suppressed by the addition of oxygen as a hydrogen atom scavenger. The rate of product formation has been shown to be pressure dependent up to about 70 nm Hg at 185, 147, and 124 nm in photolysis vessels of approximately 20 cm pathlength (75,173-176a). There is, however, disagreement about the yields at higher methanol pressures. Hagège et al. (75,174,175) report a further increase in the rate of product formation with increasing methanol pressure (up to 120 nm Hg); Herasymowych and Knight (176a) indicate that there is a noticeable decline.

In an extensive study (75,175) Hagège et al. showed that the rate of product formation is promoted not only by methanol itself but also by other gases as additives. These ranged from saturated hydrocarbons and rare gases to N_2 and CO_2. Data are given in Table 9, which shows that carbon dioxide and butane were the most powerful foreign additives to promote the photolytic decomposition of methanol. A more recent reinvestigation (176a) of this system strongly contradicts these results, and reports that carbon dioxide suppresses the rate of hydrogen formation to a residual yield of only 20 percent. The reviewers cannot judge from the published data how these discrepancies could arise.

The gas phase photolysis of methanol can be described on the basis of the reactions 22 through 25.

$$CH_3OH \xrightarrow{h\nu} CH_3O^{\bullet} + H^{\bullet} \tag{22}$$

$$\xrightarrow{h\nu} H^{\bullet} + {}^{\bullet}CH_2OH \tag{23}$$

$$\xrightarrow{h\nu} CH_2O + H_2 \tag{24}$$

$$\xrightarrow{h\nu} {}^{\bullet}CH_3 + {}^{\bullet}OH \tag{25}$$

The fate of the radicals is thought to be the same as discussed above for the liquid state. In the gas phase the H atoms produced in reactions 22 and 23 are expected to abstract hydrogen from the methyl and to some extent from the OH group. Hot hydrogen atoms might play a role. In the liquid phase hydrogen atom abstraction from the OH group is of even smaller likelihood than

TABLE 9. Gas phase photolysis of methanol. Relative efficiency of additives to raise the pressure-dependent hydrogen quantum yields (175).

Colliding partner	Relative efficiency	Colliding partner	Relative efficiency
CH_3OH	1.00	Kr	0.32
CO_2	0.82	CH_4	0.20
C_4H_{10}	0.82	N_2	0.14
Xe	0.75	SF_6	0.13
C_3H_8	0.72	Ar	0.10
C_2H_6	0.40		

in the gas phase as hydrogen bonding lowers the energy of the hydroxyl proton. In agreement with this, scarcely any D_2 is formed in the photolysis of CH_3OD in the liquid phase. Reaction 22 is believed to predominate over reaction 23. Reaction 24 contributes about 20 percent (hydrogen quenching studies), reaction 25 only 1 to 5 percent. These results agree well with those obtained for the liquid phase except that in the liquid phase the contribution of the molecular process 24 is definitely smaller.

Some very interesting features of the photolysis of methanol in the gas phase have been put forward by Hagège et al. (75,174, 175) (as mentioned before, some of their results, and hence their interpretation, have been questioned recently (176a)). The product quantum yields vary with methanol pressure. Hagège et al. rationalized their results on the basis of eq. 26 where ϕ_∞ is the quantum yield

$$\phi^{-1} = \phi_\infty^{-1} + kp^{-1} \tag{26}$$

at infinite pressure, k a constant, and p the methanol pressure. ϕ_∞ as well as k differs with the energy of the exciting photon. The equation holds for λ 185, 147, and 124 nm. The effect is thought to be due to a collision-induced predissociation of the excited methanol molecule. The effect of foreign gases seems to back this interpretation. It is noted that even molecules not containing heavy atoms are very efficient in promoting the photodecomposition of methanol. In these cases the perturbation cannot be attributed to spin-orbit coupling and is thought to be due, at least in part, to the breakdown of molecular symmetry thus allowing a symmetry-forbidden predissociation. If one allows also for a special spin-orbit term for the heavy rare gases Kr and Xe, a linear relationship holds between the transition probability and the polarizability of the added foreign gases (175).

The photolysis of 2-propanol has been studied by Herasymowych and Knight (176b) using a Hg medium-pressure arc. The effective wavelengths were in the 200 to 180 nm region. Results are reported for 30 and 80°. At the latter temperature higher pressures could be used, and in the 200 torr region the rate of product formation tended to become independent of pressure. It was thought that under these experimental conditions the effective wavelengths of 200 to 180 nm are fully absorbed by the substrate. Using N_2O as an actinometer for this region approximate quantum yields have been calculated for the products given in Table 10. Pinacol has been found to be a further product but its quantitative determination was not possible. Yield/dose plots indicate that at least CO and ethane might originate from product photolysis. CO_2 and CF_4 have been added as foreign gases in order to study the deactivation of thermally excited species. Increasing pressures of CO_2 reduce H_2 and acetone asymptotically to a smaller value, and completely inhibit acetaldehyde formation above p_{CO_2} > 800 torr whereas methane remains constant, and CO increases slightly. With CF_4, however, the acetone and acetaldehyde yields remain practically constant whereas formation of methane and CO is completely inhibited beyond p_{CF_4} > 1500 torr.

TABLE 10. Approximate quantum yields in the photolysis (λ 180-200 nm) of isopropanol vapor (200 torr at 80°) (176b).

Product	Quantum Yield
Hydrogen	0.64
Methane	0.39[a]
Acetone	0.34
Acetaldehyde	0.29
Carbon monoxide	0.15[a]
Ethane	0.08[a]

[a]Probably secondary product.

The photolysis of isopropanol in the gas phase (176b) resembles that in the liquid phase in so far as hydrogen formation is the most important process. The data given in Table 10 do not represent a good material balance and are, therefore, insufficient to draw firm conclusions. There is a quantum yield deficit of products deriving from C_3H_7O, corresponding to about 0.2 for products like acetone or pinacol. Moreover, the products methane and ethane seem partly (carbon monoxide, wholly) secondary. The results obtained in the presence of foreign gases elude interpretation, and experiments with deuterated material have not been carried out. In view of these difficulties a detailed photolysis mechanism cannot be put forward at this time.

V. PHOTOLYSIS OF ETHERS

A. Open-Chain Ethers in the Liquid Phase

Open-chain ethers were investigated in 1968 in the group of Yang (126). Preliminary results were obtained with n-butyl methyl ether, methyl n-pentyl ether, and 1,2-diethoxyethane. More recently, detailed studies were presented by the reviewers' group on diethyl ether (177), methyl n-propyl ether (88), t-butyl methyl ether (86), di-t-butyl ether (87), and 2-methoxyethanol (159), and more restricted studies (178) on ethyl n-propyl ether, ethyl isopropyl ether, isopropyl methyl ether, and n-butyl methyl ether.

The most important products are in all cases the alcohols ROH and R'OH if the ether is designated as R-O-R'. Further key products (see below) are the aldehydes in primary ethers and ketones in secondary ethers. The quantum yields of these key products and that of H_2 are tabulated in Table 11. For the quantum yields of further products that often exceed 30 in number see references given.

Excepting di-t-butyl ether (see below), the most important primary process is the homolytic C-O bond scission (reactions 27 and 28).

$$\text{R-O-R}' \xrightarrow{h\nu} \text{R}\cdot + {}^{\cdot}\text{O-R}' \qquad (27)$$

$$\text{R-O-R}' \xrightarrow{h\nu} \text{R-O}^{\cdot} + {}^{\cdot}\text{R}' \qquad (28)$$

The alkoxyl radicals ($\text{RO}^{\cdot}$ and $\text{R'O}^{\cdot}$) abstract a hydrogen atom from another ether molecule giving the alcohol (ROH and R'OH) and alkoxyalkyl radicals (e.g., reaction 29). The alcohol can also be formed

TABLE 11. UV photolysis (λ 185 nm) of liquid ethers (R-O-R'). Key products and their quantum yields.

Ref.	R-O-R'	R-OH	R'OH	R(>=O)	R'(>=O)	H_2
(177)	Et-O-Et	0.46		0.06		0.06_5
(88)	Me-O-nPr	0.16	0.70	0.004	0.13	0.019
(178)	Me-O-iPr	0.16	0.40	0.005	0.06	0.06
(178)	Me-O-nBu	0.08	0.44	0.006	0.12	0.13
(86)	Me-O-tBu	0.40_5	0.20	0.01_2	-	0.019_5
(87)	tBu-O-tBu	0.84		-		0.005
(178)	Et-O-nPr	0.31	0.28	0.07	0.02	0.04
(178)	Et-O-iPr	0.26	0.25	0.05	<0.01	0.07

$$CH_3\text{-}CH_2\text{-}O^{\bullet} + CH_3\text{-}CH_2\text{-}O\text{-}CH_2\text{-}CH_3 \rightarrow CH_3\text{-}CH_2\text{-}OH + CH_3\text{-}\dot{C}H\text{-}O\text{-}CH_2\text{-}CH_3 \qquad (29)$$

by a molecular process, an olefin being the other product (e.g., reaction 30). The other molecular process involving the C-O bond

$$CH_3\text{-}CH_2\text{-}O\text{-}CH_2\text{-}CH_3 \xrightarrow{h\nu} CH_3\text{-}CH_2\text{-}OH + CH_2 = CH_2 \qquad (30)$$

$$CH_3\text{-}CH_2\text{-}O\text{-}CH_2\text{-}CH_3 \xrightarrow{h\nu} CH_3\text{-}CHO + CH_3\text{-}CH_3 \qquad (31)$$

is the formation of a carbonyl compound and an alkane in primary and secondary ethers (e.g., reaction 31). Alcohols and carbonyl compounds are key products for the determination of the photolytic behavior of ethers because the alkoxyl radicals from reactions 27 and 28 readily abstract hydrogen atoms from neighboring molecules to give alcohols (reaction 29) but do not interact with other radicals under the conditions of these experiments (177). The carbonyl compounds cannot be formed by any process other than that exemplified in reaction 31 if the alkoxyalkyl radicals formed by H abstraction are sufficiently stable with respect to fragmentation at the experimental conditions. Fragmentation of a saturated alkoxyalkyl radical is only observed (179) if it is heavily loaded with methyl substituents (cf. the t-butoxypropyl-2 radical, reaction 32). For this reason the quantum

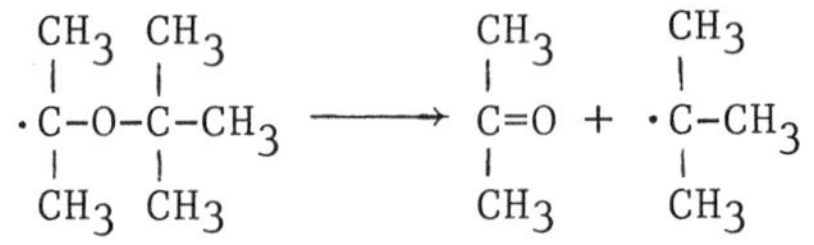

(32)

yield of the formation of alcohols and carbonyl compounds is a measure of the quantum yield of the scission C-O bond in primary and secondary ethers. In tertiary ethers only the alcohols are formed. The scission of a C-H bond is indicated by the formation of H_2, because the high abstractive power of the H atom precludes its interaction with other radicals in the liquid systems. Radical addition to olefins would become important only at high conversions of the starting material. The various modes of reaction (homolytic scission and molecular processes, e.g., reactions 27 or 28 followed by 29 and 30 as well as minor processes such as C-C bond fragmentation) can be assessed only if a complete product study has been made. Then, with the knowledge of the disproportionation/dimerization ratios of the radicals in question a detailed picture of the primary processes can be given. This has been done for diethyl ether, methyl n-propyl ether, methyl t-butyl ether, and di-t-butyl ether (Schemes 6-9). It must be

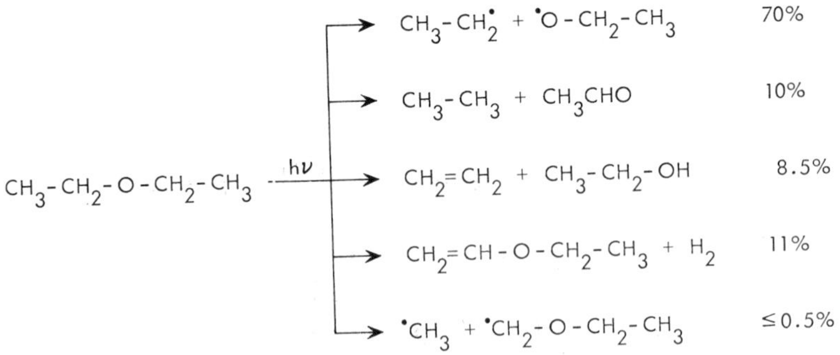

$CH_3OCH_2CH_2CH_3 \xrightarrow{h\nu}$

(1)	$CH_3^{\bullet} + {}^{\bullet}OCH_2CH_2CH_3$	(0.70)
(2)	$CH_3O^{\bullet} + {}^{\bullet}CH_2CH_2CH_3$	(0.14 to 0.16)
(3)	$CH_3OCH_2^{\bullet} + {}^{\bullet}CH_2CH_3$	($< 10^{-2}$)
(4)	$CH_3OCH_2CH_2^{\bullet} + {}^{\bullet}CH_3$	($< 10^{-2}$)
(5)	$H^{\bullet} + R^{\bullet}$	(~ 0.02) [(5) + (6)]
(6)	$H_2 + CH_3OCH{=}CHCH_3$	
(7)	$H_2 + CH_3OCH_2CH{=}CH_2$	($< 10^{-2}$)
(8)	$CH_4 + OCHCH_2CH_3$	(0.13)
(9)	$CH_4 + CH_3OCH{=}CH_2$	($\leq 10^{-2}$)
(10)	$CH_2O + CH_3CH_2CH_3$	($\leq 10^{-2}$)
(11)	$CH_3OH + CH_2{=}CHCH_3$	(≤ 0.02)
(12)	CH_3OH + cyclo-C_3H_6	($< 10^{-3}$)
(13)	$CH_3OCH_3 + C_2H_4$	($< 10^{-3}$)

kept in mind that true molecular processes and cage disproportionation reactions have not been separated in these experiments.

Carbon-carbon bond scission is rather unimportant in diethyl ether (< 0.5%) and methyl n-propyl ether (< 2%), but is enhanced somewhat in t-butyl methyl ether (≃ 4.5%) and di-t-butyl ether (7%). The reverse has been found regarding the formation of hydrogen.

The most striking phenomenon in this series of ethers is the high quantum yield of molecular decomposition in the case of di-t-butyl ether (ϕ = 0.8 ≙ 80%). Usually, the molecular processes do not exceed 20 percent. This difference has been considered to reflect the strain already present in the ground state of the di-t-butyl ether molecule.

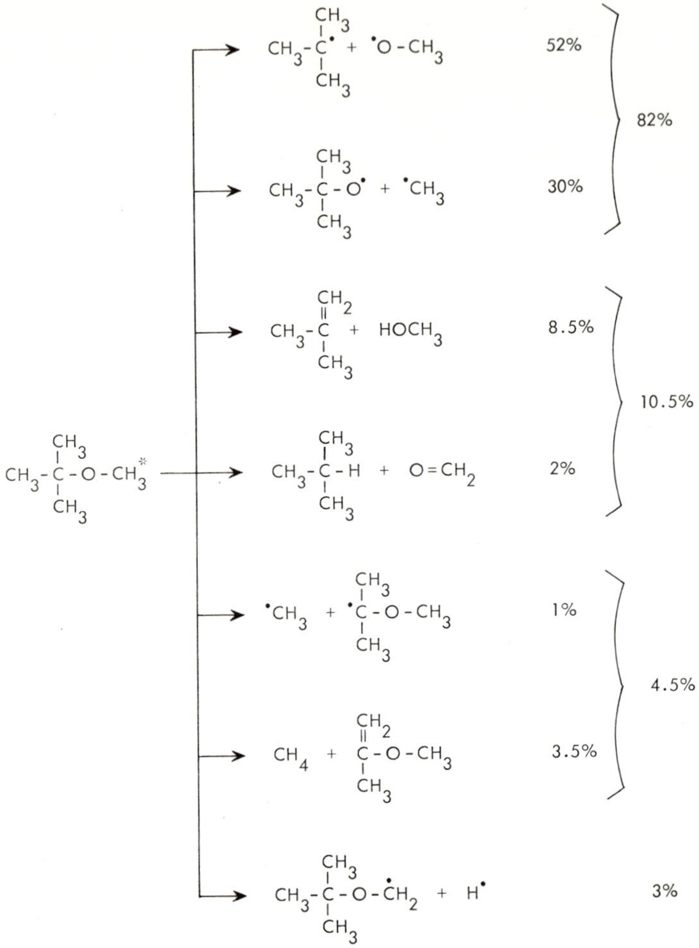

In the photochemistry of ketones, α-cleavage is one of the prominent processes (reactions 33 and 34). In asymmetric ketones the molecule is predominantly split in such a way that the larger

$$R\text{-}CO\text{-}R' \xrightarrow{h\nu} R\text{-}\dot{C}O + \cdot R' \quad (33)$$

$$\xrightarrow{h\nu} R\cdot + \dot{C}O\text{-}R' \quad (34)$$

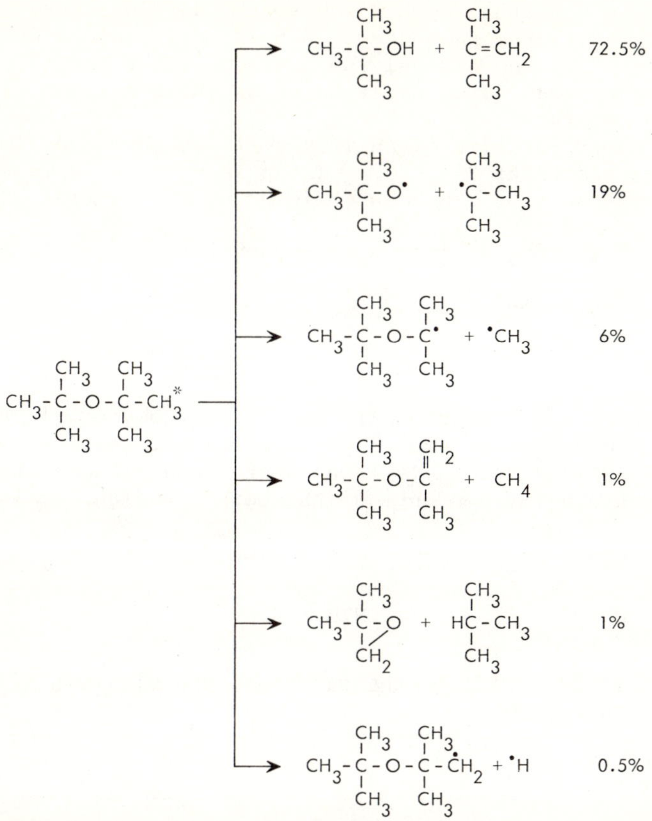

carbon chain alkyl radical, or the more highly branched alkyl radical, is formed (88). If one compares the carbonyl chromophore with the ether chromophore it is seen that except for the pair t-butyl-methyl the opposite is true.

B. Open-Chain Ethers in the Gas Phase

The photolysis of dimethyl and diethyl ethers in the gas phase was first studied by Harrison and Lake (172) using a hydrogen discharge lamp. They reported the formation of formaldehyde from dimethyl ether, and ethylene, acetaldehyde, and formaldehyde from diethyl ether.

Studies on dimethyl ether (λ 147 nm) were resumed by Meagher and Timmons (180). Products with quantum yields as reported by these authors are given in Table 12. Small amounts of dimethoxyethane and ethyl methyl ether were also observed. Photolysis of $(CD_3)_2O$ in the presence of H_2S largely yielded CD_3H. The formation of H_2 was strongly suppressed by the addition of C_2D_4 and further reduced on addition of an inert gas, C_2F_6, to an ether/C_2D_4 mixture.

In the interpretation of their results Meagher and Timmons (180) propose reactions 35 to 37. The primary process is thought to be the scission of a C-H bond followed by the thermal

$$CH_3\text{-}O\text{-}CH_3 \xrightarrow{h\nu} H^{\cdot} + {}^{\cdot}CH_2\text{-}O\text{-}CH_3^{*} \quad (35)$$

$${}^{\cdot}CH_2\text{-}O\text{-}CH_3^{*} \longrightarrow CH_2O + {}^{\cdot}CH_3 \quad (36)$$

$$\longrightarrow H_2 + CO + {}^{\cdot}CH_3 \quad (37)$$

decomposition of the vibrationally excited methoxymethyl radical. Meagher and Timmons have also considered the alternative mechanism (reactions 38-39) but given it lower preference because it would predict a pressure independence of the methyl radical yield.

$$CH_3\text{-}O\text{-}CH_3 \xrightarrow{h\nu} {}^{\cdot}CH_3 + {}^{\cdot}OCH_3^{*} \quad (38)$$

$$CH_3O^{\cdot *} \longrightarrow CH_2O + H^{\cdot} \quad (39)$$

TABLE 12. Quantum yields in the gas phase photolysis (λ 147 nm) of dimethyl ether (180).

$p(Me_2O)$/torr	11	40	95	188	302
H_2	0.402	0.428	0.349	0.338	0.330
CH_4	0.125	0.126	0.089	0.085	0.077
CO	0.110	0.100	0.049	0.055	0.060
C_2H_6	0.179	0.089	0.088	0.089	Not measured
CH_2O	0.426	0.442	0.378	0.347	0.327

However, higher products have not been measured and it can be argued that the material balance (C_2H_6 + CH_4 versus CH_2O + CO, cf. Table 12) is too poor, especially at high pressures, to draw this conclusion. It seems justified to reinterpret their results according to the mechanism represented by reactions 38 and 39. Such a mechanism would be in line with the decrease in $\phi(H_2)$ with increasing pressure and with the photolytic behavior of all other ethers studied so far.

Molecular process 40 plays a minor role in the photolysis of dimethyl ether as has been substantiated by the experiment with $(CD_3)_2O$ in the presence of H_2S. The corresponding processes

$$CH_3\text{-}O\text{-}CH_3 \xrightarrow{h\nu} CH_4 + CH_2O \tag{40}$$

in the photolysis of other methyl ethers are also of comparatively minor importance (cf. Table 11).

The photolysis of diethyl ether was studied by Johnson and Lawson (181). They showed that the distribution of the products measured is virtually the same at λ 147 and 124 nm (Table 13) although at λ 124 nm the energy absorbed is sufficient for photoionization to occur. Quantum yields are not given and some mechanistically important products such as hydrogen and formaldehyde have not been measured. It is noted that the yield of ethanol relative to propane strongly increases with increasing diethyl ether pressure suggesting that the vibrationally excited ethoxyl radical generated in the primary process 27 is thermalized at higher pressures, thus preventing reaction 41. In general the photolysis of diethyl ether in the gas phase at 147 and 124 nm

$$CH_3\text{-}CH_2\text{-}O^{\bullet} \longrightarrow {}^{\bullet}CH_3 + CH_2O \tag{41}$$

appears to follow the same lines that have been established for the liquid phase at 185 nm (177). However, in the gas phase experiments some tetrahydrofuran, and possibly also epoxybutane, is formed. Carbenes have been postulated as possible precursors. In the liquid phase no tetrahydrofuran ($\phi < 10^{-3}$) is formed at 185 nm (182).

C. Cyclic Ethers in the Liquid Phase

The fluorescent behavior, following optical excitation, of some liquid cyclic ethers, in particular p-dioxane and its solutions, has been studied by Hirayama et al. (183) and Halpern and Ware (184). Product studies on cyclic ethers are still in a preliminary state. There is a brief study by Pitts and co-workers (91) on oxetane in isoöctane and aqueous solutions. Some aspects of possible importance for photochemists working in p-dioxane as

TABLE 13. Some relative molar yields in the gas phase photolysis of diethyl ether at λ 147 and 124 nm (181).

$p(Et_2O)$ torr	147 nm		124 nm		
	10	100	2	10	100
Methane	20	20	18	19	23
Ethylene	47	42	54	48	47
Ethane	50	48	53	49	47
Propane	100	100	100	100	100
n-Butane	56	61	51	58	69
Ethyl vinyl ether	4		4	4	
Ethyl isopropyl ether	12		11	13	14
Ethyl s-butyl ether	15		14	16	18
Acetaldehyde	5		4	5	5
Ethanol	38		20	41	52

a solvent have been reported recently by Mazzocchi and Bowen (185). More detailed work on p-dioxane (186,187) and the five-membered rings (188) tetrahydrofuran, 2-methyltetrahydrofuran, and both cis- and trans-2,5-dimethyltetrahydrofuran are in progress in the reviewers' laboratory. ESR experiments on tetrahydrofuran and p-dioxane were carried out at 77 K by McIntosh and Wan (189).

Fluorescence (λ_{max} 247 nm in neat p-dioxane) occurs with a quantum yield of 0.029 (184) in the liquid state only (183). Addition of water effects a diminution and a red shift while isoöctane causes a blue shift, as well as a reduction in intensity. O_2 (183) and N_2O (187) are quenchers. It is thought that the fluorescence is from some form of excited aggregate whose composition varies with dilution, and that monomeric p-dioxane does not fluoresce (183). It has also been suggested that photodetachment of an electron might occur, and fluorescence might be a consequence of charge recombination (187).

The primary processes in the open-chain ethers so far investigated (except di-t-butyl ether) are mostly homolytic. In the cyclic compounds, the corresponding reactions lead to biradicals. Molecular reactions may play a significant role but could not yet be distinguished from biradical routes. In the cyclic ethers α-radicals appear to be formed with relatively small quantum yields compared to open-chain ethers. Such radicals are often stable at room temperature, but may also begin to undergo fragmentation reactions, depending on ring size and substituents (p-dioxanyl radical (190), α-tetrahydrofuryl radicals (191), epoxide radicals (291)). According to Benson (274), however, the isomerization of the ethylene oxide radical to $\dot{C}H_2CHO$ proceeds with an activation energy of about 10 kcal/mol so that it could compete successfully with radical combination reactions.

Ethylene and formaldehyde are the main products from oxetane (91). Preliminary quantum yield values of some products from p-dioxane (186) and the tetrahydrofurans (188) are presented in Tables 14 and 15. In the case of p-dioxane one is led to propose

TABLE 14. Quantum yields of some products (λ 185 nm) from neat liquid p-dioxane (186).

Product	ϕ	Product	ϕ
Formaldehyde	0.29	Acetaldehyde	0.03
2-Vinyloxyethanol	0.17	Hydrogen	0.03
Ethylene	0.14	2-Methyldioxolane	0.01
Oxetane	0.04	Bisdioxanyl (meso + racemic)	0.005

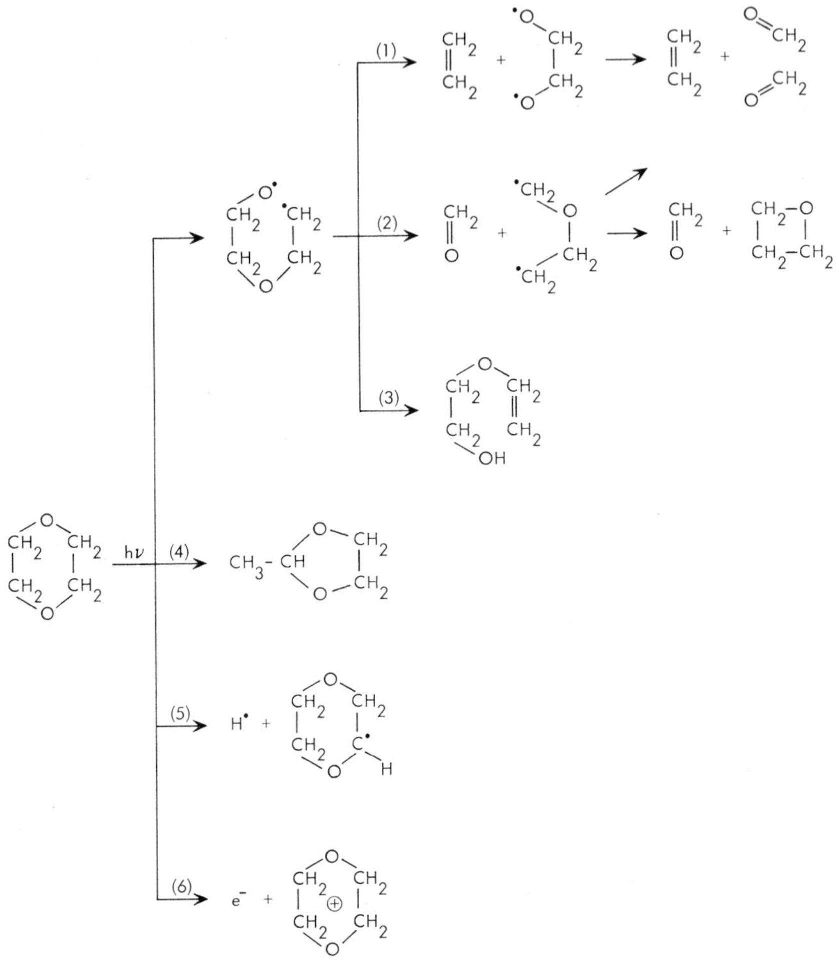

the reactions of Scheme 10. The photoionization process 6 in Scheme 10 has been included because of the high quantum yield of N_2, 0.59, in the presence of N_2O (187). N_2O is known to react efficiently with electrons according to reaction 42 but only slowly with hydrogen atoms or other radicals. In aqueous solutions $\phi(N_2)$ reaches unity. N_2O and H^+ complete for the same

$$e^- + N_2O \longrightarrow N_2 + O^- \qquad (42)$$

TABLE 15. Photolysis (λ 185 nm) of tetrahydrofuran and of some of its methyl derivatives in the liquid phase. Product quantum yields (187).

	THF	2-Methyl-THF	2,5-Dimethyl-THF		2,2,5,5-Tetra-methyl-THF
			cis	trans	
Hydrogen	0.07	0.17	0.27	0.29	0.00003
Carbon monoxide	–	–	–	–	–
Methane	0.0007	0.005	0.007	0.005	0.036
Ethylene	0.025	0.021	0.00008	0.00005	0.002
Propene	0.007	0.014	0.015	0.011	–
Cyclopropane	0.19	0.20	–	–	–
Methylcyclopropane	–	0.026	0.22	0.078	–
1,1-Dimethylcyclopropane	–	–	–	–	0.40
Butene-1	–	0.0005	0.004	0.0006	–
Iso-Butene	–	–	–	–	0.04
Formaldeyde	0.14	0.028	–	–	–
Acetaldehyde	0.008	0.21	0.22	0.09	–
Ethylene oxide	0.016	0.003	–	–	–
Acetone	–	0.015	0.008	0.04	0.40
Propylene oxide	–	0.015	0.012	0.009	–
1,1-Dimethylethylene oxide	–	–	–	–	0.04
2,3-Dihydrofuran	0.04	–	–	–	–
2,5-Dihydrofuran	–	–	–	–	–
Dehydrodimer	0.011	0.03			
Butyraldehyde	0.06	–	–	–	–
Buten-3-ol-1	0.02	–	–	–	–
Valeraldehyde	–	0.02	–	–	–
Methylpropylketone	–	0.05	–	–	–
Hexanone-(2)	–	–	0.031	0.023	–
cis-2,5-Dimethyl-THF	–	–	–	0.22	–
trans,2,5-Dimethyl-THF	–	–	0.22	–	–

precursor, which has been suggested to be the solvated electron (187). N_2O quenches the fluorescence in neat p-dioxane, and reduces the formation of some major products by about one-third. N_2O induces, however, the formation of hydroxydioxane as an important product (186). These facts seem to indicate that ionic processes, which are modified by the presence of N_2O, contribute to product formation.

It is not possible at this time to say whether process 1 in Scheme 10 occurs at all, or if all the formaldehyde and ethylene is formed via 2, the oxetane being a side product. As has been discussed above, with straight-chain ethers usually both molecular processes 30 and 31 occur and give rise to the olefin and the alcohol, and to the aldehyde and the alkane. In cyclic ethers this type of reaction leads to olefinic alcohols, or to aldehydes.

Some of the products and their quantum yields in the 185 nm photolysis of tetrahydrofuran and methylated tetrahydrofurans (188) are given in Table 15. Despite the incompleteness of the results at this stage, some of the major features are already recognizable. Principal modes of decomposition for tetrahydrofuran itself are given in Scheme 11. Route 1 is predominant. The formation of propene and acetaldehyde in processes 1 and 2 seem to point, at least to a certain extent, towards biradical intermediates. Molecular processes that directly lead to cyclopropane and ethylene oxide cannot be excluded. Methyl substitution at C-2 has a strong effect on the processes corresponding to process 1 in Scheme 11. The yield of cyclopropane is sevenfold compared to that of methylcyclopropane. Another interesting feature is observed with cis- and trans-2,5-dimethyltetrahydrofuran. The methylcyclopropane quantum yield is three times larger for the cis-compound than for the trans-compound. An unexpected result is the increase of the hydrogen yields when going from the tetrahydrofuran to the 2-methyl- and 2,5-dimethyl-substituted compounds. In the latter, hydrogen formation is the most important process recognized so far. The α-hydrogen must be involved since the hydrogen yields from 2,2,5,5-tetramethyltetrahydrofuran is virtually zero (Table 15).

D. Cyclic Ethers in the Gas Phase

The vapor phase photolysis of ethylene oxide has found more attention than that of its higher homologs. It has first been extensively studied by Gomer and Noyes (193) in the long wavelength region. Fleming et al. studied the effect of ring size on the photolysis of cyclic ethers such as ethylene oxide, oxetane, and tetrahydrofuran (90). Roquitte (194) did flash

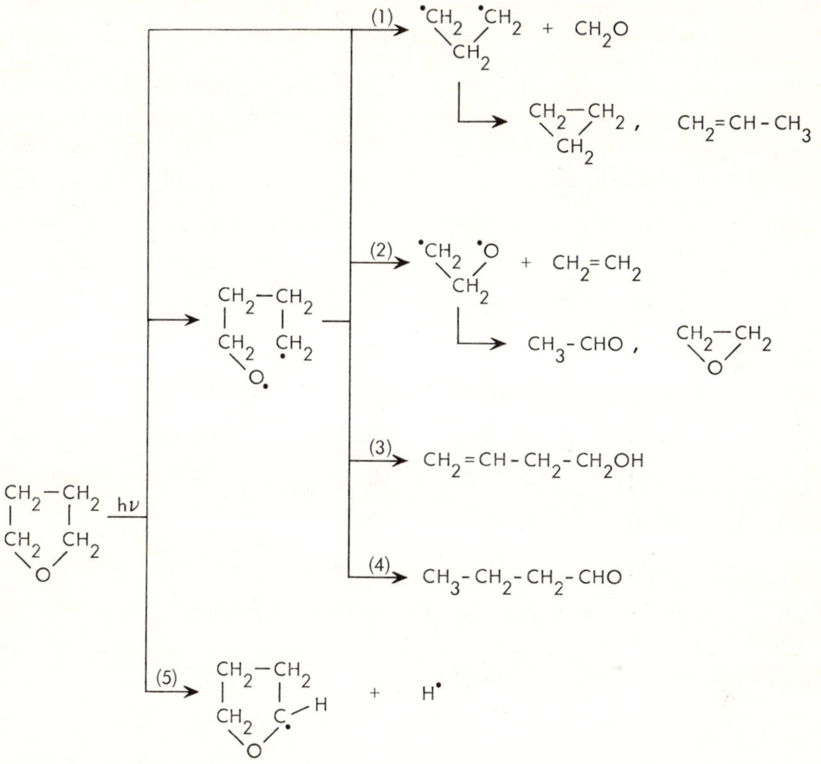

photolysis experiments. Kawasaki et al. (195) resumed the steady-state experiments over a wider wavelength range. Oxetane and 2,2-dimethyloxetane have been studied by Pitts and co-workers (91), and tetrahydrofuran by Roquitte (196,197). The photolysis of p-dioxane at 147 nm has been undertaken by Hentz and Parrish (198), and its laser-induced decomposition has been reported by Watson and Parrish (199).

The main products from the ethylene oxide photolysis at wavelengths above 170 nm are carbon monoxide and hydrogen (193, 194). Ethane and methane, and sometimes acetaldehyde, formaldehyde, and ethylene, are major products (Table 16). Ketene has also been detected (90). Ethylene shows a strong increase when the photolyzing light is below 170 nm (195). Gomer and Noyes (193) had concluded that the major primary process was reaction 1 in Scheme 12. This was confirmed by Roquitte (194), who showed a minor contribution from process 2, and recently by Kawasaki et al. (195) in their detailed study at different wavelengths

TABLE 16. Relative yields of products from the photolysis of ethylene oxide (195). Pressures near 10 torr

Product	λ147nm	λ174nm	λ≤178nm	(193)[a]	(194)[b]
CO	1.00	1.00	1.00	1.00	1.00
H_2	0.64	0.64	-	0.56	0.56
CH_3CHO	0.16	-	-	0.22	0
HCHO	1.24	-	-	0.13	0
CH_4	0.16	0.21	0.19	0.13	0.35
C_2H_2	0.12	0.03	0.02	0	0
C_2H_4	0.26	0.04	0.01	0	0.15
C_2H_6	0.28	0.40	0.32	0.43	0.13
C_3H_6	0.004	0.002	0.001	-	0
C_3H_8	0.16	0.10	0.08	-	0
i-C_4H_{10}	0.005	0.006	0.007	-	0
n-C_4H_{10}	0.04	0.006	0.007	-	0

[a]Unfiltered hydrogen discharge.
[b]Flash.

(Scheme 12). The importance and the fate of the ethylene oxide radical are not clear in this system but it seems that its concentration was minor compared to that of other radicals such as methyl since higher aldehydes and epoxides were not found. It does seem clear that acetaldehyde was not formed via $\cdot CH_2CHO$ because its formation was unaffected in the presence of nitric oxide.

The oxetane photolysis (90,91) in the 190 nm region seems largely to proceed by way of the molecular processes 43 through 45. Owing to photolysis of the carbonyl compounds there was some

$$\begin{matrix} CH_2-CH_2 \\ | \quad\quad | \\ CH_2-O \end{matrix} \xrightarrow{h\nu} CH_2O + C_2H_4 \quad (43)$$

$$\begin{matrix} CH_2-CMe_2 \\ | \quad\quad | \\ CH_2-O \end{matrix} \xrightarrow{h\nu} CH_2O + CH_2{=}CMe_2 \quad (44)$$

$$\xrightarrow{h\nu} Me_2CO + C_2H_4 \quad (45)$$

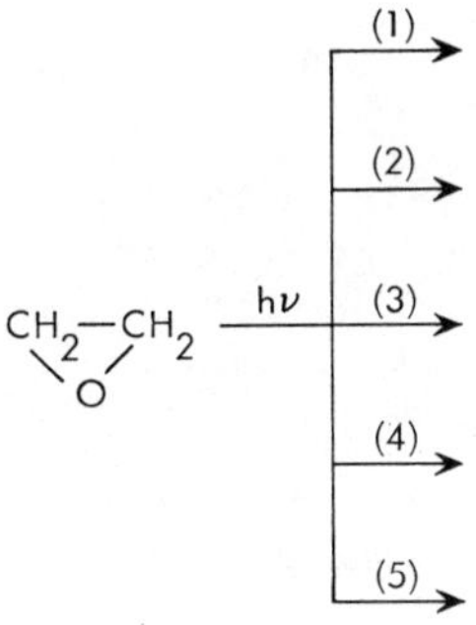

Route	Products	185 - 178nm	174nm	147nm
(1)	$^{\bullet}CH_3 + {}^{\bullet}CHO$ $(CO + {}^{\bullet}H)$	1	1	1
(2)	$O + CH_2{=}CH_2$ $(CH{=}CH + H_2)$	0.1	0.1	0.7
(3)	$H_2 + (CH_2{=}C{=}O)$	-	0.1	0.1
(4)	$CH_3\text{-}CHO$	-	-	0.2
(5)	$CH_2 + CH_2O$	-	-	0.2

formation of carbon monoxide, but all products except olefins and carbonyl compounds were minor (Table 17): route 44 and 45 appear to have similar probabilities, route 45 being somewhat favored. At photolysis temperatures above 100° a chain reaction sets in (91):

$$\begin{array}{ll} CH_2\text{-}\dot{C}H \\ | \quad\;\; | \\ CH_2\text{-}O \end{array} \longrightarrow CH_2{=}CH_2 + \dot{C}HO \qquad (46)$$

$$CHO \longrightarrow CO + H\cdot \qquad (47)$$

$$\cdot H + \begin{array}{l} CH_2\text{-}CH_2 \\ | \qquad\;\; | \\ CH_2\text{-}O \end{array} \longrightarrow H_2 + \begin{array}{l} CH_2\text{-}\dot{C}H \\ | \qquad\;\; | \\ CH_2\text{-}O \end{array} \qquad (48)$$

The photolysis of tetrahydrofuran with a medium-pressure Hg arc led to carbon monoxide, hydrogen, methane, ethane, ethylene, propylene, cyclopropane, and formaldehyde as major products (196). Quantum yields or absolute yields are not known. The yields of formaldehyde, ethylene, and propane declined with increasing pressure, those of carbon monoxide and cyclopropane increased starting from near zero at very low pressures. In the absence of a material balance mechanistic conclusions cannot be drawn except that the processes 49 to 51 are very likely important.

$$\begin{array}{c} CH_2\text{-}CH_2 \\ | \qquad\;\; | \\ CH_2 \;\; CH_2 \\ \diagdown \;\; \diagup \\ O \end{array} \xrightarrow{h\nu} \begin{array}{c} CH_2\text{-}CH_2 \\ \diagdown \;\; \diagup \\ CH_2 \end{array} + CH_2O \qquad (49)$$

$$\xrightarrow{h\nu} CH_3CH{=}CH_2 + CH_2O \quad (50)$$

$$\xrightarrow{h\nu} CH_2{=}CH_2 + CH_3CHO \quad (51)$$

TABLE 17. Products from the vapor phase photolysis of oxetanes (unfiltered Hg arc, 25°, conversion 0.3%). Rates of formation, arbitrary units (91)

Product	Oxetane	2,2-Dimethyloxetane
Ethylene	6.47	2.60
Isobutene	-	2.17
Formaldehyde	4.00	-[a]
Acetone	-	0.7[b]
Hydrogen	0.67	0.5
Carbon monoxide	0.52	0.6
Methane	0.031	0.6
Ethane	0.024	0.41

[a]Present in the same order of magnitude as isobutene.
[b]Separation incomplete.

The cyclopropane was considered to be formed in the vibrationally excited or even in the diradical state (196,197). It seems that despite low conversions near 1 percent much of the carbon monoxide is a secondary product. The same would hold for the methane. In fact, acetaldehyde has been detected beside ethylene in the tetrahydrofuran photolysis (90).

The 147 nm photolysis of 1,4-dioxane vapor is better known (198). The sum of the quantum yields of the primary processes 1 to 3 in Scheme 13 is close to unity. Product quantum yields were independent of intensity (0.18 to 2.4 x 10^{15} quanta/s), pressure (2.7 - 155 torr), and temperature (25 - 95°). This was thought to indicate that radical reactions are of minor importance (198).

The ruby laser-induced decomposition of 1,4-dioxane has been reported (199) to lead to ethylene, carbon monoxide, and hydrogen

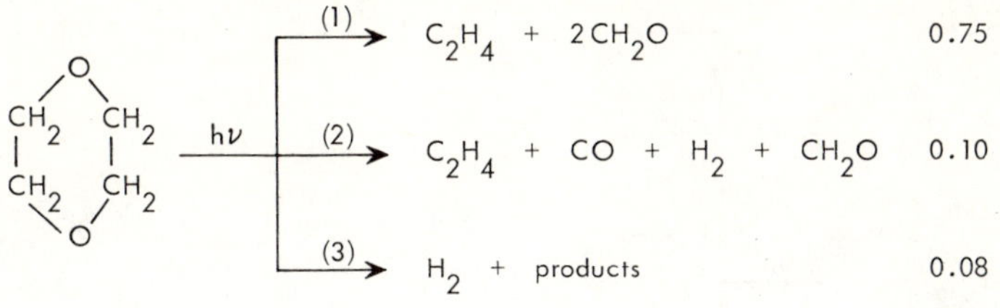

at a ratio near 1:2:2. A trace of formaldehyde was detected. Other products may have been formed that went unanalyzed. The product yield was found proportional to the square root of the energy per pulse. This behavior is unexpected and indicates a complex reaction mechanism.

VI. PHOTOLYSIS OF ACETALS

The photolysis of acetals has hitherto found scant attention. 1,3-Dioxolane has been studied by Roquitte (102) who used the full light of a medium-pressure Hg arc. The liquid phase photolysis (λ 185 nm) of formaldehyde dimethyl acetal has been undertaken in the reviewers' laboratory (103a).

On the basis of the formaldehyde dimethyl acetal results, summarized in Table 18, the quantum yields of the primary processes can be estimated (Scheme 14). It has been shown (103a) that

TABLE 18. Photolysis (185 nm) of liquid deaerated formaldehyde dimethyl acetal (15°). Products and quantum yields (103a)

Product	ϕ	Product	ϕ
H_2	0.04	$HCOOCH_3$	0.17_5
CH_4	0.22	$CH_3CH(OCH_3)_2$	0.04
C_2H_6	0.01	$CH_2(OCH_3)OC_2H_5$	0.04
CH_2O	0.15[a]	$CH_3OCH_2CH_2OCH_3$	0.02_5
CH_3OH	0.35	$CH_3OCH_2CH(OCH_3)_2$	0.07_4
CH_3OCH_3	0.13	$CH_2(OCH_3)OCH_2CH_2OCH_3$	0.05_4
$CH_3OC_2H_5$	0.02_5	Dehydrodimers	0.1

[a]Value probably too low.

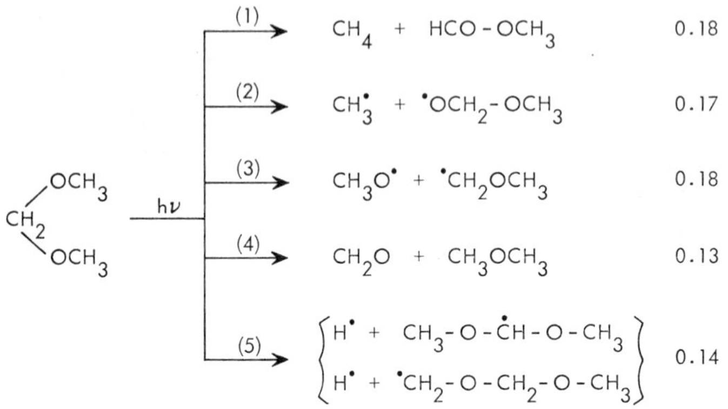

the quantum yields of all products except those which result from combination reactions of the methyl radical, are independent of the dose rate. This indicates that fragmentation (cf. 200-204) of the formaldehyde dimethyl acetal radicals is unimportant under these conditions.

As seen in Scheme 14, molecular elimination processes (including cage disproportionation reactions) account for about half of the primary decomposition. This is in contrast to the behavior of the isosteric ethers such as diethyl (177) and methyl n-propyl ethers (88), where these processes are rather less important (Schemes 6 and 7). It is not known whether the formaldehyde formed in reaction 4 (Scheme 14) incorporates the carbon atom from methylene or methyl.

The features of the gas phase 190 to 200 nm photolysis of 1,3-dioxolane (102) seem well established. The products (Table 19) give a good elemental balance. The large carbon dioxide

TABLE 19. Product yields from the photolysis (102) of 1,3-dioxolane (50 torr, active wavelength 190-200 nm, exposure 58 min)

Product	Yield (μmol)	Product	Yield (μmol)
H_2	2.21	CO_2	1.11
CO	1.53	C_2H_4	1.52
CH_4	0.27	C_4H_{10}	0.09
C_2H_6	0.08	CH_2O	1.73

and ethylene yields are remarkable. These products were only slightly quenched by oxygen whereas ethane, and presumably methane, were completely quenched. In the presence of an excess of isobutane the carbon dioxide and ethylene yields were unaffected, and the yields of the other products were very slightly affected. The conclusion was drawn (102) that the processes 52 and 53 are the most important ones.

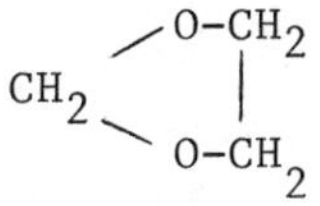

$$\xrightarrow{h\nu} CH_2{=}CH_2 + CO_2 + H_2 \quad (52)$$

$$\xrightarrow{h\nu} CH_2{=}O + CH_3 + CO + H \quad (53)$$

Some preliminary results from the reviewers' laboratory (293) may shed light on the mechanism of the ethylene and carbon dioxide formation. The liquid phase photolysis (185 nm) of 2,2-dimethyl-1,3-dioxolane affords large quantities of ethylene and some carbon dioxide. Carbon dioxide is also formed if di-t-butylperoxide is photolyzed in 2-methyl-1,3-dioxolane. Thus it seems probable that in the primary process the 1,3-dioxolanyl-2-radical is formed. The latter appears to be prone to fragmentation and further eliminates ethylene to give rise to a $O{=}CR{-}O^{\bullet}$ radical, a likely precursor of the carbon dioxide. This sequence is not possible in the open chain acetals. In agreement with this, carbon dioxide is not a product in the photolysis (185 nm) of liquid formaldehyde and acetaldehyde dimethyl acetal (103a,b).

VII. PHOTOLYSIS OF AMINES

A. Amines in the Solid Phase

1. Primary Amines. The photolysis of primary amines in the frozen state has been studied by ESR. Hadley and Volman (205) found that at 77 K neat methylamine on irradiation with 185 nm light yielded the radical $CH_3NH^{\bullet}$, which was also formed through hydrogen abstraction from methylamine by OH radicals obtained from the photodecomposition of hydrogen peroxide dissolved in the amine in the frozen state. All the evidence pointed against the presence of the radical $^{\bullet}CH_2NH_2$. Hatano et al. photolyzed methylamine and dimethylamine adsorbed on porous glass at 77 K and found the amino radicals to be primarily formed (207). In the photolysis of ethylamine, Kotake et al. (206), however, did not observe the nitrogen-centered radical but the aminoethyl radical ($CH_3{}^{\bullet}CHNH_2$). Richerzhagen and Volman (208) obtained similar results with ethylamine and n-butylamine in an adamantane matrix at room temperature. Alkylideneimino radicals were also detected. These are thought to arise from the photolysis of the α-aminoalkyl radicals (205,206,209). According to Symons (210) there is evidence of a thermal conversion of $CH_3\dot{N}H$ into $CH_2{=}N^{\bullet}$.

2. Secondary Amines. Photolysis of dimethylamine at 77 K gave rise to an ESR signal indicating the formation of $Me_2N^{\bullet}$ (205, 207). $CH_3NHCH_2^{\bullet}$ was not observed.

3. Tertiary Amines. The trimethylamine photolysis was studied by ESR at 77 K, neat and in an aqueous matrix (205). $Me_2N^{\bullet}$ and $CH_3^{\bullet}$ were observed, but not $Me_2NCH_2^{\bullet}$. The photodecomposition at 77 K of dialkylaminoalkyl radicals giving rise to an alkyl radical (and N-alkylalkylideneimine) has been observed (211).

B. Amines in the Liquid Phase

1. Primary Amines. The photolysis of neat primary amines was undertaken relatively recently by Pfordte and Leuschner (212) in 1961, but some earlier studies by Booth and Norrish (213), and later studies by Pouyet (214) and Niu (124) and Stenberg et al. (215) have been carried out in hydrocarbon solvents. Pouyet (216) investigated their photolysis in aqueous solutions.

The formation of α,α'-dehydrodimers had been recognized in the neat amines (212), but no further products were identified. In the photolysis of primary amines dissolved in hydrocarbon solvents hydrogen is the major gaseous product (213,214). In n-hexane solutions the equivalent unsaturated product is reported to be hexene-1, with complete absence of other hexenes (214). Nitrogen-containing polymers are also formed (213). Stenberg et al. (215) studied the photolysis of cyclohexylamine in cyclohexane and deduced that the only important primary process was the split into cyclohexyl and $^{\bullet}NH_2$ because they found that the amount of amine decomposed roughly balanced the bicyclohexyl yield. This is an odd result, even if almost exclusively the C-N bond was cleaved. One expects, besides bicyclohexyl, cyclohexene if cyclohexyl radicals play a part in this photolysis. Their disproportionation/combination ratio is about 1.1 (217). It is conceivable that an intermolecular photoelimination of ammonia might take place in this system. Evidence for similar reactions yielding small molecules other than H_2 has been described (148,149). Booth and Norrish (213), and Pouyet (214) had, however, recognized that H atom detachment at the amino group, or from the methylene group α to it, must be the main process in alkane solution of primary amines.

2. Secondary Amines. Neat secondary amines have been photolyzed by Pfordte and Leuschner (212) and Allan and Swan (218). They observed from diethylamine the α,α'-dehydrodimer (212), and also N-ethyl-but-2-enylideneimine, N,N'-diethyl-2,3-diaminobutane, 1,3-diethyl-2,4,5-trimethyl-imidazolidine, and tetraethylhydrazine (218).

In hydrocarbon solvents, the photolysis of ethylmethylamine has been studied by Booth and Norrish (213), that of dihexylamines by Niu (124), and dimethylamine by Hancock and Dickinson (219). Whereas ethylmethylamine yielded hydrogen, an unsaturated compound, methane, and a nitrogen-containing polymer, the dihexylamines afforded N-alkyl-alkylideneimines (the main products), alkylalkanes, and alkylamines. Conversions were carried to near maximum yields of the N-alkyl-alkylideneimines (124). Dimethylamine (219) yielded methylamine, bis(dimethylamino)methane, small amounts of N,N-dimethylamino-N'-methylaminomethane, and tetramethylhydrazine, but no methane or N,N'-dimethyldiaminoethane. Analysis for hydrogen was not carried out. Product yields were linear with time. The first three products were explained by a transamination mechanism operating through the primarily formed N-methylmethyleneimine (219):

$$Me_2NH \xrightarrow{h\nu} Me_2N^{\bullet} + H^{\bullet} \quad (54)$$

$$2Me_2N^{\bullet} \xrightarrow{h\nu} Me_2NH + MeN{=}CH_2 \quad (55)$$

$$MeN{=}CH_2 + Me_2NH \xrightarrow{h\nu} MeNH\text{-}CH_2\text{-}NMe_2 \quad (56)$$

$$MeNH\text{-}CH_2\text{-}NMe_2 + Me_2NH \xrightarrow{h\nu} Me_2N\text{-}CH_2\text{-}NMe_2 + MeNH_2 \quad (57)$$

3. Tertiary Amines. The photolysis of triethylamine has been studied by Pfordte and Leuschner (212) and by Allan and Swan (218). They identified the dehydrodimer N,N,N',N'-tetraethyl-2,3-diaminobutane (meso and racemic) and tetraethylhydrazine as well as N,N'-diethyl-2,3-diaminobutane (218).

Fluorescence of tertiary amines has been observed by Phillips and co-workers (220), Tsumbomura and co-workers (112), and Halpern and co-workers (221-223,294). Fluorescence is reported from tertiary, but not secondary or primary, amines (220), presumably because of the ease with which the amino group can lose an H atom in these compounds (220). Fluorescence has been observed in hydrocarbon solution (112) as well as the vapor phase, with comparable quantum yields (222). For 1,4-diaza-bicyclo-[2,2,2]octane the fluorescence quantum yields are 0.9 (vapor) and

0.5 (10^{-4}M in n-hexane)(221). Fluorescence is quenched by oxygen, perhaps because of charge-transfer complex formation (112). Aqeuous solutions of tertiary amines do not fluoresce (112). Tertiary amines form excimers; in the case of bifunctional di-tertiary diamines there is evidence of "intramolecular" excimer formation (222,223). A progressive red shift of the fluorescence band is observed (112) when solvents of increasing polarity are used.

C. Amines in the Gas Phase

1. Primary Amines. Early studies on the photolysis of primary amines in the gas phase by Bamford (224) in 1939 were resumed by Taylor and co-workers (47,225), and Michael and Noyes (226) using the wavelength region around 200 nm. Studies at λ 147 nm were undertaken by Magenheimer et al. (227), and at λ 124 nm by Kawasaki and Tanaka (228). Flash photolysis experiments were carried out by Black and Porter (229) and by Dodonova and Terenin (230).

The most extensively studied primary amine by far is methylamine (47,224-230), and only little information is available on other primary amines ethylamine (230), n-butylamine (224) and n-amylamine (224). Products and their quantum yields from methylamine as reported by Michael and Noyes (226) are given in Table 20.

TABLE 20. Photolysis (λ 224-194 nm) of methylamine in the gas phase (226) (100 torr, room temperature). Quantum yields are based on the hydrogen bromide actinometer.

	H_2	NH_3	C_2H_5N	Me_2NH	CH_4	C_2H_6	Me_2N_2	N_2
φ	0.89	0.89	0.22	0.13	0.042	0.024	0.019	0.011

The primary processes have been suggested to be:

$$CH_3NH_2 \xrightarrow{h\nu} CH_3NH^{\bullet} + H^{\bullet} \quad (58)$$

$$\xrightarrow{h\nu} H^{\bullet} + {}^{\bullet}CH_2NH_2 \quad (59)$$

$$\xrightarrow{h\nu} {}^{\bullet}CH_3 + {}^{\bullet}NH_2 \quad (60)$$

$$\xrightarrow{h\nu} H_2 + :CHNH_2 \qquad (61)$$

$$\xrightarrow{h\nu} CH_3N: + H_2 \qquad (62)$$

$$\xrightarrow{h\nu} H_2 + CH_2{=}NH \qquad (63)$$

$$\xrightarrow{h\nu} CH_4 + :NH \qquad (64)$$

The hydrogen yield is strongly suppressed by the addition of propene (47,225), suggesting that H atoms are important precursors in the formation of hydrogen. The H atoms abstract hydrogen giving rise to further carbon- and nitrogen-centered radicals. There is still uncertainty whether these abstraction reactions preferentially give rise to carbon- or nitrogen-centered radicals. Arrhenius parameters for H abstraction at C and at N by CH_3 from methylamine were determined by Gray and Thynne (231). The activation energy as well as the frequency factor ($cm^3\ mol^{-1}s^{-1}$) at C were higher than at N (E_c = 8.7 kcal/mol, E_N = 5.7 kcal/mol; log A_C = 11, log A_N = 9.5). Gray and Jones (232) found that at 150° H abstraction by CH_3 from ethylamine took place to 67 percent at methylene, to 29 percent at amine, and to 4 percent at methyl. Thus it would seem that at room temperature both N-centered and C-centered radicals should be formed at similar rates. There could also be intermolecular rearrangement. ESR evidence for rearrangement of N-centered into α-carbon-centered radicals has been presented by Symons (210). Another indication that the carbon-centered radicals are more abundant in such systems is the high G-value of 3.4 for ethylenediamine in the liquid phase radiolysis of methylamine given by Smithies and Whitworth (233), who also found that G(N,N'-dimethylhydrazine) must be smaller than 0.05. In aqueous solution, alkylamino radicals undergo a fast, possibly proton-catalyzed rearrangement into the corresponding α-aminoalkyl radicals (234).

Ammonia is formed in a yield about equal to hydrogen. Part of it may have the amino radical (reaction 60) identified by flash photolysis (229,230), as precursor, but other routes to ammonia are probably more important (reactions 65-68). The combination of the methylamino radical [6] with the aminomethyl radical [7] leads to [8]. The structure of [8] resembles that of the hemiacetals which are known to be unstable so that [8] is involved in the equilibrium 67. An alternative route (reaction 68), which gives rise to ammonia and an alkylideneimine such as [10],

$$CH_3\dot{N}H + {}^{\bullet}CH_2NH_2 \longrightarrow CH_3NH\text{-}CH_2NH_2 \quad (65)$$

[6] [7] [8]

$$2CH_3\dot{N}H({}^{\bullet}CH_2NH_2) \longrightarrow CH_3NH_2 + CH_2\text{=}NH \quad (66)$$

[9]

$$[8] \rightleftharpoons CH_2\text{=}NH + CH_3NH_3 \quad (67)$$

[9]

$$[8] \rightleftharpoons NH_3 + CH_3\text{-}N\text{=}CH_2 \quad (68)$$

[10]

has been established by Baum et al. (235). N-alkylalkylideneimines have been found in the photolysis of other primary amines (124,215). Methyleneimine [9] can also be formed in the disproportionation reaction 66. These reactions, especially 67 and 68, are not expected to be instantaneous. The observation (225) that ammonia continued to be evolved after irradiation may find its explanation herein.

The absence of ammonia reported by Smithies and Whitworth (233) in their study of the γ-radiolysis of neat liquid methylamine apparently contradicts this view. McAlonie (236), however, studying the gas phase γ-radiolysis of methylamine, found ammonia to be among the main products.

By mass spectrometry the fairly important product C_2H_5N (Table 20) has been found which was considered to be ethyleneimine (226). It is difficult to see how this would be formed. It seems more likely that the product was in fact N-methylmethyleneimine [10]. It can be expected that the mass spectra of both compounds are quite similar if the analogy to other molecules isosteric to it is any guide. The MS of cyclopropane and ethylene oxide resemble fairly closely those of propylene and acetaldehyde, respectively.

Carbene and nitrene compounds might rearrange to imines or aminoolefines, react with the starting material by insertion, or react with each other. It has been suggested (226) that azomethane might be formed via $2CH_3N{:} \longrightarrow CH_3\text{-}N\text{=}N\text{-}CH_3$. The alkylideneimines as well as the N-alkylalkylideneimines can polymerize. The trimer of N-methylmethyleneimine, 1,3,5-trimethyl-1,3,5-triazacyclohexane, was observed by Wetmore and

Taylor (47) and Watson and Darwent (237). Higher polymers, more viscous liquids, have been observed as a rule in these photolyses.

The determination of the relative importance of the individual primary photochemical processes is complicated because the nitrogenous radicals and products can further react in a variety of ways. However, it is clear that at wavelengths of 147 nm and above, H atom detachment (processes 58 and 59) is the main source of H_2 as proved by the fact that its formation is strongly reduced in the presence of scavengers. It further seems that at wavelengths above 194 nm process 58 outranks process 59 (226) while the reverse may be true at 147 nm (227).

The extent of reaction 60 appears to be the most uncertain. It has been observed that in methylamine $\phi(NH_3)$ far exceeds $\phi(CH_4) + 2\phi(C_2H_6)$, the sum being taken to represent $\phi(CH_3)$ (226, 227). However, it must be expected that a considerable portion of the methyl radicals would react with the radicals most abundant in this system forming ethylamine and dimethylamine (266). An induction period has been found for the latter, and it has been concluded to be a secondary product while the former was not reported. It is therefore possible that process 60 might be more important than has been hitherto assumed. With the higher homologs the ratio $\phi(H_2)/\phi(NH_3)$ is larger than in methylamine. In n-butylamine it is 1.9, and 2.5 in n-amylamine (224). These results indicate that hydrogen and ammonia formation, rather than being consecutive processes as had been suggested (226) are independent.

Processes 61 through 63 appear relatively more important at shorter wavelengths. While Michael and Noyes (226) working in the range from 194 to 244 nm rated them at less than 10 percent in methylamine, Magenheimer et al. (227) concluded that they add up to about 20 percent at 147 nm. At still higher energies (24 nm) the appearance of the species NH has been observed by Kawasaki and Tanaka (228) (process 64). There, only 80 percent of the original methane formation could be quenched by oxygen.

2. Secondary Amines. Studies on the gas phase photolysis of secondary amines have been restricted to dimethylamine. Experiments were carried out in the long-wavelength region by Bamford (224), at 147 nm by Magenheimer (238), and at 124 nm by Kawasaki and Tanaka (228). Photoionization measurements at 147 nm were performed by Magenheimer with dimethylamine, as well as by Salomon and Scala (239) with di-n-propylamine and diisopropylamine.

The photolysis of dimethylamine is even less clear than that of methylamine. A complete and thorough analysis has not yet been attempted. Nitrogen-containing products have not been determined. The most prominent gaseous product appears to be hydrogen followed by methane and ethane. A viscous liquid of low volatility formed in the long-wavelength region (224), but no ammonia. At 124 nm methane and ethane were formed in about equal yields in the absence of oxygen, but while ethane was almost completely suppressed in the presence of oxygen, methane was merely reduced to one-third of its previous yield (228). It has been suggested that the residual methane is eliminated via a molecular process (228). Photoionization occurs already at 147 nm. Values reported are $\phi = 0.03$ for dimethylamine (238), $\phi = 0.096$ for di-n-propylamine, and $\phi = 0.15$ for diisopropylamine (239).

3. Tertiary Amines. Studies on trimethylamine were performed in 1939 by Bamford (224) in the long-wavelength region. They were resumed by Gesser et al. (240). Some results at 147 nm were reported by Magenheimer (238). A study on the photolysis of triethylamine has been published by Kozak and Gesser (241). Photoionization measurements at 147 nm were made on trimethylamine by Magenheimer (238), and on both trimethylamine and triethylamine by Salomon and Scala (239).

There is agreement that hydrogen is a lesser product than ethane and methane combined (224,240). The same holds for triethylamine (241). Some rates of product formation from the latter are shown in Table 21. Addition of NO strongly diminished the rates of formation of all these products except hydrogen, which was quenched by about 20 percent only (241). In the absence of a full material balance it is unprofitable to propose a detailed reaction scheme. The evidence is that C-N bond scission predominates over C-H bond scission. A detectable amount

TABLE 21. Photolysis of triethylamine in the gas phase (241); rates of product formation (arbitrary units) at different temperatures.

T(^oC)	H_2	CH_4	C_2H_4	C_2H_6	C_3H_8	C_4H_{10}
28	9.0	1.8	3.0	16.8	5.0	1.8
254	11.5	35.5	12.2	150	13.7	5.6

of C-C splitting (in triethylamine) also occurs, even at room temperature. Chain reactions seem to become significant at higher temperatures (Table 21) if one assumes, in the absence (241) of quantum yield values, a quantum yield of triethylamine decomposition between 0.1 and unity at room temperature. At 147 nm, photoionization is a major primary process in tertiary amines. Some ionization quantum yields have been reported: trimethylamine 0.35 (238), 0.38 (239); triethylamine 0.46 (239).

4. Cyclic Amines. In the series of cyclic amines the gas phase photolysis of ethyleneimine at both 147 and 124 nm has found the detailed attention of Kawasaki and co-workers (242,243,228). Only little information is available on 1,3,5-trimethyl-1,3,5-triazacyclohexane, which had been studied briefly by Wetmore and Taylor (47).

In the photolysis of ethyleneimine (242,243,228), the formation of hydrogen, ethylene, ethane, methane, acetylene, propane, and n-butane under varying conditions has been measured. The formation of ammonia and of a dehydrodimer has been shown only qualitatively. A complete material balance is lacking, and the primary processes discussed are restricted to those given in Scheme 15. Assuming unity for the quantum yields of the sum of primary processes there is a deficit of 20 percent in the 147 nm photolysis and a deficit of 62 percent in the 124 nm photolysis. This is largely due to processes giving rise to hydrogen. Process 1 in Scheme 15 leads to NH and ethylene. The species NH have been identified in the 124 nm photolysis. Fluorescence measurements show (243) that NH is released in its $C^1\pi$ state if ethyleneimine is photolyzed at 124 nm, but no

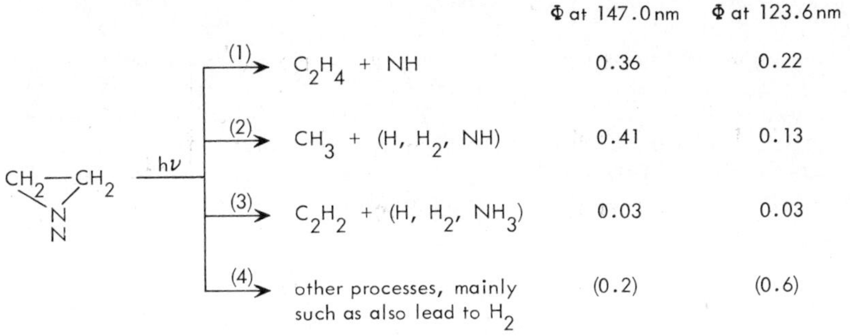

emission has been observed at 147 nm excitation. Calculation shows (243) that the excess energy at 147 nm for process 69 is only 0.1 eV, possibly too low for it to occur. The excess

$$C_2H_4NH^* \text{ (147 nm)} \longrightarrow C_2H_4(X^1A_{1g}) + NH\ (c^1\pi) \qquad (69)$$

energy in the 124 nm photolysis is about 1.6 eV. No emission from the $NH(A^3\pi)$ state has been observed in the direct photolysis in accordance with the spin conservation rule (for results on Kr- and Xe-sensitized reactions see below).

The formation of the methyl radical (reaction 2 in Scheme 15) is believed to be caused by the primary split of a C-N bond, followed by a hydrogen shift and further fragmentation. This mechanism is backed by the observation that addition of NH to ethylene yields methyl radicals (244). CH_3 formation could also be a secondary process, reaction 70 (245-247). The precursor of some of the acetylene is thought to be an excited ethylene

$$CH_2\text{-}CH_2(\text{-}\dot{N}\text{-}) \longrightarrow {}^{\bullet}CH_3 + HCN \qquad (70)$$

molecule from process 1 (Scheme 15). This is concluded (242) from the pronounced pressure dependence of the acetylene yield relative to that of ethylene.

VIII. Hg-PHOTOSENSITIZED DECOMPOSITION

In recent years it has been shown that the primary event in the Hg-sensitized decomposition of many hydrogen-bearing compounds, in particular paraffins and H_2, can be considered the process of H abstraction by the excited Hg. There is evidence that the excited Hg atom forms complexes with alcohols (248), amines (249), and alkanes (250). This has been concluded from fluorescence observation by Phillips and co-workers (248,249), and Gunning and co-workers (250). The transient molecule HgH has been detected spectroscopically in such a system by Vikis and LeRoy (251). The sequence 71 could be of importance in the

$$Hg^* + RH \longrightarrow (Hg\ RH)^* \longrightarrow {}^{\bullet}R + HgH \longrightarrow {}^{\bullet}R + H^{\bullet} + Hg \qquad (71)$$

Hg-sensitized photolysis of compounds such as open-chain saturated ethers as well, where the C-H bond is exclusively cleaved (252). Energy transfer, however, may operate in other cases. For example, in the Hg-sensitized photolysis of ethyl vinyl (253) and methyl vinyl (254,255) ethers the main process is the split of the alkyl C-O bond.

A. Alcohols

Most of the work in this field has been done in the gas phase. Some experiments were also carried out in the liquid phase by Juvet and Turner (256), and at 77 K by Bremer et al. (257), with results similar to those obtained in the gas phase. Gas phase studies with methanol as a substrate were carried out by Phibbs and Darwent (258) and later resumed by Pottie et al. (259) as well as Knight and Gunning (260a,b). Subsequently the latter also investigated ethanol (260c), isopropanol (260d,e), and t-butanol (260f). Brief studies on ethanol (261a) and isopropanol (261b) were carried out by Kato and Cvetanović. The Hg-sensitized luminescence of alcohols was investigated by Phillips and co-workers (248) (luminescence of alcohols can also be sensitized by Cd (292,295); λ 326 nm ≙ 87.7 kcal/mol).

In the series of primary and secondary alcohols (258-260e, 261) the formation of hydrogen is the major process. The dehydrodimers were recognized as the main corresponding products although they were not measured quantitatively except in study (261a), whereas the carbonyl compounds appeared of lesser importance (cf. Table 22). On irradiation with an intermittent light source (light period $< 10^{-4}$s, dark period $> 10^{-1}$s), $\phi(H_2)$ approaches unity (26b,c,d). In the presence of nitric oxide alkyl nitrites have been found as products in all cases investigated (260a,b,c,e). CD_3OD was formed from a mixture of CD_3OH and CH_3OH (260a). At very low pressures of CD_3OH ($\sim 5\times10^{-3}$torr) and in the presence of CD_3 radicals from $Hg(CD_3)_2$ photolysis led to the formation of CD_3OCD_3 but not C_2D_5OH (259). Alkoxyl radicals are also prominent in the photolysis of glassy Hg-containing alcohols at 77 K (257). These facts are in apparent conflict with the finding (258) that the hydrogen obtained from CH_3OD contained only 7 percent HD while the rest was H_2.

The formation of alkyl nitrites, CD_3OD, and dimethyl ether as well as the appearance of the alkoxyl radicals indicates the importance of the primary process 72, whereas the low HD content of the hydrogen formed from CH_3OD seems to favor reaction 73, which

$$RR'CHOH + Hg^* \longrightarrow RR'CHO^{\cdot} + H^{\cdot} + Hg \qquad (72)$$

$$RR'CHOH + Hg^* \longrightarrow RR'\dot{C}OH + H^{\cdot} + Hg \qquad (73)$$

TABLE 22. Hg-sensitized decomposition of saturated alcohols in the gas phase at room temperature. Quantum yields.

	Methanol			Ethanol		Isopropanol			t-Butanol	
Reference	(258)[a]	(260a)[b]	(260b)[c]	(260c)[b]	(261a)[e]	(260d)[d]	(260d)[c]	(261b)[b]	(260f)[b]	(260f)[c]
H_2	0.43	0.46	∿1.0	0.53 >0.95[c]	0.77	0.63 (0.72)	1.0	0.77	0.045	>0.1
CO	0.016			0.025	0.026			0.005		<0.01
CH_4	0.010			0.002	0.007	0.008(0)		<0.01	0.015	>0.05
C_2H_6		0.005		0.006		0.01 (0)			0.03	
CH_2O	0.020	0.01								
CH_3CHO				0.08	0.24	0.011(0)		<0.005		
CH_3OCH_3		0.006								
CH_3COCH_3						0.22 (0.26)	0.4	0.60	0.09	
$C(CH_3)_4$									0.007	
$CH(CH_3)_3$									0.004	
$CH_3OC(CH_3)_3$									0.003	
dehydrodimers	present	present		present	0.42	present		present	not identified	

[a]Exposure time 2 hr.
[b]Extrapolated to zero exposure time.
[c]Intermittent illumination, light periods $\sim 10^{-4}$s, dark periods $\sim 10^{-1}$s.
[d]Exposure time 5 min; values in parentheses from extrapolation to zero exposure time.
[e]Initial yield.

is considered not unlikely in ethanol (261a). The most important primary process in competition with luminescence in alcohols is claimed to be α hydrogen abstraction except with methanol (248). Ethanol is a much better quencher of Hg^* than methanol, and this is thought to be so on account of the former having secondary H atoms (262).

Except for the case of t-butanol (260), $\phi(H_2)$ approaches unity at intermittent light conditions. This means that in steady illumination experiments a considerable part of the H atoms disappear in combination reactions. The high radical density (at low hydrogen donor concentration) is reached by the high absorption coefficient of the Hg vapor. Some minor products are thought to originate from "atomic cracking" reactions 74.

$$RO^{\bullet} + H^{\bullet} \longrightarrow ROH^{*} \longrightarrow R^{\bullet} + \dot{O}H \qquad (74)$$

t-Butanol behaves differently from the primary and secondary alcohols. At steady illumination, acetone with $\phi = 0.09$ had the highest quantum yield, twice as large as that of hydrogen. At intermittent illumination, the $\phi(H_2)$ grew with increasing tendency and reached about 0.1 under the conditions of the shortest light periods possible (Table 22). The differences to the primary and secondary alcohols can be explained if one assumes that here also the primary process is the production of H atoms and alkoxyl radicals. Some of the t-butoxyl radicals will disappear via unimolecular decomposition into acetone and methyl. The tertiary alcohol, in contrast to the primary and secondary ones, is relatively inert to abstractive attack with the result that radical concentrations are high because the rate constant for recombination of t-butoxyl radicals is low. $k_{rec} = 10^{8.3}\ M^{-1}\ s^{-1}$ has been estimated (263). In consequence of this, more H atoms disappear in combination reactions than in other alcohols, and $\phi(H_2)$ is lower.

B. Ethers

The Hg-sensitized photolysis of several ethers in the liquid phase was briefly studied by Tsao (264). The gas phase photosensitized decomposition of ethers has found much more interest. In Steacie's group before 1950 there were investigated diethyl ether (265), ethylene oxide (266), and dimethyl ether (267). Cvetanović and co-workers studied ethylene oxide (268) and trans-2,3-epoxybutane (269). Subsequent work on dimethyl ether was done by Pottie et al. (259), Takezaki et al. (270), Loucks and Laidler (271), and Payette et al. (252). The Hg-sensitized photolysis of ethyl methyl ether was undertaken by Filseth (272).

Most authors agree that the initial chemical event in the Hg-sensitized photolysis of saturated straight-chain ethers in the decomposition of the ether molecule into the 1-alkoxylalkyl radical and the H atom, for example, reaction 75. In the initial stages of the photolysis the main products are hydrogen and the

$$Hg^* + CH_3OCH_2CH_3 \longrightarrow H^\bullet + CH_3O\dot{C}HCH_3 + Hg \qquad (75)$$

dehydrodimers (Table 23). It appeared that the initial $\phi(-Et_2O)$ (265) and $\phi(-Me_2O)$ (267) were close to two.

TABLE 23. Quantum yields in the Hg-sensitized photolysis of some straight-chain ethers.

	Et_2O^a	Me_2O^b	Me_2O^c	Me_2O^d	$EtOMe^e$
H_2	0.48	0.81	4.65	1.0	0.55
Dehydrodimer	0.48	-	2.53	1.0	0.54
CO	0.018	0	0.048		-
CH_4	0.005	-	1.59		0.003
C_2H_4	0.013 (C₂H₄ + C₂H₆)	-			0.015
C_2H_6			0.116		0.002
Aldehyde	0.024	-	-		0.02
$CH_3CH_2OCH_3$			0.84		
CH_2=CH-OCH_3					0.13
$(CH_3)_2$CH-OCH_3					0.025

[a] 150 torr, exposure 10 min, 35°, dose rate 1.7 x 10^{18} quanta/s (265).
[b] 110 torr, exposure 1 hr, 25°, dose rate 4.5 x 10^{15} quanta/s (267).
[c] Yields, arbitrary units; 75 torr, exposure 600 s, 200° (271).
[d] 200 torr, exposure 120 min, room temperature, dose rate 1.2 x 10^{14} quanta/s (252).
[e] 2 torr, exposure 20 s, 55°, dose rate 1 x 10^{16} quanta/s (272).

Under conditions of relatively high radical density, that is, at low substrate pressures and high light intensities, and at elevated temperatures, products other than hydrogen and dehydrodimers become important (Table 23). Methane, ethane, and ethyl methyl ether were found (271) from dimethyl ether at 200° and above, but there was no evidence for CH_3 production at 30°; if present such products would not have escaped detection. This result was confirmed (252) when it was definitively shown (that at room temperature, 200 torr, and very low intensity the relationship $\phi(-Me_2O) = 2\phi(H_2) = 2\phi(\text{dehydrodimers}) = 2.0$ is exactly obeyed, whereby it was certain that H$^{\bullet}$ is the precursor of all the hydrogen since this is completely quenched by olefin (272). The decrease of the hydrogen quantum yield at lower pressures and higher light intensities is due to the recombination of H atoms with radicals whose reaction then competes more successfully with H abstraction. The results of Pottie et al. (259), which showed considerable C-O bond cleavage to occur, can be explained on this basis inasmuch as this process was not brought about by λ 185 which accompanies λ 254. In their system (p_{Me_2O} 10^{-2} torr, p_{He} 8 torr, high-intensity low-pressure Hg lamp), reactions 76 and 77 would probably occur. An apparent second-order rate constant k_{77} of about 1×10^{10} l $mol^{-1}s^{-1}$ in the

$$H^{\bullet} + {}^{\bullet}CH_2OCH_3 \longrightarrow CH_3OCH_3^{*} \longrightarrow \dot{C}H_3 + {}^{\bullet}OCH_3 \quad (76)$$

$$H^{\bullet} + \dot{C}H_3 + M \longrightarrow CH_4 + M \quad (77)$$

presence of 6.7 torr He has been given (273).

While the mechanism of the Hg-sensitized photolysis of linear saturated ethers thus seems well established, the picture appears less clear for ethylene oxide and its homologs. Table 24 shows some product quantum yields. It is concluded from Cvetanović's detailed study (268) that reactions 78 through 81 are probably important. H atoms must play a large part in the

$$(CH_2)_2O + Hg^{*} \longrightarrow (CH_2)_2O^{*} + Hg \quad (78)$$

$$(CH_2)_2O^{*} \longrightarrow {}^{\bullet}CH_3 + {}^{\bullet}CHO \quad (79)$$

Hg-sensitized decomposition of ethylene oxide since hydrogen formation is strongly reduced in the presence of olefins. Two routes to H atoms were proposed, reactions 80 and 81.

$$(CH_2)_2O + Hg^{*} \longrightarrow \dot{C}_2H_3O + H^{\bullet} + Hg \quad (80)$$

$$^{\bullet}CHO \longrightarrow CO + H^{\bullet} \quad (81)$$

TABLE 24. Quantum yields of products from the Hg-sensitized photolysis (25°) of some epoxides.

	$(CH_2)_2O$ (266)	$(CH_2)_2O$ (268)			$Me_2(CH)_2O$ (269)
	ϕ^a	ϕ^b	ϕ^c	ϕ^d	ϕ^e
CO	0.24	0.29	0.13	0.14	0.07
H_2	0.18	0.22	0.04	0.03	0.02
CH_4	0.04	0.04	0.01	0.02	0.015
C_2H_4	} 0.05 (C_2H_4 + C_2H_6)	trace	-	0.04	0.05
C_2H_6		0.04	0.04	0.02	0.11
C_3H_8		0.04	0.04	0.001	0.03
CH_2CO		<0.01	0.02	-	
Carbonyl Compounds	0.15	0.17	0.24	0.18	0.18

[a]100 torr, exposure 1 hr, dose rate 1.6×10^{16} quanta/s.
[b]100 torr, exposure 30 min, dose rate 4.9×10^{15} quanta/s, 2.8 torr.
[c]100 torr, exposure 90 min, dose rate 4.9×10^{15} quanta/s, 2.8 torr ethylene.
[d]100 torr, exposure 100 min, dose rate 3.7×10^{15} quanta/s, 2.8 torr butene-1.
[e]60 torr, low conversion, dose rate 3.5×10^{15} quanta/s.

Rearrangement of the oxiranyl radical (274) might lead to $\cdot CH_2CHO$ which would explain the formation of propionaldehyde (268).

Methyl radicals played an important role in the more complicated trans-2,3-epoxybutane system (269) and the primary process 82 was postulated. The α-methyloxiranyl radical might be

```
CH3-CH-CH-CH3                 ·CH-CH-CH3
     \ /        + Hg* --->        \ /      + ·CH3 + Hg        (82)
      O                            O
```

stabilized which would explain a pronounced decrease of the carbonyl product compounds with increasing pressure from

ϕ = 0.36 at 10 torr to 0.18 at 60 torr. The dimer of this radical could not be determined, however. An interesting result was the detection of minor amounts of *cis*-2,3-epoxybutane.

C. Amines

The Hg-sensitized photolysis of methylamine was studied by Wetmore and Taylor (47) and Watson and Darwent (237). The behavior of some primary and secondary amines was investigated by Baum et al. (235). Hg-sensitized luminescence of primary, but not of secondary and tertiary amines was found by Phillips and co-workers (249). The Cd-sensitized luminescence ($Cd^* \triangleq$ 326 nm $\triangleq$ 87.7 kcal/mol) of primary, secondary, and tertiary amines has been studied by Yamamoto and Sato, and found to occur via a metal/amine complex (275). There has been considerable interest in the sensitized photolysis of ethyleneimine. Its Hg-sensitized photolysis has been done by Luner and Gesser (245), by Klemm (276), and by Jones and Lossing (246). Kawasaki et al. carried out its Kr- and Xe-photosensitized decomposition (242, 243).

The data available suggest that the Hg-sensitized photolysis of acyclic aliphatic amines (47,237,235) yields the same products that are found in the direct photolysis (see above) but perhaps in different proportion. Hydrogen, ammonia, and a cyclic trimer of N-methylmethyleneimine were obtained from methylamine (237). Baum et al. (235) suggested that the Hg-sensitized photolysis of primary and secondary amines is of preparative value for the synthesis of N-alkylalkylideneimines.

CH_3ND_2 yielded 33 percent H_2, 65 percent HD and 2 percent D_2. The corresponding values for CD_3NH_2 were 30 percent H_2, 70 percent HD and < 1 percent D_2 (237), indicating that in the primary event hydrogen atoms are detached (reactions 83 and 84) from both C and N even though isotopic effects with respect to hydrogen abstraction by Hg^* and hydrogen atoms complicate the picture. The rate of hydrogen formation in these systems was substantially reduced in the presence of olefins (235). The

$$CH_3NH_2 + Hg^* \dashrightarrow CH_3\dot{N}H + Hg + H^\bullet \quad (83)$$

$$\dashrightarrow \dot{C}H_2NH_2 + Hg + H^\bullet \quad (84)$$

primary processes 83 and 84 indicate that linear aliphatic amines resemble both alcohols and ethers in that the hetero radical (as in the former) as well as the α-carbon radical (as in the latter) is formed. Abstraction of α hydrogen is claimed

TABLE 25. Product yields in the sensitized photolysis of ethyleneimine.

	$\phi^a(Hg^*)$	$R^b(Hg^*)$	$R^c(Hg^*)$	$\phi^d(Hg^*)$	$R^e(Kr^*)$	$R^f(Xe^*)$
H_2	0.50	0.43	1	0.4		
CH_4	0.012	0.03		0.02	0.54	0.36
C_2H_2	0.002				0.31	0.13
C_2H_4	0.52	1	1	0.4	1	1
C_2H_6	0.009				0.77	0.66
C_3H_8	0.003				0.20	0.07
C_4H_8	Present					
N_2	0.13	0.12	0.63	0.12		
NH_3	Present	0.09		0.27		
HCN	Present	0.08		0		
CH_3CN	-	0.08		0		
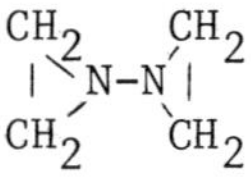	Present			Present	Present	Present
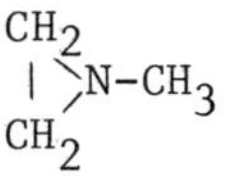				Present		
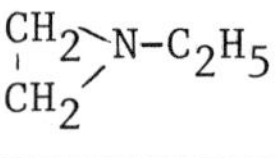				Present		

[a]123 torr, 27°, conversion 3% (245).
[b]Relative yields, flow system, 7.5 x 10^{-3} torr, plus 12 torr He, 55°, intense low-pressure Hg arc (246).
[c]Relative yields, static system, 2 torr, 25° (246).
[d]Primary quantum yields, 100 torr (276).
[e]Kr-sensitized, relative yields, 0.7 torr, 2.7 torr Kr (242).
[f]Xe-sensitized, relative yields, 0.9 torr, 0.8 torr Xe (242).

to be in competition with the Hg-sensitized luminescence of primary amines, except methylamine (249).

The situation is more complex in the case of ethyleneimine. Experimental results are summarized in Table 25. There is evidence that the main product ethylene is formed in a molecular-type elimination reaction (1 in Scheme 16). Primary processes

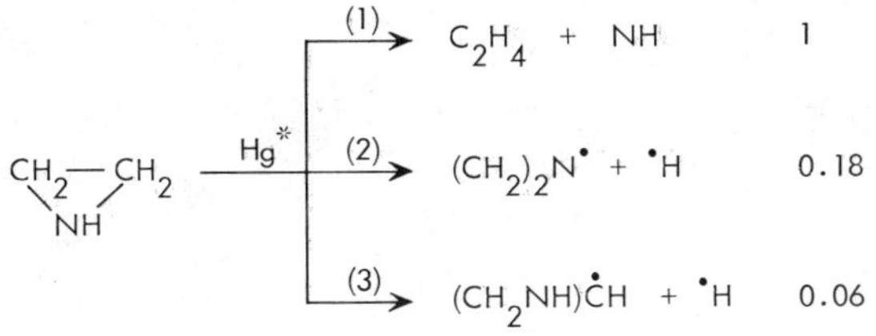

such as 2 and 3 in Scheme 16 must occur in this system. The quantum yields of hydrogen and nitrogen were quenched in the presence of ethylene (269). Most of the hydrogen from $(CD_2)_2NH$ was H_2 (Table 26). Material considered to represent dehydro-dimer (245,276,242) was detected. N-methyl- and N-ethylethylene-imine were identified (276). The intermediacy of the radical $(CH_2NH)\dot{C}H$ could also be established by trapping it with CH_3 and identifying the ensuing 2-methylethyleneimine (246). Methane was completely quenchable by oxygen (242).

A complete quantitative mechanism cannot be constructed with confidence on the basis of the products since not all of them were determined. A variety of secondary processes is likely. Hydrogen abstraction is expected to involve mainly the N-bound H atom. The activation energy for abstraction by methyl is given as 4.8 kcal/mol (277). Reaction 85 has been found (277) to yield ethylene and nitrogen in the radical-sensitized decomposition of ethyleneimine. Reaction 70 (see above) could explain the production of hydrocyanic acid.

TABLE 26. Isotopic composition (%) of various products from the Hg-sensitized photolysis of $(CD_2)_2NH$ (276).

Hydrogen		Methane		Ethylene	
H_2	67	CD_2H_2	63	C_2D_3H	2
HD	33	CD_3H	33	C_2D_4	98
D_2	<1	CD_4	3		

$$2(CH_2)_2N^{\cdot} \longrightarrow 2C_2H_4 + N_2 \quad (85)$$

Data from the Kr- and Xe-sensitized photolysis (242) are shown in Table 25. Although they are incomplete they indicate that the rare gas-sensitized photolysis of ethyleneimine resembles more the direct photolysis at the rare gas lines and less the Hg-sensitized one.

IX. PHOTOLYSIS OF O_2-CT COMPLEXES

In recent years a certain amount of experimental material on the photolysis of O_2-CT complexes from ethers and amines has become available whereas no such work seems to have been done during this time on alcohols. This discussion is therefore restricted to the former two classes of compounds.

A. Ether-O_2-CT-Complexes

The photolysis of oxygenated diethyl ether has been studied in 1955 by Horner and Jurgeleit (278) was later further investigated by Stenberg and Kulevsky (279,280), and more recently resumed by the reviewers' group (122), and Tsubomura and co-workers (281). There are studies on the tetrahydrofuran-O_2 system by Stenberg and Kulevsky (280,282), on oxygenated polyethylene glycol by Reimann, Hollatz, and Eckert (283), and some unpublished material from the reviewers' laboratory (284,285).

Most extensively investigated has been the photolysis of the diethyl ether-O_2 CT-complex (122) carried out with λ 254 at lower temperature (-30°) where oxygen is more soluble than at room temperature, and thus stronger absorbance through the CT-complex minimized photolysis of the products. Special care was taken to keep conversions low. A satisfactory material balance was reached. Products and their quantum yields are listed in Table 27. Tsubomura et al. (281) also reported quantum yields for some of the products such as ethylacetate (ϕ = 0.49), ethyl formate (ϕ = 0.23), acetaldehyde (ϕ = 0.17), and ethanol and/or methanol (ϕ = 0.43). They also undertook ESR experiments at 77 K but the radical observed ($^{\cdot}CH_3$) was thought not to originate in the primary photoprocesses.

The reviewers' interpretation is through electron transfer from the ether to the oxygen molecule (reaction 86) as the primary process:

TABLE 27. Quantum yields of the major products in the photolysis (λ 254 nm) of O_2-saturated diethyl ether at -30° (122).

Product	ϕ
Peroxide (total)	0.28
Hydrogen peroxide	0.24
Diethyl ether-α-hydroperoxide	0.04
Ethyl acetate	0.26
Acetaldehyde	0.18
Ethanol	0.18
Ethyl formate	0.04
Methanol	0.035
Formaldehyde	0.005
Ethyl vinyl ether	0.013

$$\text{Et-O-Et}\cdots O_2 \xrightarrow{h\nu} \text{Et-O-Et}^{\cdot +} + O_2^{-} \qquad (86)$$

$$\text{Et-O-Et}^{\cdot +} + O_2^{-} \longrightarrow CH_2\text{-}\dot{C}H\text{-O-Et} + HO_2^{\cdot} \qquad (87)$$

$$CH_3\text{-}\dot{C}H\text{-O-Et} + HO_2^{\cdot} \longrightarrow CH_3\text{-}\underset{\substack{|\\ OOH}}{CH}\text{-O-Et} \qquad (88)$$

$$CH_3\text{-}\dot{C}H\text{-O-Et} + O_2 \longrightarrow CH_3\text{-}\underset{\substack{|\\ O_2^{\cdot}}}{CH}\text{-OEt} \qquad (89)$$

$$2HO_2^{\cdot} \longrightarrow H_2O_2 + O_2 \qquad (90)$$

$$2CH_3\text{-}\underset{\substack{|\cdot \\ O_2}}{CH}\text{-}O\text{-}Et \longrightarrow CH_3\text{-}CO\text{-}OEt + O_2 + CH_3\text{-}\underset{\substack{| \\ OH}}{CH}\text{-}OEt \quad (91)$$

$$CH_3\text{-}\underset{\substack{|\cdot \\ O_2}}{CH}\text{-}O\text{-}Et + HO_2^{\cdot} \longrightarrow CH_3\text{-}CO\text{-}OEt + H_2O + O_2 \quad (92)$$

$$2CH_3\text{-}\underset{\substack{|\cdot \\ O_2}}{CH}\text{-}O\text{-}Et \longrightarrow 2\,^{\cdot}CH_3 + 2\ HCO\text{-}OEt + O_2 \quad (93)$$

Proton transfer from the radical cation to the O_2^- gives the ethoxyethyl and $HO_2^{\cdot}$ radicals. Cage combination of these two radicals gives the ethoxyethyl hydroperoxide (reaction 88). This hydroperoxide is only a minor product. With high probability the two radicals separate and the ethoxyethyl radical is scavenged by oxygen to give the ethoxyethylperoxyl radical (reaction 89). Disproportionation reactions of the $HO_2^{\cdot}$ radicals give hydrogen peroxide (reaction 90). The disproportionation reaction of the ethoxyethylperoxyl radicals yields ethyl acetate and acetaldehyde ethyl hemiacetal which decomposes into the products acetaldehyde and ethanol. More ethyl acetate is formed through disproportionation of $HO_2^{\cdot}$ with the ethoxyethylperoxyl radicals (reaction 92). The minor products ethyl formate, formaldehyde and methanol are thought to arise from a fragmentation reaction 93. The methyl radicals are scavenged by oxygen giving rise to methylperoxyl radicals, the likely peecursors of the products methanol and formaldehyde. Another fragmentation reaction yields ethyl vinyl ether. Except for the primary photochemical event, the mechanism described follows the same lines as traced by Ingold and co-workers (286) regarding the fate of the peroxyl radicals derived from diethyl ether in another system. It is calculated that the quantum yield of the primary processes leading to products is 0.5.

In a preliminary study (284) on the photolysis of the t-butyl methyl ether-O_2 CT-complex some products have been identified and their quantum yields determined. They are peroxidic compounds ($\phi = 0.15$), t-butyl formate ($\phi = 0.21$), t-butanol ($\phi = 0.035$), 2-methoxy-2-methylpropionaldehyde ($\phi = 0.03$), formaldehyde ($\phi = 0.04$), water ($\phi = 0.3$), and carbon dioxide ($\phi = 0.007$). It was noted (284) that deprotonation of the t-butyl methyl ether radical cation, although it largely occurs at the methyl group next to the oxygen (product, t-butyl formate), is also possible at the β-position (product, 2-methoxy-2-methylpropionaldehyde).

The study on the photolysis of the tetrahydrofuran-O_2 CT-complex presented by Stenberg and Kulevsky (280) must be considered as preliminary. More recently it has been found that initially hydrogen peroxide is formed as a major part of the peroxidic compounds (285). Hence the conclusions drawn previously (280) would have to be revised.

Oxygen saturated polyethylene glycol (283) shows an absorption band in the ultraviolet which is ascribed to a CT-absorption. Photolysis induces the formation of peroxidic compounds. Glycolaldehyde is an important end product.

B. Amine-O_2 CT-Complexes

Stratford observed that the rate of amine oxydation was enhanced by the action of ultraviolet light, and some products were isolated (287). Studies were taken up again by Niu (124), Kulevsky, Niu and Stenberg (288), and by Tsubomura and his group (123). In the series of tertiary amines the formation of N-oxides (287,124) as well as of the corresponding secondary amines (287,123) were observed. Acetaldehyde was a product in the case of triethylamine (123). However, there remained much unidentified material (287,123). In the photolysis of triethylamine the formation of Et_2NO was detected by ESR (123). In the photolysis of the cyclohexylamine-O_2 CT-complex (124,288) cyclohexanone oxime, N-cyclohexylcyclohexylideneimine, ammonia, water, and peroxidic products were observed. Owing to the lack of quantitati·e measurements, material balances cannot be established. Further mechanistic discussion would, therefore, seem unjustified at this time.

REFERENCES

(1) E.-M. Raoult, Compt. Rend., 73, 1049 (1871).

(2) U. Kreusler, Ber. Deut. Chem. Ges. 8, 93 (1875).

(3) H. Gillot, Bull. Acad. Roy. Belg. 40, 863 (1901); Chem. Zentralbl., 1901 I, 377.

(4) (a) H. Bierry, V. Henri, and A. Ranc, Compt. Rend. 151, 316 (1910); (b) ibid., 152, 1629 (1911); (c) ibid. 155, 1151 (1912).

(5) (a) D. Berthelot and H. Gaudechon, Compt. Rend., 151, 395 (1910); (b) ibid., 155, 401 (1912); (c) ibid., 155, 831 (1912); (d) ibid., 155, 1016 (1912); (e) ibid., 155, 1153 (1912); (f) ibid., 156, 707 (1913).

(6) A. Guntz and J. Minguin, Compt. Rend., 152, 372 (1911).

(7) C. Neuberg, Biochem. Z., 39, 158 (1912); Chem. Zentralbl., 1912 I, 1617.

(8) H. Euler and E. Lindberg, Biochem. Z., 39, 410 (1912); Chem. Zentralbl., 1912 II, 1989.

(9) Y. Dalström, Arkiv Kemi, Min. och. Geol., 4 No. 30 (1912); Chem. Zentralbl., 1912 II, 1013.

(10) A. Ranc and R. Wurmser, Ind. Chim., 7, 429 (1920).

(11) P. Holtz, Naunyn-Schmiedebergs Arch. exp. Pathol. Pharmakol., 182, 141 (1936); Chem. Zentralbl., 1936 II, 3440.

(12) A.L. Bernoulli and R. Cantieni, Helv. Chim. Acta, 15, 119 (1932).

(13) R. Cantieni, Helv. Chim. Acta, 17, 1528 (1934); ibid., 18, 473, 933 (1935).

(14) T.C. Laurent and E.M. Wertheim, Acta Chim. Scand., 6, 678 (1952).

(15) T.C. Laurent, J. Am. Chem. Soc., 78, 1875 (1956).

(16) G.O. Phillips and G.J. Moody, J. Chem. Soc., 1960, 3398.

(17) G.O. Phillips, Advan. Carbohydrate Chem., 18, 9 (1963) (Review).

(18) G.O. Phillips and T. Rickards, J. Chem. Soc. B, 1969, 455.

(19) P.J. Baugh and G.O. Phillips, "Cellulose and Cellulose Derivatives", N. Bikales and L. Segal, Eds., Vol. 5, Part 5, Wiley, 1971, p. 1047 (Review).

(20) A. Bos, J. Polym. Sci., 12, 2283 (1974).

(21) (a) D. Berthelot and H. Gaudechon, Compt. Rend., 151, 478 (1910); (b) ibid., 151, 1349 (1910); (c) ibid., 153, 383 (1911); (d) ibid., 156, 68 (1913).

(22) M. Volmar, Compt. Rend., 182, 1607 (1926).

(23) J.R. Bates and H.S. Taylor, J. Am. Chem. Soc., 49, 2438 (1927).

(24) F. Paneth and W. Hofeditz, Ber. Deut. Chem. Ges., 62, 1335 (1929).

(25) F.O. Rice and K. Herzfeld, J. Am. Chem. Soc., 56, 284 (1934).

(26) K.F. Bonhoeffer, Z. Elektrochem., 40, 425 (1934).

(27) F. Patat, Anz. Akad. Wiss. Wien, math.-nat. Kl., 1933, 93; Chem. Zentralbl., 1933 II, 1975.

(28) F. Patat, Z. Elektrochem., 41, 494 (1935).

(29) O. Mabuchi, J. Chem. Soc. Japan, 63, 948 (1942); C. A., 41, 3372g (1947).

(30) A. Terenin and H. Neujmin, Nature, 134, 255 (1934).

(31) H. Neujmin and A. Terenin, Acta physicochim. URSS, 5, 465 (1936); J. Chem. Phys., 3, 436 (1935).

(32) L. Farkas and Y. Hirshberg, J. Am. Chem. Soc., 59, 2450 (1937).

(33) L. Farkas, Y. Hirshberg, and L. Sandler, J. Am. Chem. Soc., 61, 3393 (1939).

(34) G. Leuschner and K. Pfordte, Liebigs Ann. Chem., 619, 1 (1958).

(35) K. Pfordte and G. Leuschner, Strahlentherapie, 111, 417 (1960).

(36) K. Pfordte and G. Leuschner, Atompraxis, 6, 50 (1960).

(37) H. Bierry, V. Henri, and A. Ranc, Compt. Rend., 152, 535 (1911).

(38) V. Henri and A. Ranc, Compt. Rend., 154, 1261 (1912).

(39) G.O. Phillips and P. Barber, J. Chem. Soc., 1963, 3990.

(40) K. Pfordte and G. Leuschner, Liebigs Ann. Chem., 643, 2 (1961).

(41) K. Pfordte, Liebigs Ann. Chem., 625, 30 (1959).

(42) D. Berthelot and H. Gaudechon, Compt. Rend., 152, 377 (1911).

(43) H.J. Emeléus and H.S. Taylor, J. Am. Chem. Soc., 53, 3371 (1931).

(44) H.J. Emeléus and L.J. Jolley, J. Chem. Soc., 1935, 1612.

(45) J. Franck, Z. Phys., 9, 259 (1922).

(46) G. Cario and J. Franck, Z. Phys., 11, 161 (1922).

(47) O.C. Wetmore and H.A. Taylor, J. Chem. Phys., 12, 61 (1944).

(48) A. Richardson and E.C. Fortey, Trans. Chem. Soc., 69, 1349 (1896).

(49) L. Mathieu, Bull. Assoc. Chim. Sucre Dist., 22, 1283 (1905); Chem. Zentralbl., 1905 II, 782.

(50) H.D. Gibbs, Philippine J. Sci., 7 A, 57 (1912); Chem. Zentralbl., 1912 II, 2044.

(51) G. Ciamician and P. Silber, Ber. Deut. Chem. Ges., 46, 3894 (1913).

(52) N.R. Dhar and R.P. Sanyal, J. Phys. Chem., 29, 926 (1925).

(53) H.S. Taylor and A.J. Gould, J. Phys. Chem., 37, 367 (1933).

(54) R. Cantieni, Ber. Deut. Chem. Ges., 69, 1101, 1386, 1796 (1936).

(55) N.A. Milas, U.S. Pat. 2115, 1, 206-207; Chem. Zentralbl., 1938 II, 768.

(56) C.E. Redemann, J. Am. Chem. Soc., 64, 3049 (1942).

(57) W.R. Dunstan and T.S. Dymond, Trans. Chem. Soc., 57, 574, 988 (1890).

(58) A. Richardson and E.C. Fortey, Trans. Chem. Soc., 59, 51 (1891); ibid., 69, 1352 (1896).

(59) A.M. Clover, J. Am. Chem. Soc., 44, 1107 (1922).

(60) A.M. Clover, J. Am. Chem. Soc., 46, 419 (1923).

(61) N.A. Milas, J. Am. Chem. Soc., 53, 227 (1931).

(62) R. Neu, Angew. Chem., 45, 519 (1932).

(63) G. Radulesco, Ann. Office Nat. Combust. Liq., 13, 1071 (1938); Chem. Zentralbl., 1939 I, 4456.

(64) C.W. Tasker and C.B. Purves, J. Am. Chem. Soc., 71, 1018 (1949).

(65) G. Ciamician and P. Silber, Ber. Deut. Chem. Ges., 48, 183 (1915).

(66) R. Cantieni, Z. Wiss. Photogr. Photophys. Photochem., 36, 119 (1937).

(67) N.A. Milas, J. Phys. Chem., 33, 1204 (1929).

(68) A. Bach, Compt. Rend., 124, 951 (1897).

(69) J.G. Calvert and J.N. Pitts, Jr., "Photochemistry," Wiley, New York, 1966.

(70) H. Tsumbomura, K. Kimura, K. Kaya, J. Tanaka, and S. Nagakura, Bull. Chem. Soc. Japan, 37, 417 (1964).

(71) M.B. Robin and N.A. Kuebler, J. Electron Spectrosc., 1, 13 (1972).

(72) A.J. Harrison, B.J. Cederholm, and M.A. Terwilliger, J. Chem. Phys. 30, 355 (1959).

(73) H. Kaiser, Doctoral thesis, Universität München (1970).

(74) D.R. Salahub and C. Sandorfy, Chem. Phys. Lett., 8, 71 (1971).

(75) J. Hagège, P.C. Roberge and C. Vermeil, J. Chim. Phys., 65, 641 (1968); Ber. Bunsenges. Phys. Chem., 72, 138 (1968).

(76) J.C. Weeks, G.M.A.C. Meaburn, and S. Gordon, Radiat. Res., 19, 559 (1963).

(77) J. Barrett, A.L. Mansell, and M.F. Fox, J. Chem. Soc. (B), 1971, 173.

(78) D. Sänger, Doctoral thesis, Universität Karsruhe (1970).

(79) M. Ogawa and G.R. Cook, J. Chem. Phys., 28, 747 (1958).

(80) K. Watanabe, T. Nakayama, and J. Mottl, J. Quant. Spectrosc. Radiat. Transfer, 2, 369 (1962).

(81) W. Kaye and R. Poulsen, Nature, 193, 675 (1962).

(82) D. Schulte-Frohlinde, D. Sänger, and C. von Sonntag, Z. Naturforsch., 27b, 205 (1972).

(83) H.-P. Schuchmann, C. von Sonntag, and D. Schulte-Frohlinde, J. Photochem., 4, 63 (1975).

(84) J. Doucet, P. Sauvageau, and C. Sandorfy, Chem. Phys. Lett., 17, 316 (1972).

(85) A.J. Harrison and D.R.W. Price, J. Chem. Phys., 30, 357 (1959).

(86) H.-P. Schuchmann and C. von Sonntag, Tetrahedron, 29, 1811 (1973).

(87) H.-P. Schuchmann and C. von Sonntag, Tetrahedron 29, 3351 (1973).

(88) H.-P. Schuchmann and C. von Sonntag, Z. Naturforsch., 30b, 399 (1975).

(89) H.-P. Schuchmann and C. von Sonntag, unpublished results.

(90) G. Fleming, M.M. Anderson, A.J. Harrison, and L.W. Pickett, J. Chem. Phys., 30, 351 (1959).

(91) J.D. Margerum, J.N. Pitts, Jr., J.R. Rutgers, and S. Searles, J. Am. Chem. Soc., 81, 1549 (1959).

(92) L.W. Pickett, N.J. Hoeflich, and T.-C. Liu, J. Am. Chem. Soc., 73, 4865 (1951).

(93) R. McDiarmid, J. Chem. Phys., 60, 4091 (1974).

(94) G.J. Hernandez, J. Chem. Phys., 38, 1644 (1963).

(95) R. McDiarmid, J. Chem. Phys., 60, 3340 (1974).

(96) G.J. Hernandez, J. Chem. Phys., 39, 1355 (1963).

(97) P.A. Clark, Doctoral thesis, University of Massachusetts (1967); Diss. Abstr., 28 B, 130 (1967/68).

(98) G.J. Hernandez, J. Phys. Chem., 38, 2233 (1963).

(99) R. Davidson, J. Hög, P.A. Warsop, and J.A.B. Whiteside, J. Chem. Soc. Faraday Trans. II, 68, 1652 (1972).

(100) G.D. Gray, J. Hög, P.A. Warsop, and J.A.B. Whiteside, Trans. Faraday Soc., 66, 2130 (1970).

(101) G.J. Hernandez and A.B.F. Duncan, J. Chem. Phys., 36, 1504 (1962).

(102) B.C. Roquitte, J. Phys. Chem., 70, 2863 (1966).

(103) (a) H.-P. Schuchmann and C. von Sonntag, J. Chem. Soc. Perkin Trans. II, 1976, 1408. (b) Z. Naturforsch., in print.

(104) P. Naderwitz, H.-P. Schuchmann and C. von Sonntag, Z. Naturforsch., in print.

(105) E. Tannenbaum, E.M. Coffin, and A.J. Harrison, J. Phys. Chem., 21, 311 (1953).

(106) A.M. Halpern and T. Gartman, J. Am. Chem. Soc., 96, 1393 (1974).

(107) A.M. Halpern, J.L. Roebber, and K. Weiss, J. Chem. Phys., 49, 1384 (1963).

(108) T.M. McKinney, Spectrochim. Acta, 25A, 501 (1969).

(109) A.M. Halpern, J. Am. Chem. Soc., 96, 7655 (1974).

(110) Y. Hamada, A.Y. Hirakawa, and M. Tsuboi, J. Mol. Spectrosc., 47, 440 (1973).

(111) A.M. Halpern, Mol. Photochem., 5, 517 (1973).

(112) Y. Muto, N. Nakato, and H. Tsubomura, Chem. Phys. Lett., 9, 597 (1971).

(113) M. Tsuboi, A.Y. Hirakawa, and H. Kawashima, J. Mol. Spectrosc., 29, 216 (1969).

(114) A. Hirakawa, H. Miyazaki, and M. Tsuboi, Bull. Chem. Soc. Japan, 45, 757 (1972).

(115) D.P. Stevenson, J. Am. Chem. Soc., 84, 2849 (1962).

(116) (a) R.S. Mulliken and W.B. Person, "Molecular Complexes", Interscience, New York, 1969; (b) Electron Donor-Acceptor Complexes and Charge Transfer Spectra, in "Physical Chemistry: An Advanced Treatise", (D. Henderson, Ed.), Vol. 3, Academic Press, New York, 1969.

(117) D.F. Evans, J. Chem. Soc. 1953, 345.

(118) R.S. Mulliken, J. Chem. Phys., 61, 20 (1964).

(119) A.V. Munck and J.F. Scott, Nature, 177, 587 (1956).

(120) H. Tsubomura and R.S. Mulliken, J. Am. Chem. Soc., 82, 5966 (1960).

(121) J. Jortner and U. Sokolov, J. Phys. Chem., 65, 1633 (1961).

(122) C. von Sonntag, K. Neuwald, H.-P. Schuchmann, F. Weeke, and E. Janssen, J. Chem. Soc. Perkin Trans. II, 1975, 171.

(123) H. Tsubomura, T. Yagishita, and H. Toi, Bull. Chem. Soc. Japan, 46, 3051 (1973).

(124) C.-H. Niu, Doctoral thesis, University of North Dakota (1972); Diss. Abstr., 33 B, 4199 (1972/73).

(125) N.C. Yang, D.P.C. Tang, Do-Minh Thap, and J.S. Sallo, J. Am. Chem. Soc., 88, 2851 (1966).

(126) S.S. Yang, Doctoral thesis, University of Chicago (1968).

(127) C. von Sonntag, Tetrahedron, 25, 5853 (1969).

(128) C. von Sonntag, Z. Phys. Chem. N. F., 69, 292 (1970).

(129) C. von Sonntag, Z. Naturforsch., 27b, 41 (1972).

(130) C. von Sonntag and D. Schulte-Frohlinde, Z. Physik. Chem. N. F., 55, 329 (1967).

(131) C. von Sonntag, Tetrahedron, 24, 117 (1968).

(132) C. von Sonntag, Int. J. Radiat. Phys. Chem., 1, 33 (1969).

(133) E. Tschuikow-Roux and J.R. McNesby, Trans. Faraday Soc., 62, 2158 (1966).

(134) S.G. Lias, G.J. Collin, R.E. Rebbert, and P. Ausloos, J. Chem. Phys., 52, 1841 (1970).

(135) D. Sänger and C. von Sonntag, Z. Naturforsch., 25b, 1491 (1970).

(136) D. Sänger and C. von Sonntag, Tetrahedron, 26, 5489 (1970).

(137) C. von Sonntag and D. Sänger, Tetrahedron Lett., 43, 4515 (1968).

(138) F.S. Dainton, I.V. Janovsky, and G.A. Salmon, J. Chem. Soc. Chem. Commun., 1969, 335.

(139) D.H. Ellison, G.A. Salmon, and F. Wilkinson, Proc. Roy. Soc. (London) A, 328, 23 (1972).

(140) H. Seki, R. Nagai and M. Imamura, Bull. Chem. Soc. Japan, 41, 2877 (1968).

(141) W.A. Seddon and A.O. Allen, J. Phys. Chem., 71, 1914 (1967).

(142) M.H. Awan, S.A. Chaudri, and Farhataziz, The Nucleus, 8, 87 (1971).

(143) L. Gilles and J. Sutton, J. Chim. Phys., 67, 128 (1970).

(144) C. von Sonntag, Doctoral thesis, Universität Karlsruhe (1964).

(145) A. Henne and H. Fischer, EUCHEM Research Conference on Useful Preparative Aspects of Photochemistry, Gent (1975). Preprints.

(146) C. von Sonntag, Fortschritte Chem. Forsch., 13, 333 (1969).

(147) B. Blank, A. Henne, G.P. Laroff, and H. Fischer, Pure Appl. Chem., 41, 475 (1975).

(148) D. Bryce-Smith, A. Gilbert and P.J. Twitchett, J. Chem. Soc. Chem. Commun., 1973, 457.

(149) H.-P. Schuchmann, C. von Sonntag, and D. Schulte-Frohlinde, J. Photochem., 3, 267 (1974/75).

(150) U. Sokolov and G. Stein, J. Chem. Phys., 44, 3329 (1966).

(151) H.J. van der Linde and C. von Sonntag, Photochem. Photobiol., 13, 147 (1971).

(152) R. Livingston and H. Zeldes, J. Am. Chem. Soc., 88, 4333 (1966).

(153) N.M. Bazhin, E.V. Kuznetsov, N.N. Bubnov, and V.V. Voevodskii, Kinet. Katal., 7, 732 (1966).

(154) A.L. Buley, R.O.C. Norman, and R.J. Pritchett, J. Chem. Soc. (B), 1966, 849.

(155) F. Seidler and C. von Sonntag, Z. Naturforsch., 24b, 780 (1969).

(156) C. von Sonntag and E. Thoms, Z. Naturforsch., 25b, 1405 (1970).

(157) C.E. Burchill and K.M. Perron, Can. J. Chem., 49, 2382 (1971).

(158) P.J. Venter, H.J. van der Linde, and R.A. Basson, J. Chem. Soc. Chem. Commun., 1972, 187.

(159) R. Ford, H.-P. Schuchmann, and C. von Sonntag, J. Chem. Soc. Perkin II, 1975, 1338.

(160) L. Farkas and O.H. Wansbrough-Jones, Z. Physik. Chem., B18, 124 (1932).

(161) E.O. Wilg and G.B. Kistiakowsky, J. Am. Chem. Soc., 54, 1806 (1932).

(162) A. Bernas, M. Bodard, and D. Saghattchian, J. Chim. Phys., 62, 1418 (1965).

(163) J. Hagège and C. Vermeil, J. Chim. Phys., 62, 403 (1965).

(164) N. Getoff and G.O. Schenck, Photochem. Photobiol., 8, 167 (1968).

(165) W.A. Noyes, Jr. and P.J. Leighton, "The Photochemistry of Gases", Reinhold, New York, 1941, p. 370.

(166) F.S. Dainton and P. Fowles, Proc. Roy. Soc. London A, 287, 295 (1965).

(167) J. Barrett and J.H. Baxendale, Trans. Faraday Soc., 56, 37 (1960).

(168) J. Barrett, M.F. Fox, and A.L. Mansell, J. Chem. Soc. (A), 1967, 483.

(169) F. Weeke, E. Bastian, and G. Schomburg, Chromatographia, 7, 163 (1974).

(170) E.E. Tucker and E.D. Becker, J. Phys. Chem., 77, 1783 (1973).

(171) R. Wolff, Angew. Chem., 67, 89 (1955).

(172) A.J. Harrison and J.S. Lake, J. Phys. Chem., 63, 1489 (1959).

(173) R.P. Porter and W.A. Noyes, Jr., J. Am. Chem. Soc., 81, 2307 (1959).

(174) J. Hagège, S. Leach, and C. Vermeil, J. Chim. Phys., 62, 736 (1965).

(175) J. Hagège, P.C. Roberge, and C. Vermeil, Trans. Faraday Soc., 64, 3288 (1968).

(176) (a) O.S. Herasymowych and A.R. Knight, Can. J. Chem., 51, 147 (1973); (b) ibid., 50, 2217 (1972).

(177) H.-P. Schuchmann, C. von Sonntag, and G. Schomburg, Tetrahedron, 28, 4333 (1972).

(178) H.-P. Schuchmann, F. Mark, and C. von Sonntag, To be published.

(179) S. Steenken, H.-P. Schuchmann, and C. von Sonntag, J. Phys. Chem., 79, 763 (1975).

(180) J.F. Meagher and R.B. Timmons, J. Chem. Phys., 57, 3175 (1972).

(181) C.A.F. Johnson and W.M.C. Lawson, J. Chem. Soc. Perkin II, 1974, 353.

(182) H.-P. Schuchmann and C. von Sonntag, Unpublished results.

(183) F. Hirayama, C.W. Lawson, and S. Lipsky, J. Phys. Chem., 74, 2411 (1970).

(184) A.M. Halpern and W.E. Ware, J. Phys. Chem., 74, 2413 (1970).

(185) P.H. Mazzocchi and M.W. Bowen, J. Org. Chem., 40, 2689 (1975).

(186) C. von Sonntag, H.-P. Schuchmann, and H. Bandmann, Unpublished results.

(187) C. von Sonntag and H. Bandmann, J. Phys. Chem., 78, 2181 (1974).

(188) N. Kizilkilic, H.-P. Schuchmann, and C. von Sonntag, Unpublished results.

(189) A.R. McIntosh and J.K.S. Wan, Mol. Photochem., 4, 417 (1972).

(190) S. Steenken, Unpublished results.

(191) T.J. Wallace and R.J. Gritter, J. Org. Chem., 26, 5256 (1961).

(192) G. Behrens and D. Schulte-Frohlinde, Angew. Chem., 85, 993 (1973); Angew. Chem. (Int. Ed.), 12, 932 (1973).

(193) R. Gomer and W.A. Noyes, J. Am. Chem. Soc., 72, 101 (1950).

(194) B.C. Roquitte, J. Phys. Chem., 70, 2699 (1966).

(195) M. Kawasaki, T. Ibuki, M. Iwasaki, and Y. Takezaki, J. Chem. Phys., 59, 2075 (1973).

(196) B.C. Roquitte, J. Phys. Chem., 70, 1334 (1966).

(197) B.C. Roquitte, J. Am. Chem. Soc., 91, 7664 (1969).

(198) R. Hentz and C.F. Parrish, J. Phys. Chem., 75, 3899 (1971).

(199) E. Watson and C.F. Parrish, J. Chem. Phys., 54, 1427 (1971).

(200) M.J. Perkins and B.P. Roberts, J. Chem. Soc. Perkin Trans. II, 1975, 77.

(201) L.P. Kuhn and C. Wellmann, J. Org. Chem., 22, 774 (1957).

(202) (a) E.S. Huyser and D.T. Wang, J. Org. Chem., 29, 2720 (1964). (b) E.S. Huyser, Ibid., 25, 1820 (1960).

(203) A.L.J. Beckwith and P.K. Tindal, Aust. J. Chem., 24, 2099 (1971).

(204) C. Bernasconi and G. Descotes, Compt. Rend. C, 280, 469 (1975).

(205) S.G. Hadley and D.H. Volman, J. Am. Chem. Soc., 85, 1035 (1967).

(206) Y. Kotake, M. Ono, and K. Kuwata, Bull. Chem. Soc. Japan, 45, 110 (1972).

(207) K. Hatano, N. Shimamoto, T. Katsu, and Y. Fuyita, Bull. Cham. Soc. Japan, 47, 4 (1974).

(208) T. Richerzhagen and D.H. Volman, J. Am. Chem. Soc., 93, 2062 (1971).

(209) D.E. Wood, R.V. Lloyd, and D.W. Pratt, J. Am. Chem. Soc., 92, 4115 (1970).

(210) M.C.R. Symons, Tetrahedron, 29, 615 (1973).

(211) G.V. Pukhalskaya, A.G. Kotov, and S. Ya. Pshezhetskii, Dokl. Akad. Nauk SSSR, 171, 1380 (1966); High Energy Chem., 1, 471 (1967).

(212) K. Pfordte and G. Leuschner, Liebigs Ann. Chem., 646, 25 (1961).

(213) G.H. Booth and R.G.W. Norrish, J. Chem. Soc., 1952, 188.

(214) B. Pouyet, Bull. Soc. Chim. France, 1964, 2582; Compt. Rend., 258, 2317 (1964).

(215) V.I. Stenberg, N. Kulevsky, and C.-H. Niu, J. Org. Chem., 38, 1227 (1973).

(216) B. Pouyet, Bull. Soc. Chim. France, 1965, 90.

(217) M.J. Gibian and R.C. Corley, Chem. Rev., 73, 441 (1973).

(218) L.T. Allan and G.A. Swan, J. Chem. Soc., 1965, 4822.

(219) K.G. Hancock and D.A. Dickinson, J. Org. Chem., 40, 969 (1975).

(220) C.G. Freeman, M.J. McEwan, R.F.C. Claridge, and L.F. Phillips, Chem. Phys. Lett., 8, 77 (1971).

(221) A.M. Halpern and R.M. Danziger, Chem. Phys. Lett., 16, 72 (1972).

(222) A.M. Halpern, J. Am. Chem. Soc., 96, 4392 (1974).

(223) A.M. Halpern and P.P. Chan, J. Am. Chem. Soc., 97, 2971 (1975).

(224) C.H. Bamford, J. Chem. Soc., 1939, 17.

(225) C.I. Johnson and H.A. Taylor, J. Chem. Phys., 19, 613 (1951).

(226) J.V. Michael and W.A. Noyes, Jr., J. Am. Chem. Soc., 85, 1228 (1963).

(227) J.J. Magenheimer, R.E. Varnerin, and R.B. Timmons, J. Phys. Chem., 73, 3904 (1969).

(228) M. Kawasaki and I. Tanaka, J. Phys. Chem., 78, 1784 (1974).

(229) G. Black and G. Porter, Proc. Roy. Soc. (London) A, 266, 191 (1962).

(230) N. Ya. Dodonova and A.N. Terenin, Bull. Acad. Sci. USSR, Phys. Ser., 27, 1064 (1963).

(231) P. Gray and J.C.J. Thynne, Trans. Faraday Soc., 59, 2275 (1963).

(232) P. Gray and A. Jones, Trans. Faraday Soc., 62, 112 (1966).

(233) D. Smithies and A. Whitworth, J. Chem. Soc. (A), 1969, 1987.

(234) N.H. Anderson and R.O.C. Norman, J. Chem. Soc. (B), 1971, 993.

(235) A.A. Baum, L.A. Karnischky, D. McLeod, Jr., and P.H. Kasai, J. Am. Chem. Soc., 95, 617 (1973).

(236) G.E. McAlonie, Doctoral thesis; Diss. Abstr., B 27, 783 (1966).

(237) J.S. Watson and B. deB. Darwent, J. Chem. Phys., 20, 1041 (1952).

(238) J.J. Magenheimer, Doctoral thesis, Washington, 1971; Diss. Abstr., B32, 1485 (1971/72).

(239) D. Salomon and A.A. Scala, J. Chem. Phys., 62, 1469 (1975).

(240) H. Gesser, J.T. Mullhaupt, and J.E. Griffiths, J. Am. Chem. Soc., 79, 4834 (1957).

(241) P.J. Kozak and H. Gesser, J. Chem. Soc., 1960, 448.

(242) M. Kawasaki, M. Iwasaki, I. Ibuki, and Y. Takezaki, J. Chem. Phys., 59, 6321 (1973).

(243) M. Kawasaki, M. Iwasaki, and I. Tanaka, J. Chem. Phys., 59, 6328 (1973).

(244) D.W. Carwell, R.S. Berry, and W. Lwowski, J. Am. Chem. Soc., 88, 544 (1966).

(245) C. Luner and H. Gesser, J. Phys. Chem., 62, 1148 (1958).

(246) A. Jones and F.P. Lossing, Can. J. Chem., 45, 1693 (1967).

(247) B.G. Dzantiev, A.T. Koritskii, R.G. Kostyanovskii, I.I. Mustafaev, Yu. M. Rumyantsev, and A.V. Shishkov, Khim. Vysokikh Energii, 8, 552 (1974).

(248) (a) C.G. Freeman, M.J. McEwan, R.F.C. Clardige, and L.F. Phillips, Trans. Faraday Soc., 67, 67 (1971); (b) ibid., 67, 2567 (1971).

(249) (a) R.H. Newman, C.G. Freeman, M.J. McEwan, R.F.C. Claridge, and L.F. Phillips, Trans. Faraday Soc., 67, 1360 (1971); (b) ibid., 67, 3247 (1971).

(250) H.E. Gunning, J.M. Campbell, H.S. Sandhu, and O.P. Strausz, J. Am. Chem. Soc., 95, 746 (1973).

(251) A.C. Vikis and D.J. LeRoy, Can. J. Chem., 50, 595 (1972).

(252) R. Payette, M. Bertrand, and Y. Rousseau, Can. J. Chem., 46, 2693 (1968).

(253) E. Murad, J. Am. Chem. Soc., 83, 1327 (1961).

(254) R.V. Morris and S.V. Filseth, Can. J. Chem., 48, 924 (1970).

(255) J. Castonguay and Y. Rousseau, Can. J. Chem., 49, 2125 (1971).

(256) R.S. Juvet, Jr., and L.P. Turner, Anal. Chem., 37, 1464 (1965).

(257) N. Bremer, B.J. Brown, G.H. Morin, and J.E. Willard, J. Phys. Chem., 79, 2187 (1975).

(258) M.K. Phibbs and M. deB. Darwent, J. Chem. Phys., 18, 495 (1950).

(259) R.F. Pottie, A. Harrison, and F.P. Lossing, Can. J. Chem., 39, 102 (1961).

(260) (a) A.R. Knight and H.E. Gunning, Can. J. Chem., 39, 1231 (1961); (b) ibid., 39, 2251 (1961); (c) ibid., 39, 2466 (1961); (d) ibid., 40, 1134 (1962); (e) ibid., 41, 763 (1963); (f) ibid., 41, 2849 (1963).

(261) (a) A. Kato and R.J. Cvetanović, Can. J. Chem., 45, 1845 (1967); (b) ibid., 46, 235 (1968).

(262) A.J. Yarwood, O.P. Strausz, and H.E. Gunning, J. Chem. Phys., 41, 1705 (1964).

(263) S.W. Benson and H.E. O'Neal, "Kinetic Data on Gas Phase Unimolecular Reactions," NSRDS-NBS 21, U.S. Dept. Commerce, Washington, 1970, p. 430.

(264) J.C.Y. Tsao, J. Chinese Chem. Soc. (Taipei), 16, 152 (1969).

(265) B. deB. Darwent, E.W.R. Steacie, and A.F. van Winckel, J. Chem. Phys., 14, 551 (1946).

(266) M.K. Phibbs, B. deB. Darwent, and E.W.R. Steacie, J. Chem. Phys., 16, 39 (1948).

(267) R.A. Marcus, B. deB. Darwent, and E.W.R. Steacie, J. Chem. Phys., 16, 987 (1948).

(268) R.J. Cvetanović, Can. J. Chem., 33, 1684 (1955).

(269) R.J. Cvetanović and L.C. Doyle, Can. J. Chem., 35, 605 (1957).

(270) Y. Takezaki, S. Mori, and H. Kawasaki, Bull. Chem. Soc. Japan, 39, 1643 (1966).

(271) L.F. Loucks and K.J. Laidler, Can. J. Chem., 45, 2763 (1967).

(272) S.V. Filseth, J. Phys. Chem., 73, 793 (1969).

(273) A.F. Dodonov, G.K. Lavrovskaya, and V.L. Tal'rose, Kinet. Katal., 10, 14 (1969).

(274) S.W. Benson, J. Chem. Phys., 40, 105 (1964).

(275) S. Yamamoto and S. Sato, Bull. Chem. Soc. Japan, 48, 1382 (1975).

(276) R.F. Klemm, Can. J. Chem., 45, 1936 (1966).

(277) R.K. Brinton and D.H. Volman, J. Chem. Phys., 20, 25 (1952).

(278) L. Horner and W. Jurgeleit, Liebigs Ann. Chem., 591, 138 (1955).

(279) V.I. Stenberg, R.D. Olson, C.T. Wang, and N. Kulevsky, J. Org. Chem., 32, 3227 (1967).

(280) N. Kulevsky, C.T. Wang, and V.I. Stenberg, J. Org. Chem., 34, 1345 (1969).

(281) K. Maeda, A. Nakane, and H. Tsubomura, Bull. Chem. Soc. Japan, 48, 2448 (1975).

(282) V.I. Stenberg, C.T. Wang, and N. Kulevsky, J. Org. Chem., 35, 1774 (1970).

(283) (a) I. Reimann, G. Hollatz, and T. Eckert, Arch. Pharmaz., 307, 321 (1974); (b) ibid., 307, 328 (1974).

(284) C. von Sonntag, H.-P. Schuchmann, and J. Jung, Unpublished results.

(285) C. von Sonntag, H.-P. Schuchmann, and N. Kizilkilic, Unpublished results.

(286) D. Lindsay, J.A. Howard, E.C. Horswill, L. Iton, K.U. Ingold, T. Cobbley, and A. Li, Can. J. Chem., 51, 870 (1973).

(287) M.J.W. Stratford, U.S. Dept. Com., Office Tech. Serv. PB Rept. 144,250 (1959); C.A., 55, 19740 b (1961).

(288) N. Kulevsky, C.-H. Niu, and V.I. Stenberg, J. Org. Chem. 38, 1154 (1973).

(289) (a) P.J. Sullivan and W.S. Koski, J. Am. Chem. Soc., 84, 1 (1962); (b) ibid., 86, 159 (1964).

(290) M.E. Jacox and D.E. Milligan, J. Mol. Spectros., 47, 148 (1973).

(291) R.J. Gritter and E.C. Sabatino, J. Org. Chem., 29, 1965 (1964).

(292) S. Yamamoto, K. Tanaka, and S. Sato, Bull. Chem. Soc. Japan, 48, 2172 (1975).

(293) C. von Sonntag, H.-P. Schuchmann, and E. Cetinkaya, Unpublished results.

(294) A.M. Halpern and D.K. Wong, Chem. Phys. Lett., 37, 416 (1976).

(295) S. Tsunashima, K. Morita, and S. Sato, Chem. Lett., 1976, 453.

Excitation and Deexcitation of Benzene

R. B. CUNDALL

Department of Chemistry and Applied Chemistry
University of Salford, United Kingdom

and

D. A. ROBINSON

Department of Chemistry, University of Botswana
Lesotho, and Swaziland, Gaborone, Botswana

and

L. C. PEREIRA

Department of Physics, University of Minho
Portugal

CONTENTS

I. INTRODUCTION

A large amount of spectroscopic data are now available for benzene, providing much information on photophysics and photochemistry properties of the lowest excited states of this molecule. Nevertheless the nature of the various radiative and nonradiative transitions involved in the deexcitation of the first excited states is far from being completely understood.

This work reviews the photophysics and relevant photochemistry of benzene in gas, liquid, and solid phases, with special emphasis on the experimental advances obtained by spectroscopic techniques in the past few years. Throughout, the symbolism used is that shown in the following reaction scheme:

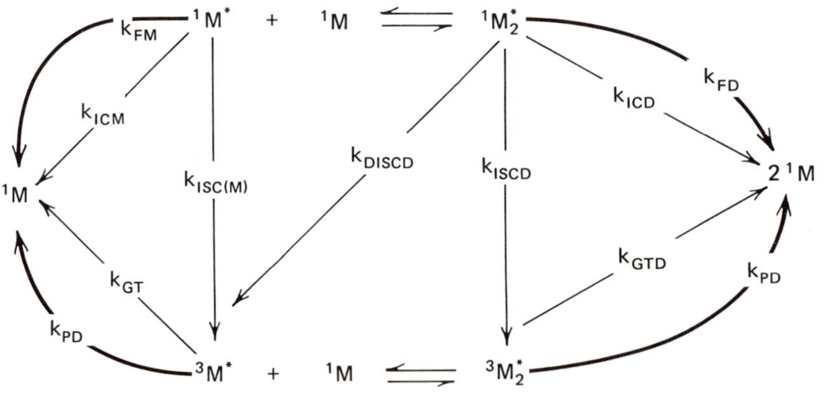

II. THE NATURE OF THE LOWER EXCITED STATES OF BENZENE

The near-ultraviolet spectral characteristics, and photophysical decay processes of benzene are generally described in terms of transitions between the various π molecular orbitals formed by linear combination of the carbon 2p atomic orbitals which have symmetries b_{2g}, e_{2g}(two), e_{1g}(two), and a_{2u} within the

D_{6h} point group. Though the absolute energies and energy spacings of the orbitals are not known precisely, and are, to a degree, dependent upon the complexity of the calculatory method, the symmetries and ordering of energies generated by the simpler theories remain valid.

The ground state benzene configuration is $(a_{2u})^2$, $(e_{1g})^4$. Singly excited configurations are produced by promotion of an electron from one of the two degenerate e_{1g} orbitals to one of the degenerate e_{2u} orbitals (1). The predicted degeneracy of the states so produced (B_{2u}, B_{1u}, E_{1u}) is removed by electron interaction. Calculations (2) have shown that the ordering by energy of the transitions in which spin conservation is observed should be ${}^1E_{1u} > {}^1B_{1u} > {}^1B_{2u}$.

Experimental observations in the gas phase (3), solution (4), low-temperature glasses (5), rare gas matrices (6), and the crystal phase (7) have all confirmed that the lowest excited singlet state has symmetry, B_{2u}, and that a transition from the ground to this state occurs around 260 nm. Band intensity is very low (ε at λ (max) $\sim$ 250 (8))* as befits a symmetry forbidden, but vibrationally induced transition. Two more intense bands are observed on the high energy side of this ${}^1B_{2u} \leftarrow A_{1g}$ absorption; a medium-intensity band ($\varepsilon \sim$ 9000 (8)) around 205 nm and a high-intensity band ($\varepsilon \sim$ 70,000 (8)) around 190 nm.

The 205 nm band is widely believed to be the ${}^1B_{1u} \leftarrow {}^1A_{1g}$ transition though some controversy is still attached to this assignment. Several recent semi-empirical (9-11), and _ab initio_ (11) calculations place an E_{2g} state, arising from a doubly excited π-π^{**} transition below the B_{1u} state arising from a single π-π^* excitation; supporting an early assignment (13) of the 205 nm band to a ${}^1E_{2g} \leftarrow {}^1A_{1g}$ transition. A doubly excited configuration may be of comparable energy to a singly excited configuration due to the sensitivity of the energy of the E_{2g} configuration to electron interaction. Contrary evidence comes from another _ab initio_ calculation (14) which places the ${}^1E_{2g}$ state between the ${}^1E_{1u}$ and ${}^1B_{1u}$ π-π^* states. Still other calculations, semi-empirical (15) and _ab initio_ (16) have placed the elusive ${}^1E_{2g}$ state above ${}^1E_{1u}$. It is clear from this confusing situation that even the most sophisticated computations presently available, incorporating multiple excited configurations and multicenter corrections, are unsatisfactory in assigning even relative energies to the doubly excited π-π^{**} states. This is apparently due to the extreme sensitivity of E_{2g} state energies to the values chosen for the resonance integrals (17).

---

*Extinction coefficient units are M^{-1} cm^{-1} throughout the article.

The common assignment of the 205 nm band to the $^1B_{1u} \leftarrow {}^1A_{1g}$ transition is consistent with the expected ordering of state energies for those configurations arising from $e_{2u} \leftarrow e_{1g}$ excitation, and with the assigned position of $^1B_{2u}$. Despite early inconclusive reports (18), recent evidence from vibrational analyses of low temperature absorption data in rare gas matrices and films (19) and of reflection spectra (20) confirms the $^1B_{1u}$ assignment. However, electron impact spectra (21) under high resolution suggests that two electronic transitions occur in the 205 nm region; and this has been confirmed recently by absorption studies on benzene in perfluoro-n-hexane (22) and rare gas matrices (22,23), at low temperature. In these environments the shift in the $^1E_{2g} \leftarrow {}^1A_{1g}$ transition is greater than that of the $^1B_{1u} \leftarrow {}^1A_{1g}$ transition due to differences in the charge distribution of the configurations and allows separate observation of the two bands. Available experimental evidence favors the energy sequence deduced by Koutecky et al. (9):

$$^1E^+_{2g} > {}^1E^+_{1u} > {}^1B^+_{1u} > {}^1E^-_{2g} > {}^1B^-_{2u} > {}^1A_{1g}$$

The high-intensity band ($\varepsilon \sim 70{,}000$) around 190 nm is usually assigned to the $^1E_{1u} \leftarrow {}^1A_{1g}$ symmetry-allowed π-π^* transition, though some contribution to the intensity may come from the superposition of the first member of one of the Rydberg series. The π-π^* assignment has been confirmed (24) from a vibrational analysis of the low-temperature spectrum of benzene in a rare gas matrix. Recent studies using multiphoton ionization spectroscopy have been considered to provide some indications of a $^1E_{1g}$ state as an alternative to assignments for the elusive $^1E_{2g}$ state (25). This conclusion awaits confirmation or otherwise by further experiments.

The relevant energies associated with each state are, to varying extents affected by the molecular environment, as discussed later. Table 1 contains approximate Franck-Condon maxima for the transitions from the $^1A_{1g}$ state.

In addition to transitions in the π manifold discussed above, some attention has been given recently to possible σ-π^* states in the ultraviolet nonradiative transitions from the $^1B_{2u}$ state. Callomon et al. (27) have postulated the transitions from higher vibrational levels of $^1B_{2u}$ to a $^3E_{2u}(\sigma$-$\pi^*)$ state to explain the onset of diffuseness in the high resolution absorption spectrum of vapor phase benzene in the 5.0 eV region. *Ab initio* (12,28) calculations have predicted the presence of at least one σ state (e_{2g}) lying between two filled π-molecular orbitals. Photoionization (29,30) experiments, and other evidence (31) appear to agree with the conclusion that a σ-π^* transition may be

TABLE 1. Approximate Franck-Condon maxima for transitions from $^1A_{1g}$ ground state to the lower lying π-electronic states

$^1E^+_{2g}$	$^1A_{1g}$	∿ 7.25 eV[a]
$^1E^+_{1u}$	$^1A_{1g}$	∿ 6.75 eV
$^1B^+_{1u}$	$^1A_{1g}$	∿ 6.08 eV
$^1E^-_{2g}$	$^1A_{1g}$	∿ 5.9 eV
$^1B_{2u}$	$^1A_{1g}$	∿ 4.88 eV

[a]The location of the second E_{2g} state, $^1E^+_{2g}$ has been suggested by Birks (26), from an analysis of published data on $S_x \leftarrow {}^1B_{1u}$ absorption experiments.

expected in the ultraviolet region conventionally associated with $\pi-\pi^*$ transitions. A recent CNDO -CI plus SO calculation by Minaev (32) has predicted a $^1A_{2u}(\sigma-\pi^*) \leftarrow {}^1A_{1g}$ transition at 5.11 or 5.98 eV depending upon the parameters chosen. Much of the earlier work, theoretical and practical, in the neglected consideration of σ-electron excitation has been reviewed by Burnelle and Kranepool (33). While most of the photophysical decay parameters of benzene that we shall discuss may be explained in terms of the lowest $\pi-\pi^*$ singlet and triplet states, the mechanism of radiationless $S_1 \rightarrow S_o$ decay may involve the elusive $^1E_{2g}$ and $\sigma-\pi^*$ states discussed above and their participation cannot be disregarded.

Since electron spin may invert during excitation, $\pi-\pi^*$ triplet states, of the same symmetries as the previously described excited singlet states should exist. Again, electron interaction removes the degeneracy of these states, though the precise ordering of state energies is still a matter of some dispute. The ranking now generally accepted (34) is:

$$^3B_{2u} > {}^3E_{1u} > {}^3B_{1u}$$

a sequence for which there is considerable experimental and theoretical justification (35).

There is no interference in the lower $\pi-\pi^*$ manifold from the $^3E_{2g}$ states; both recent calculatory (12,16,17,36) and experimental (37) evidence place the lowest energy $^3E_{2g}$ configuration above the $^3B_{2u}$ state. Again, the relevant energies of the various states are dependent, to a degree, on the molecular environment, and in this case are more difficult to determine experimentally because of the multiplicity as well as symmetry forbidden nature of transitions from the ground state.

The selective data on Franck-Condon maxima for transitions from $^1A_{1g}$, recorded in Table 2, were obtained from electron impact (38) and $T_x \leftarrow {}^3B_{1u}$ transient absorption studies (39). They are considered at this stage to illustrate the relative positions of the triplet states vis-à-vis the singlet states discussed earlier.

TABLE 2. Approximate Franck-Condon maxima for transitions from $^1A_{1g}$ ground state to the lower lying triplet π-electronic excited states.[a]

$^3E_{2g}$	$^1A_{1g}$	∿ 6.55 eV
$^3B_{2u}$	$^1A_{1g}$	∿ 5.60 eV
$^3E_{1u}$	$^1A_{1g}$	∿ 4.75 eV
$^3B_{1u}$	$^1A_{1g}$	∿ 3.95 eV

[a]Data from refs. 38 and 39.

No information is available concerning $\sigma-\pi^*$ transitions, though from the approximate singlet assignments and presuming validity of Hund's Rule a $\sigma-\pi^*$ state or states may be expected to be within the energies of the lower $\pi-\pi^*$ manifold. It is interesting to speculate that the radiationless $T_1 \rightarrow S_o$ decay, explained with such difficulty in theoretical analyses of the benzene problem, may involve a transition to a previously unconsidered $^3(\sigma-\pi^*)$ state.

III. TRANSITION BETWEEN THE LOWER ENERGY STATES

Interstate transitions are generally considered to be governed principally by the familiar symmetry and multiplicity selection rules. The manner in which these rules affect the major excitation and decay routes of benzene are discussed below.

A. Excitation from the $^1A_{1g}$ State

Within the singlet manifold of the lower energy transitions only the $^1E_{1u} \leftarrow {}^1A_{1g}$ excitation is symmetry allowed, though transition to any of the $^1B_{1u}$, $^1E_{2g}$, and $^1B_{2u}$ states can be promoted by coupling of the "pure" electronic states with vibrations of symmetry (b_{2g}, e_{2g}), (e_{2u}, b_{1u}, b_{2u}, e_{1u}), and (e_{2g}, b_{1g}), respectively. As mentioned above, all four transitions may be observed under certain circumstances, although $^1E_{1u} \leftarrow {}^1A_{1g}$ is, as expected, the most intense. The selection rules for double photon excitation predict that the modes of b_{1u}, b_{2u}, e_{2u} and e_{1u} symmetry should induce the absorption (40). Studies carried out by Hochstrasser, Sung, and Wessel (41) on a single crystal of benzene at 2^oK show that a certain number of vibrations of b_{1u}, b_{2u}, e_{1u}, and e_{2u} symmetry can be assigned in the two photon absorption spectrum. Gas phase two photon absorption spectra for benzene were reported recently by Wunsch, Neusser, and Schlag (42), who have found that the most powerful mode is of b_{2u} symmetry which is the extremely weak the ν_{19} mode of e_{1u} symmetry. Definitive experiments on the excited states of benzene using this promising technique are awaited.

For excitation into the triplet manifold, the transition $^3E_{1u} \leftarrow {}^1A_{1g}$ has first-order spin-orbit allowedness, while the $^3B_{1u} \leftarrow {}^1A_{1g}$ transition requires vibronic coupling via e_{2g}, b_{1g}, or D_{2g} modes or combination tones. Absorption studies (39) on oxygen free crystalline benzene have revealed a weak $T_2 \leftarrow S_o$ transition, but no $T_1 \leftarrow S_o$ transition, confirming the relative allowedness of the two transitions. Furthermore, an enhancement of the $T_1 \leftarrow S_o$ absorption induced by high pressures of oxygen was recently reported for benzene and for a large number of simple benzene derivatives (43).

No experimental evidence is available for the $^3B_{2u} \leftarrow {}^1A_{1g}$ transition which is both spin and symmetry forbidden in the first order, and should occur in the same spectral region as the medium-intensity $^1B_{1u} \leftarrow {}^1A_{1g}$ transition.

B. Emission to the $^1A_{1g}$ State

Kasha's Rule (44) requires that emission should occur from the lowest $^1B_{2u}$ and $^3B_{1u}$ states in benzene. This is essentially the case though a predictable very weak emission has been reported recently from higher states of oxygen-quenched benzene in gaseous (45) and solution (46) phases. The major fluorescence and phosphorescence processes for benzene are thus forbidden, and the radiative decay probabilities may be influenced by the environment as will be discussed below.

C. Transitions Between Excited Singlet and Triplet States

Intersystem crossing from the $^1B_{2u}$ envelope of vibronic states is thought to occur from the lowest vibrational level, or perhaps the lowest two or three such levels. Given the lower relative energies of $^3E_{1u}$ and $^3B_{1u}$ states, transitions to each of these configurations from $^1B_{2u}$, are possible. There is first-order spin-orbit allowedness in that to the $^3B_{1u}$ state, but vibronic coupling is required for the transition to $^3E_{1u}$. Experimental evidence for intersystem crossing from higher electronic states is lacking, but there is no reason to neglect this route in the consideration of the decay processes of benzene molecules that have been so excited.

IV. EXCITED STATE GEOMETRY

The ground state of the benzene molecule is perfectly hexagonal and the molecular wave functions may be used as bases for irreducible representations of the D_{6h} point group. Transitions between the molecular energy states are usually excited in terms of the symmetry restrictions of this point group. The forbidden nature of the lower energy electronic transition has led to considerable discussion about the geometry of the excited states involved, since even slight deviations from regularity lead to reduction in symmetry, and require consequent modification of the selection rules. Despite early indications (47) that the phosphorescence spectrum may be explained using regular hexagonal symmetry, and recent evidence (48) the D_{6h} symmetry is maintained for $^3B_{1u}$ benzene in cyclohexane at 77°K, there is much evidence to show that the molecule is distorted when in the $^3B_{1u}$ state. Thirty years ago, Lewis and Kasha (49) presented evidence for a molecular contraction along the 1-4 axis, from analysis of phosphorescence data. Further results from gas phase $^3B_{1u} \leftarrow {}^1A_{1g}$ oxygen-perturbed absorption spectra (50) led to a D_{2h} assignment of the upper state. van der Waals and co-workers (51-55) have shown by ESR studies that the vibronic coupling in the triplet manifold between $^3B_{1u}$ and $^3E_{1u}$ via the e_{2g} vibrational mode plays the dominant role for the phosphorescence process and that this leads to a non hexagonal phosphorescent state distorted to a quinoidal form. Gwait et al. (56), using phosphorescence microwave double resonance, have studied the zero field origin of the

0,0 band of phosphorescence of benzene-h_6 in benzene-d_6 at 1.6^oK. The results obtained are consistent with a $^3B_{1u}$ assignment for a distorted D_{2h} benzene phosphorescent state and the lowest triplet of toluene is also shown to have an expanded quinoidally distorted ring by the calculations of Haaland and Nieman (57). Evidence for a D_{2h} assignment of the lowest triplet state in solid phase was also provided by other authors (58,59).

Due to the highly forbidden nature of the $^3B_{1u} \leftarrow {}^1A_{1g}$ absorption, observations are only possible in perturbed systems such as in the presence of oxygen or in solid matrices, and it has been suggested that the symmetry distortions that have been deduced may arise from the influence of the molecular environment, rather than a genuine geometric property of the isolated molecule. While this possibility cannot be overlooked, theoretical considerations support the view that the $^3B_{1u}$ state is inherently distorted. Davaquet (60) has shown that the vibrations that stabilize the free $^3B_{1u}$ molecules are precisely those that would cause a slight distortion similar to that suggested for the solid phase triplet molecule. Fujimura, Yanaguchi, and Nakajima (61) have calculated the $^3B_{1u}$ state to have quinoidal D_{2h} symmetry in agreement with an earlier theoretical analysis (62). These authors predict that the $^1B_{1u}$ state will also be distorted with antiquinoidal D_{2h} symmetry, and maintain that the lowest excited singlet state $^1B_{2u}$ will retain D_{6h} symmetry, albeit with a slightly enlarged ring. This is not in agreement with conclusions of Nieman and Tinti (58) who have interpreted experimental results on the absorption, fluorescence, and phosphorescence of several isotopic benzenes in terms of a distortion of both the $^1B_{2u}$ and $^3B_{1u}$ states.

It is worthy of note that the photochemical production of Dewar benzene from the $^1B_{1u}$ state (63) and from the $^3B_{1u}$ (64,65) is consistent with the quinoidal deformations suggested for these states.

A distortion of the $^1B_{2u}$ state of benzene to D_{3d} has also been proposed (66) to account for inconsistencies in the interpretation of the ultraviolet spectrum of pure crystalline benzene at low temperatures. Such distortions may be the result of the solid medium or, more probably, the spectral inconsistencies may be due to state splitting by the environment.

V. THE INFLUENCE OF THE ENVIRONMENT ON THE ELECTRONIC STATES

The influence of different environments on molecular states has been studied, theoretically and practically, from changes in the absorption spectra. Such effects as frequency shifts, peak width changes, and variation of oscillator strength for various transitions are measures of the relative effects of different environments on the state involved.

The first absorption band of benzene shifts slightly to the red ($\sim$ 200 - 300 cm^{-1}) with the change between vapor phase and solution, and further shifts dependent upon the nature of the solvent occur. The extent of such red shifts may be related qualitatively to medium refractive index and dielectric constant. Several theoretical approaches to the problems of spectral shifts have been attempted (67), but in general these are more useful for polar molecules, and to allowed transitions where shifts are of greater magnitude. Suffice to say that in benzene there is no discontinuity in the trend of shifts with refractive index, through gas, liquid, and solid phases, other than indicating that the forces of interaction between this molecule and the environment are weak. The crystalline phase is exceptional, since the "strict" selection rules for transition are broken down by interaction with the crystal field, which leads to factor group splitting (68) and the appearance of four (69) components of the 0-0 band, all of which are shifted.

Much experimental evidence is available on the small solvent shifts which are measureable but, since little effect on the photophysical decay processes is expected, no review of this work will be attempted; the reader is referred to the work of Eastman and Rehfeld (70) who report data for benzene in the vapor phase and in 37 different solvents.

Changes in oscillator strength are expected since this is a function of the refractive index of the medium (71), though the effect is much smaller than theoretically expected in the case of benzene. To a first approximation the oscillator strength for the ${}^1B_{2u} \leftarrow {}^1A_{1g}$ transition is independent of molecular environment (70).

A typical environmental effect on $S_1 \leftrightarrow S_0$ transitions concerns the vibrational structure of the spectra. Loss of structure in solution relative to vapor and solid phases is the result of specific solute-solvent interactions and as such, is dependent upon the dielectric properties of the solvent. Lawson, Hirayama, and Lipsky (72) have reported observations on the absorption and fluorescence spectra of benzene in vapor phase and various solvents. Their data are consistent with the trend of increased broadening of bands and loss of structure with solvent polarity. Perfluorinated solvents exert the least effect, the spectra retaining much of the fine structure of the vapor phase. Figure 1 shows the data of Lipsky and co-workers (72) for vapor phase benzene, and for solutions in hexane and perfluorohexane.

The influence of environmental perturbations on the degree of allowedness of the symmetry-forbidden 0-0 band has been

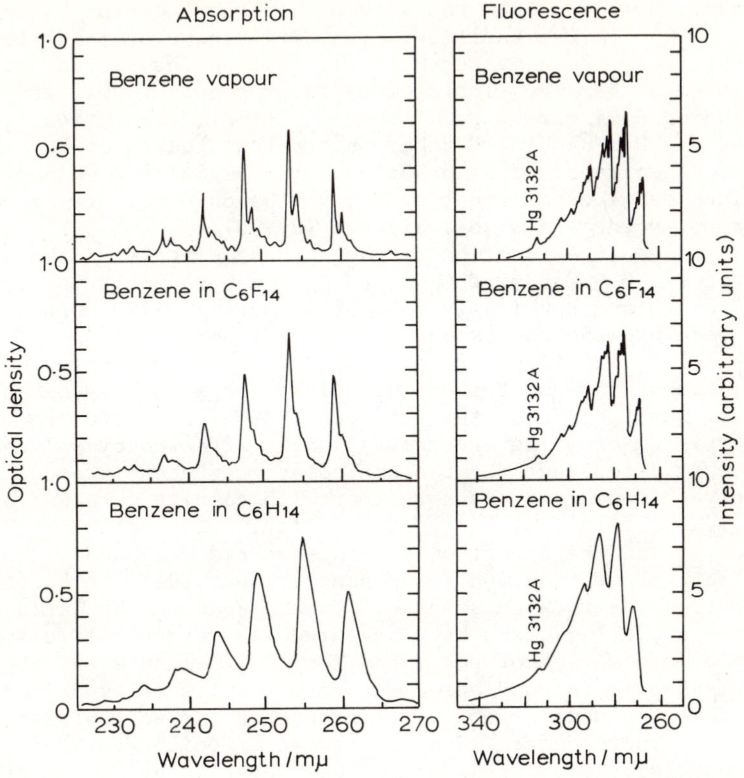

FIGURE 1. Absorption and emission spectra for benzene in (a) vapor phase, (b) perfluorohexane, (c) hexane. (Lawson et al.[72]).

thoroughly investigated. The first study on the effect was carried out by Ham (73) using benzene dissolved in hydrocarbon glasses with addition of carbon tetrachloride in various proportions. Bayliss and Hulme (74) also found a new absorption band appearing in benzene dissolved in carbon tetrachloride, chloroform and water, among other solvents, and a more detailed study was later carried out by Kanda (75) and by Bayliss and Cant (76). Platt (77) in a review concerned with the location of the second triplet state of benzene called these Ham bands and suggested that they arise from the spin forbidden $^3E_{1u} \leftarrow {}^1A_{1g}$ transition, a view shared by other authors (78,79). Robinson (80) proposed that the Ham bands in carbon tetrachloride solid matrices are due to intensity borrowed from the solvent and have nothing to do with the $^3E_{1u} \leftarrow {}^1A_{1g}$ transition. The correctness of this view was conclusively shown by Leach et al. (5), who have resolved the 0-0 band of benzene fluorescence in presence of CCl_4 at 77°K, finding a mirror image of the $^1B_{2u} \leftarrow {}^1A_{1g}$ absorption including the Ham bands. Further absorption studies on Ham bands have been reported by other authors (82-86) and evidence for environmental enhancement of the 0-0 bands of fluorescence and phorphorescence is also available (87-90).

A detailed investigation of the solvent effects on the intensity of the 0-0 bands of C_6H_6 and C_6D_6 was recently carried out by Cundall and Pereira (91-93) by measuring fully corrected fluorescence and phosphorescence spectra, fluorescence and phosphorescence lifetimes, triplet yields, and electronic absorption spectra at various temperatures in a large variety of solvents. It was found that the rate constants for fluorescence and intersystem crossing are only slightly affected by solvent while the rate constant for the nonradiative process from S_1 to S_o is solvent dependent, in agreement with results obtained by Eastman and Rehfeld (70). A good correlation was established between k_{ICM} and the extent of solvent-induced 0-0 band intensity in both absorption and fluorescence bands (91). A solvent dependence on the rate constant k_{ICM} was also found by Luria, Ofran, and Stein (94), who have measured fluorescence lifetimes of benzene in a series of polar and nonpolar solvents and the trend obtained is in good agreement with results of Cundall and Pereira (91). A similar correlation between the solvents enhancing the fluorescence 0-0 band and the efficiency of the internal conversion from $^1E_{1u}$ to $^1B_{2u}$ was reported by Lawson et al. (72). This shows that solvents which give a higher probability of the nonradiative transition from S_3 to S_1 also increase the nonradiative transition from S_1 to S_o. In Fig. 2 the fully corrected fluorescence spectrum of benzene solutions in methanol at 77°K and the absorption spectrum at 233°K are represented (91); a good mirror relationship is displayed, including 0-0 bands enhanced bands by solvent perturbation. It was also established by these authors that in the presence of chloroform and carbon tetrachloride the symmetry-forbidden 0-0 bands are enhanced in both fluorescence and phosphorescence, while xenon enhances these bands only in phosphorescence. These effects are illustrated at 77°K in Figs. 3 and 4, respectively, for benzene-d_6 in methylcyclohexane + chloroform and for benzene-h_6 in ethanol and in ethanol + xenon. The extent of enhancement was given for fluorescence by the ratio Fo/F_1 in which Fo represents the peak

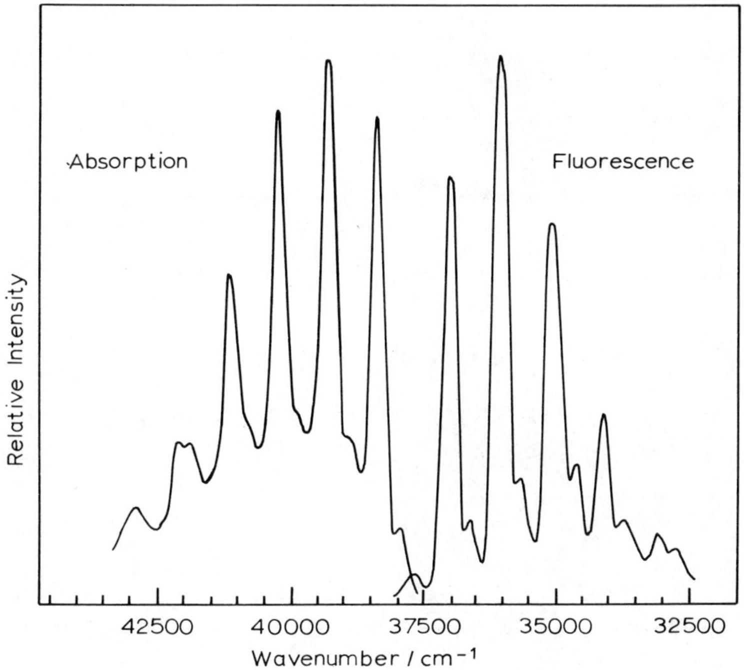

FIGURE 2. Absorption and fully corrected fluorescence spectra of benzene at 233°K and 77°K, respectively.[92]

height of the 0-0 band and F_1 the peak height of the vibronic allowed band at the wavenumber $\bar{\nu}_{o-o}$-606 cm^{-1}, and for phosphorescence by the ratio Po/P_2, where Po is the peak height of the 0-0 band and P_2 the corresponding peak intensity at about $\bar{\nu}_{o-o}$-1500 cm^{-1}, as shown in Fig. 4. Some of the results obtained from this type of measurement are indicated in Tables 3 and 4. From phosphorescence lifetimes, (τ_P), it was shown by the above authors that a decrease of τ_P and increase of the ratio of phosphorescence to fluorescence yield, (χ), parallels the enhancement of the 0-0 phosphorescence band in most of the solvents investigated. The results were interpreted in terms of a distortion of both the $^1B_{2u}$ and $^3B_{1u}$ states induced by solvent perturbation, distortion in the $^3B_{1u}$ state being greater than in the $^1B_{2u}$ state, in agreement with results reported by Nieman and Tinti (58).

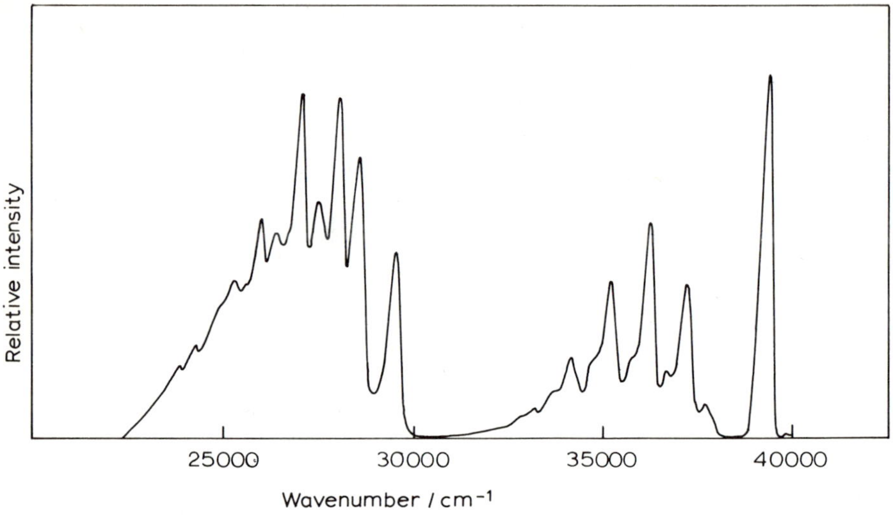

FIGURE 3. Fully corrected spectra (fluorescence and phosphorescence) of benzene in methylcyclohexane + 10% chloroform at 77°K,[92] first-excitation band.

Observations on transitions between S_o and the lowest triplet state are difficult in nonrigid media and it is not possible to say with certainty that this transition is similarly affected by environment.

Nonradiative processes may occur other than through π-π^* configurations. Other types of state may, as a result of their charge distributions, be more susceptible to interaction with the environment than the emitting π,π^* states.

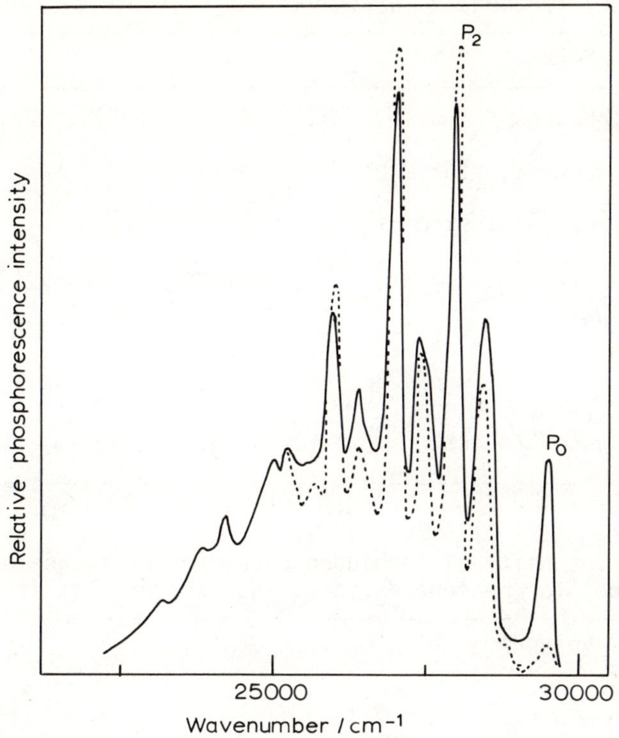

FIGURE 4. Fully corrected phosphorescence spectra of benzene in ethanol (----) and in ethanol + xenon (——) at 77°K.[92]

VI. THE $S_1 \leftrightarrow S_0$ RADIATIVE TRANSITION

The $^1B_{2u} \leftarrow {}^1A_{1g}$ absorption transition has been well studied and is discussed in the previous sections. Theory predicts an integrated absorption intensity independent of temperature. Below room temperature, Eastman and Rehfeld (70) found a similar absorption spectrum for benzene in 2-methyl-1-propanol at 25°C and at -140°C, that at low temperature being sharper and better resolved. Above room temperature there is a rather more marked effect on peak intensity (95) accompanied by peak broadening and the expected increase in hot band intensity at 260 nm. This is

TABLE 3. Intensity of forbidden bands (fluorescence and phosphorescence) for benzene-h_6 in various solvents at 77°K (91-93).

Solvent	F_0/F_1	P_0/P_2
Methylcyclohexane	0.050	0.074
Methylcyclohexane + 10% $CHCl_3$	0.155	0.530
Methylcyclohexane + 7% C Cl_4	0.094	0.280
Ethanol	0.076	0.070
Methanol	0.077	0.060
Acetonitrile	0.110	-
Ethanol + xenon (∿0.8M)	-	0.380

TABLE 4. Intensity of forbidden bands (fluorescence and phosphorescence) for benzene-d_6 in various solvents at 77°K (91-93).

Solvent	F_0/F_1	P_0/P2
Methylcyclohexane	0.072	0.040
Methylcyclohexane + 10% CHC_3	0.234	0.470
Methylcyclohexane + 20% $CHCl_3$	0.260	0.440
Methylcyclohexane + 10% CCl_4	0.146	0.260
Ethanol	0.082	0.047
Methanol	0.076	0.060
Acetonitrile	0.100	-
Ethanol + xenon (∿0.8H)	-	0.26

well illustrated in Figs. 5 and 6. The changes are brought about by variations in energy distribution within the system, and the integrated intensity is approximately constant (70,94). At very high temperatures the spectrum becomes very broad as observed in shock wave experiments (96).

The $^1B_{2u} \leftarrow {}^1A_{1g}$ symmetry forbidden emission transition observed in all three phases, results from vibrational perturbation of the ground state. The fluorescence spectrum follows the changes in resolution, shape, and frequency with environment, observed in the absorption spectrum. It is, in all cases, a fair mirror image of the absorption spectrum, demonstrating that emission occurs from the electronic state created by absorption in the 260 nm region. Fluorescence intensity varies considerably with environment (see Tables 5 - 11), though as will be shown, this results from media effects on competitive processes, rather than on the probability of emission.

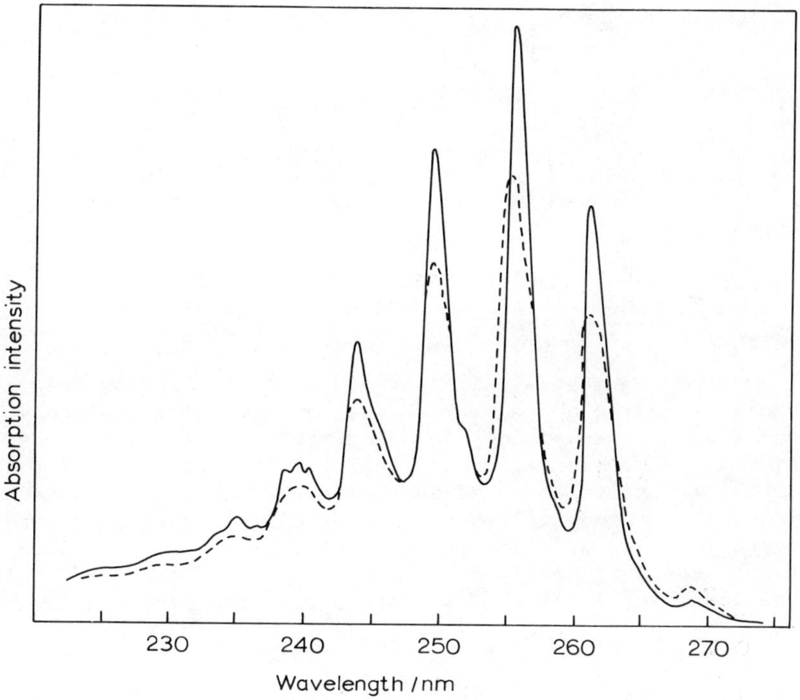

FIGURE 5. Variation in absorption spectrum (first band) with temperature for benzene in cyclohexane, 11°C (——) 75°C (----) (Robinson[95]).

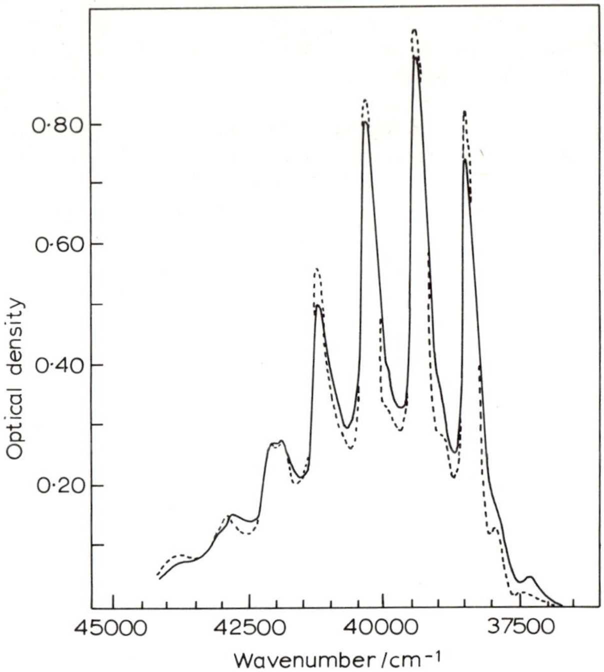

FIGURE 6. Absorption spectra of benzene in methanol at room temperature (——) and at $-40^{o}C$ (----)[92].

One of the most significant advances in spectroscopic techniques in recent years has given data on single vibronic level fluorescence yields and associated emission lifetimes of individual vibronic levels. Details of the experimental technique and of the results achieved have been reviewed by Parmenter (97). Parmenter and Schuyler (98,99) succeeded in recording fluorescence spectra from low pressure (<0.2 torr) benzene vapor excited with a band pass sufficiently small ($\sim$ 40 cm^{-1}) to ensure that single vibrational levels of the $^{1}B_{2u}$ state were created. In the lower levels the spectra recorded were sharp and structured, as would be expected in a system in which collisional redistribution of energy prior to fluorescence does not occur. At higher energy excitation the emission spectra become more diffuse, and the fluorescence intensity drops to zero when levels of excess energy

TABLE 5. Monomer fluorescence yields in the gas phase.

ØF	Isomer	Temp. (°C)	Conditions/Remarks	Ref.
0.21	C_6H_6	30	2 torrs, 254 nm excitation	(100)
0.32	C_6H_6	30	0.077-0.12 torr λ_{exc} 254 nm	(101)
0.39	C_6H_6	30	0.01-0.4 torr λ_{exc} 254 nm	(102)
0.22	C_6H_6		For zeroth level	(98)
0.27	C_6H_6		For 1st e_{2g} band low-pressure vapor	(98)
0.18	C_6H_6			(103)
0.39	C_6H_6		$P \geq 10^{-2}$ torr λ_{exc} 254 nm	(104)
0.18	C_6H_6	25	$P > 10$ torrs λ_{exc} 254 nm	(105)
0.31	C_6H_6	30	P 10-32 torrs λ_{exc} 264 nm	(100)
0.23	C_6D_6	25	P 10 torrs λ_{exc} 254 nm	(105)
0.53	C_6D_6	30	P 4-15 torrs λ_{exc} 264 nm	(100)
0.26	C_6D_6	30	P 5.2 torrs λ_{exc} 254 nm	(100)
0.35	C_6D_6		Low-pressure-single level fluorescence, zeroth level	(106)

> 2500 cm^{-1} are excited. Quantum yields for the emission from the various levels were estimated from the established high pressure yield of 0.18. Data are given in Tables 5 and 6.

Further quantum yield measurements (111) and emission lifetimes (111-113,115) have also been reported which demonstrate that the radiative transition probability from a vibronic state is determined by the combination of vibrational modes that have been excited. The transition rates that may be of interest in comparing data from all three phases are, of course, those from the lower levels of S_1, since relaxation in higher-pressure gases and condensed media will lead to these states. The data of Spears and Rice (111) show that the emission probability from the levels 0, $6^1 16^1$ and 1^1 are $2.2 \times 10^6 s^{-1}$, $3.4 \times 10^6 s^{-1}$,

TABLE 6. Monomer fluorescence lifetimes in the gas phase.

τ_F/ns	Isomer	Conditions	Ref.
60	C_6H_6	High pressure (5-40 mm) λ_{exc} 230-280nm	(107)
80	C_6H_6	High pressure λ_{exc} 259 nm	(108)
77	C_6H_6	λ_{exc} 260, extrapolated to zero pressure	(109)
80±10	C_6H_6	3-10 torrs, λ_{exc} 254 nm	(110)
100	C_6G_6	-Zeroth level } low pressure	(111)
79	C_6H_6	1st e_{2g} level } low pressure	
90	C_6H_6	-Zeroth level } low pressure	(112)
80	C_6H_6	-1st e_{2g} level } low pressure	(112)
72	C_6H_6	High pressure λ_{exc} 247-266 nm	(113)
164	C_6D_6	Zeroth level } low pressure	(113)
125	C_6D_6	-1st e_{2g} level } low pressure	(113)
92	C_6D_6	High pressure limit λ_{exc} 247-266	(113)
200	C_6D_6	Zeroth level	(113)
95	C_6D_6	High pressure (5-40 torrs) λ_{exc} 230-280	(106)
72	C_6H_6	50 torrs indep. of λ_{exc} 245-270 nm	(114)
92	C_6D_6	50 torrs indep. of λ_{exc} 245-270 nm	(114)

$3.4 \times 10^6 s^{-1}$, and $2.5 \times 10^6 s^{-1}$, respectively. Data from the comparable transitions (106) in C_6D_6 give the values $1.75 \times 10^6 s^{-1}$, $2.9 \times 10^6 s^{-1}$, $3.3 \times 10^6 s^{-1}$, and $3.0 \times 10^6 s^{-1}$, respectively. In benzene-d_6 the sharp drop in quantum yield with energy in excess of zero point level for the transition is noted (106) in a similar region to that for C_6H_6, about 3500 cm^{-1} above the 0-0 band.

With increase in benzene (or added gas) pressure, the rate of collisional redistribution of excitation energy becomes

TABLE 7. Rate constants for fluorescence of benzene-h_6 and benzene-d_6, as a function of temperature assuming a non-quenchable internal conversion process.

	T(°C)	$k_F \times 10^6 (s^{-1})$
Benzene-h_6	25	2.76
	40	2.79
	50	2.84
	60	2.93
	70	3.06
	80	3.24
	100	3.72
Benzene-d_6	25	2.85
	40	3.09
	50	3.30
	60	3.51
	70	3.78
	80	4.42
	100	4.98

competitive with the fluorescence rate to an extent which depends upon the total pressure of the system. Quantum yield and emission lifetime measurements reflect the nonequilibrium distribution of excited molecules in the medium-pressure range; for example, for excitation of C_6H_6 at 254 nm, yields of (100) 0.21 (for 2 mm pressure) and a negative dependence upon increasing gas pressure have been reported. A similar dependence upon gas pressure is evident for C_6D_6; again for excitation at 254 nm, yields of 0.33 (100), 0.28 (100), 0.26 (100), and 0.23 (116) have been reported for pressures of 1.2 3.0, 5.2, and >10 torrs, respectively. Also the work of Poole (100) has demonstrated that the yield is a function of excitation wavelength, dropping to zero in both benzene and the per-deutero derivative at excitation wavelengths of 240 nm and below.

TABLE 8. Monomer fluorescence yields for benzene in solution.

ϕ^{o}_{FM}	Isomer	Temp. (°C)	Solvent	Remarks	Ref.
0.060	C_6H_6	25	Hexane	Relative to toluene	(119)
0.052	C_6H_6	25	Me-cyclohexane		(119)
0.061	C_6H_6	25	Cyclohexane		(70)
0.046	C_6H_6	25	Me-cyclohexane		(70)
0.058	C_6H_6	25	Hexane		(70,120)
0.060	C_6H_6	25	Hexane		(121)
0.061	C_6H_6	25	Cyclohexane		(122)
0.053	C_6H_6	AMB	Hexane		(123)
0.050	C_6H_6	20	Hexane		(124)
0.071	C_6H_6	25	Me-cyclohexane	Relative to ϕ^{o}_{FM} (DPA)=1.0	(125)
0.070	C_6H_6	RMT	Cyclohexane		(126)
0.066	C_6D_6	25	Cyclohexane		(126)
0.042	C_6D_6	24	Me-cyclohexane		(127)
0.042	C_6H_6	RMT	Ethanol		(128)
0.040	C_6H_6	25	Ethanol		(119)
0.025	C_6H_6	25	Acetonitrile		(119)
0.029	C_6H_6	25	Methanol		(119)
0.027	C_6H_6	25	Methanol		(70)
0.033	C_6H_6	25	Ethane		(70)
0.025	C_6H_6	25	Acetonitrile		(119)
0.006	C_6H_6	25	Water		(129,130)
0.0058	C_6H_6	25	Water		(70)
0.065	C_6H_6	25	Per-fluoro-n-hexane		(91)

TABLE 9. Monomer fluorescence emission lifetimes for benzene in solution.

τ_{FM} (ns)	Isomer	Temp. (^oC)	Solvent	Ref.
29	C_6H_6	RMT	Cyclohexane	(126)
24	C_6H_6	20	Hexane	(131)
26	C_6H_6	RMT	Cyclohexane	(132)
27.5	C_6H_6	24	Hexane	(133)
29.5	C_6H_6	25	Cyclohexane	(134)
32	C_6H_6	20	n-Nonane	(135)
27.9	C_6H_6	25	Me-cyclohexane	(136)
45	C_6H_6	RMT	Hexane	(137)
34	C_6H_6	25	Cyclohexane	(119)
30	C_6H_6	25	Me-cyclohexane	(119)
33.6	C_6H_6	25	Hexane	(138)
26.6	C_6D_6	25	Cyclohexane	(125)
33	C_6D_6	22	Cyclohexane	(139)
28	C_6H_6	25	Cyclohexane	(139)
30	C_6D_6	25	Me-cyclohexane	(127)
31	C_6H_6	RMT	Ethanol	(121)
26	C_6H_6	25	Ethanol	(119)
20	C_6H_6	25	Methane	(119)
2.2	C_6H_6	25	Water	(129)
15	C_6H_6	25	Acetonitrile	(119)

TABLE 10. Fluorescence yields for benzene in the solid phase.

ØFM	Isomer	Temp. (°K)	Solvent	Remarks	Ref.
0.21	C_6H_6	77			(148)
0.26	C_6H_6	77	EPA	± 10%	(149)
0.20	C_6H_6	77	EPA		(150)
0.22	C_6H_6	173	Benzene		(151)
0.29	C_6H_6	77	Crystal		(152)
0.22	C_6H_6	77	EOA	Ether/isooctane/ alcohol	(144)
0.19	C_6H_6	77	EPA		(153-155)
0.27	C_6D_6	77	EPA		(153-155)
0.21	C_6H_6		n-Hexane		(120)
0.18	C_6H_6		Methanol		(120)
0.17	C_6H_6	30	30/70 iso but/isopent	Limiting low-temp. values	(120)
0.21	C_6H_6		EPA		(120)

TABLE 11. Fluorescence emission lifetimes for benzene in the solid phase.

τ_F (ns)	Temp. (°K)	Solvent	Remarks	Ref.
84	173	Benzene	No phosphorescence detected	(151)
120±6	77	Me-cyclohexane		(93)
115±5	77	Ethanol		(93)

It is considered that pressures of 10 torr or greater are required for complete collisional deactivation to an equilibrium distribution of energy between levels prior to emission. Under these conditions, the fluorescence yield appears independent of further increases in pressure, and achieves a value of 0.18 (100,103,105,116) for all first singlet band excitation wavelengths in excess of 247 nm.

Fluorescence yields for C_6H_6 and C_6D_6 decrease (100,105) with increasing temperature; the data of Dunnicliff (105) are presented in Fig. 7 for 2.5 x 10^{-4}M and 7.1 x 10^{-4}M benzene between 20 and 140°C.

Emission lifetimes for benzene with vapor phase are subject to similar variations with pressure and excitation wavelength as are fluorescence yields. Data are collected in Table 6 and can be seen to show considerable variation with source and experimental technique. Recent measurements, using a single-proton counting technique (114) have shown the emission lifetimes of high pressure C_6H_6 and C_6D_6 to be 77 and 92 ns, respectively, at 25°C. The temperature dependence of the fluorescence lifetime is also shown in Fig. 7.

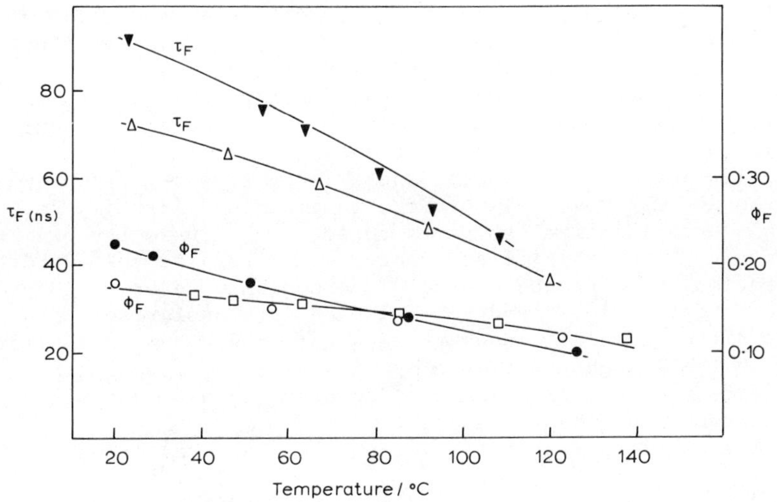

FIGURE 7. The variation in $\emptyset_F$ and τ_F for vapor-phase benzene as functions of temperature (Dunnicliff[105] and Lockwood[114]). The filled points are for C_6D_6. The open points are for C_6H_6.

At high pressures gaseous system most closely resemble the situation in condensed media, and it is instructive to determine the radiative rate constant for this system, and its variation with temperature. Assuming that internal conversion from S_1 is a nonquenchable process which occurs prior to vibrational relaxation from a nonequilibrated state and using results of ϕ_F and ϕ_T determined by Cundall and Dunnicliff (105) and values of τ_F due to Lockwood (114), the calculated k_F values are presented in Table 7 for both C_6H_6 and C_6D_6. The similarity of rate constants for the protonated and deuterated molecules indicates that the large differences in yield and lifetime for the two isomers are the result of an isotope effect on a nonradiative transition.

In the medium pressure region, it has been suggested that fluorescence in the collisionally equilibrated gas phase system results from two (117) or three (118) vibrational levels of S_1. Whether this be the case, and if emission involves a multistep cascade through vibrational levels, it seems likely from considerations of the levels initially populated, the rates of vibrational deactivation, and the Boltzmann distribution of levels in the equilibrated system, that emission is dominated by transitions from the zeroth level. This is borne out by comparison of high-pressure values of k_F with zeroth-level transition probability discussed previously; $2.8 \times 10^{-6} s^{-1}$ and $2.2 \times 10^6 s^{-1}$, respectively, for C_6H_6 at 25°C; though the discrepancy is greater for C_6D_6 for which the values are $2.9 \times 10^6 s^{-1}$ and $1.7(5) \times 10^6 s^{-1}$.

In solution, the quantum yields and lifetimes for fluorescence emission for benzene are dependent upon solvent (see Tables 8 and 9, and more detailed data in reference 70) and temperature (140), but yield has been shown (141-145) to be independent of excitation wavelength within the first absorption band (λ exc > 230 nm). The observed variations arising from different exciting sources and experimental methods, in yield and lifetime for similar and even identical solvents reflect the difficulty experienced in obtaining reliable fluorescence data. Despite this unfortunate limitation, some trends may be discerned. In general, quantum yields and lifetimes tend to be larger in nonpolar solvents (91, 119), although in all cases these quantities are much smaller than in the gas phase. There appears to be no significant isotope effect on either parameter, contrary to expectation from a comparison with gas phase data (107,112).

The oscillator strength for the $S_1 \leftarrow S_0$ transition is almost independent of solvent (70,74,94), and it might thus be expected that the effect of solvent on the radiative probability would be slight. Cundall and Pereira (91,119) have reported measurements of k_{FM} in polar solvents, nonpolar solvents, and in particular perfluoro-n-hexane (in which solvent the similarity of spectral characteristics to those in vapor phase show that the solute-solvent interactions are minimal). It is found that although slight variations in k_{FM} with solvent occur, the variations are not accounted for by considerations of solvent polarity, and it may well be that k_{FM} is independent of solvent; lying within the range 1.5 - 1.8 x $10^6 s^{-1}$. The variation of ϕ_F and τ_F with solvent must, therefore, be the result of a medium effect on the nonradiative decay process. The difference between mean monomer transition probability in a range of solvents, and values from high-pressure and zeroth-level gas phase data is significant: 1.5 - 1.8 x $10^6 s^{-1}$ compared with 2.2 x $10^6 s^{-1}$ and 2.8 x $10^6 s^{-1}$. Application of the disputed refractive index, n^2, correction further increases the divergence between these data; while the theoretical value for k_{FM} (2.5 x $10^6 s^{-1}$) derived using the Strickler-Berg equation, neglecting n^2 is in close agreement with the gas phase data. The phase effect may be the result of uncertainties in measurement of various parameters, or, more probably interactions in the condensed phase which are not explained by the n^2 relationship. Reported data on C_6D_6 are less extensive than for C_6H_6 and contradictory. The data of Berlman (126) for ϕ_F and τ_F indicate that $k_F(C_6D_6)$ is larger than that for the perprotonated species. The data used by Sandros (146) and Helman (139) provide conclusions intermediate between those of Berlman (126), and those of Cundall, Olgivie, and Robinson (127), who reported a value for $k_F(C_6D_6)$ lower than that of C_6H_6. Rice and co-workers noted the similar trend in radiative lifetimes of the zeroth vibrational level in fluorescence (106,111), and attributed the effect to a greater purity of the electronic levels in perdeuterobenzene, which is also mirrored by changes in oscillator strength with isotopic substitution (86).

It was noted earlier that there was a small temperature effect on k_F in the gas phase. A similar effect is also observed in solution (94,119). Although in nonpolar solvents the effect is relatively small (119,136,147), it is accentuated in more polar environments (119). The effect is smaller for C_6D_6.

Reliable fluorescence quantum yields and lifetimes in the solid phase are difficult to determine, and accurate data are sparse (see Tables 10 and 11). It is well established that fluorescence and lifetimes of dilute solutions of benzene increase with decreasing temperature (8,119). In a recent study of similar molecules (156-158), it was shown that the increase in both ϕ_F and τ_F is smooth, and that both these quantities tend towards a limiting value at low temperature. An analysis of the temperature dependence of the two parameters for benzene in a variety of solvents (119) has indicated that the limiting values for both should be reached at a temperature well above 77°K, since the controlling variable is the radiationless Channel III process(es) which has a solvent dependent activation energy characteristically in excess of 0.2 eV. The term "Channel III" has been introduced to describe the process or processes other than fluoresence or triplet state formation which lead to the disappearance of $^1B_{2u}$ state (see Section VIII). Measurements at 77°K are thought to provide the limiting low-temperature values for the two fluorescence parameters.

Eastman (120) has reported limiting yields of 0.21, 0.18, 0.17, and 0.21 for benzene in n-hexane, methanol, 30/70 isobutanol/isopentane, and EPA, respectively. The yields are seen to be solvent dependent and determination of k_F requires emission lifetimes in identical media. Such are not available though limiting values of 120 and 115 ns have found (93) for benzene in methylcyclohexane and ethanol, respectively, at 77°K. Making the crude assumption that media effects are due to differences in solvent polarity we can calculate radiative rate constants of 1.75 x $10^6 s^{-1}$ for hydrocarbon solvents and 1.6 x $10^6 s^{-1}$ for alcohol solvents.

The similarity of these values to those mentioned earlier for room temperature solution work leads to the conclusions that the radiative rate constant is independent of temperature (below ambient) and that the difference between gas phase and condensed phase data is a real effect resulting from some molecular interaction not accounted for by the n^2 correction which conventional theory requires.

VII. THE $S_1 \rightarrow$ TRIPLET RADIATIONLESS TRANSITION

From analysis of fluorescence yields and lifetime measurements on emission from single vibronic levels (98,99,111-113,115) "rate constants" for <u>total</u> radiationless decay may be determined. These are also dependent upon vibrational level involved. Spears and Rice (111) have reported nonradiative lifetimes of 128, 108, 104, 105 ns, respectively, for the levels considered previously. The fluorescence "cut off" region, noted in these reports, suggests, in agreement with several other indications discussed in Section VIII, that radiationless $S_1 \rightarrow S_0$ decay proceeds via a highly vibrationally energetic level of S_1. The report of Rice and co-workers (106) on benzene-d_6 in this type of experiment shows that the nonradiative transition from S_1 is affected by deuteration to a greater extent than the fluorescence transition; the nonradiative lifetimes of the four comparable levels being 565, 345, 300, and 340 ns, respectively.

Triplet quantum yields for benzene vapor in collisionally deactivating systems have been estimated by a number of techniques (Table 12), involving sensitized isomeration, decomposition, or

TABLE 12. The S_1 triplet transition for benzene in the gas phase.

ϕ_T	Isomer	Temp. (°C)	Conditions	Technique	Ref.
0.72	C_6H_6	Room	2-19 torrs λ_{exc}254nm	Butene-2	(159)
0.72	C_6H_6	Room	7-26 torrs λ_{exc}252-67nm	Butene-2	(160)
0.71	C_6H_6	Room	10 torrs λ_{exc}252nm	Ketene	(161)
0.71	C_6H_6	Room	Cyclobutanone	Cyclo-butanone	(162)
0.79	C_6H_6	Room		Pyrazine	(163)
0.63	C_6D_6	Room	2-19 torrs λ_{exc}254 nm	Butene-2	(159)
0.71	C_6H_6	Room	> 10 torrs λ_{exc}=254 nm	Butene-2	(105)
0.70	C_6H_6	40	> 10 torrs λ_{exc}=254 nm	Butene-2	(105)
0.68	C_6H_6	60	> 10 torrs λ_{exc}=254 nm	Butene-2	(105)
0.65	C_6H_6	80	> 10 torrs λ_{exc}=254 nm	Butene-2	(105)
0.63	C_6H_6	100	> 10 torrs λ_{exc}=254 nm	Butene-2	(105)
0.61	C_6D_6	Room	> 10 torrs λ_{exc}=254 nm	Butene-2	(105)
0.52	C_6D_6	50	> 10 torrs λ_{exc}=254 nm	Butene-2	(105)
0.46	C_6D_6	70	> 10 torrs λ_{exc}=254 nm	Butene-2	(105)
0.37	C_6D_6	100	> 10 torrs λ_{exc}=254 nm	Butene-2	(105)

luminescence of additives. It is to be expected that the overall intersystem crossing efficiency is dependent upon the extent of collisional deactivation and redistribution of energy following excitation. The presence of appreciable quantities of added gas, which may arise because of the presence of the triplet monitor, and the use of benzene pressures in excess of 10 torrs ensure that the yields of triplets reported for gas phase systems result from intersystem crossing from thermally equilibrated range of levels in S_1. Of the many triplet monitors used, it is widely believed that butene-2 effectively quenches only the triplet state.

Dunnicliff (115) has shown that the olefin also quenches singlet S_1 benzene, but only with a bimolecular rate constant of 2.2 x 10^8 $M^{-1}s^{-1}$, so at the olefin concentrations commonly used for triplet scavenging, singlet state quenching is negligible compared with that of the triplet state.

Noyes and Harter (160) obtained evidence of a decrease in triplet yield with decrease of excitation wavelength. A limiting high-pressure yield of 0.71 at room temperature after excitation at 254 nm is now well established. The observed isotope effect is large; with corresponding high-pressure limit values of 0.63 (159) and 0.61 (105) reported for C_6D_6, under the same conditions. Using the emission lifetime measurements of Lockwood (114), intersystem crossing rate constants (for heavy and light benzene) of 7.9 x 10^6s^{-1} and 11 x 10^6s^{-1} are obtained assuming a nonquenchable internal conversion also occurs. For this transition, the crude assumption that the high-pressure rate constant is dominated by the contribution from the zeroth level may not be valid. As discussed earlier, transition may be to either or both of the $^3E_{1u}$ and $^3B_{1u}$ states. Transition to $^3B_{1u}$ has first-order spin-orbit allowedness, but the rate constant may be reduced by the small Franck-Condon overlap factor resulting from the 1.1 eV energy gap between states. Transition to $^3E_{1u}$ requires vibronic coupling. The positions of the three states in C_6D_6 relative to the equivalent states in C_6H_6 are not accurately known, though spectroscopic evidence (39,164,165) suggests that the energy spacings are slightly increased (200 cm^{-1}) by deuteration. In addition, deuterium substitution reduces the energy separation of the vibrational levels (166-169). There is some confusion concerning whether deuterium atom substitution affects (170) the radiationless $T_1 \rightarrow S_o$ transition or otherwise. Because of this uncertainty, it is probably unacceptable to seek similarities between the measured rate constant in high-pressure gases and in the nonradiative decay rates of the zeroth vibrational levels of S_1 states of C_6H_6 and C_6D_6.

Considerable data are available for triplet yields of benzene in dilute solutions of different solvents (see Table 13). In the main, two techniques have been used; sensitized phosphorescence of biacetyl, sensitized cis-trans isomerization of butene-2, octene-2, and stilbene. All yield comparable results. In saturated hydrocarbon solvents at room temperature, the triplet yield for C_6H_6 is found to be about 0.24 ± 0.01. There is a solvent dependence of this quantity, the yield dropping to 0.15 in ethanol, 0.13 in methanol, and 0.09 in acetonitrile (91). In determining the effect of environment on the rate constant controlling intersystem crossing, values for emission lifetimes in the various systems are needed. These are, as mentioned previously, often unreliable. Cundall and Pereira (91) have reported consistent data which give the variation in k_{ISCM} (= ϕ_T/τ_F) with solvent. It is found that, though differences between solvents occur, there is no consistent trend with solvent polarity, and we can consider k_{ISCM} to be independent of solvent within the range of approximately $7(\pm1) \times 10^6 s^{-1}$. Further, there appears to be little isotopic effect on k_{ISCM} since a rate constant of $8.7 \times 10^6 s^{-1}$ has been reported for C_6D_6 (127) in methyl-cyclohexane at room temperature compared with $8.0 \times 10^6 s^{-1}$ for the perprotonated solute (91).

The variation of k_{ISCM} with temperature for C_6H_6 in cyclohexane (182) is more marked than that for C_6D_6 (127) similar to the trends noted in gas phase studies (105). It is erronous to fit this temperature dependence to a simple Arrhenius function, since the yield of phosphorescence at low temperature shows a substantial triplet production in the solid phase. More realistically, the room temperature process may be viewed as the sum of two intersystem crossing steps to the two lowest triplet states. One process being temperature independent ($\rightarrow {}^3B_{1u}$?) and the other temperature dependent ($\rightarrow {}^3E_{1u}$?).

Estimation of triplet yields from benzene in the solid phase presently requires several approximations, since the available triplet monitoring methods do not function in solid media. Pereira (93) has shown that the quantum deficit $1-(\phi_F + \phi_T)$ may be assigned in benzene although not necessarily to substituted benzene derivatives, to a single process, and that $(\phi_F + \phi_T)$ approaches unity with decreasing temperature. In this way, the deficit $(1 - \phi_F)$ may be assigned to ϕ_T at low temperatures. Using the emission yields and limiting low temperature lifetime data from Tables 10 and 11, together with the solvent effect assumption noted previously, we may calculate rate constants of 6.6×10^6 and $7.1 \times 10^6 s^{-1}$ for intersystem crossing in hydrocarbon and polar media, respectively.

It is seen that the rate constants for intersystem crossing are broadly, that is, within ±10 percent, independent of molecular environment. We must conclude that the significant differences noted in fluorescence and triplet yields and lifetimes, with phase, must be the result of an effect on the "hidden" $S_1 \rightarrow S_o$ decay process.

VIII. THE $S_1 \rightarrow S_0$ RADIATIONLESS TRANSITION

Evidence that decay of the first excited singlet state may occur via a nonradiative process which depends on the vibrational energy and probably does not involve the lower triplet states of the π manifold comes from various types of experiment:

TABLE 13. Triplet yields in the solution phase.

ϕ_T	Isomer	Temp. (°C)	Conditions	Monitor	Ref.
0.24	C_6H_6	29	Pure benzene	Piperylene	(172)
0.57	C_6H_6	25	Pure benzene	Butene-2	(173)
0.60	C_6H_6	25	Pure benzene	Octene-2	(174)
0.56	C_6H_6	24	Pure benzene	Butene-2	(175)
0.58	C_6H_6	25	Pure benzene	Stilbene	(176)
0.71	C_6H_6	25	Pure benzene	Octene	(177)
0.24	C_6H_6	24	0.11 2M cyclohexane	Butene-2	(147)
0.25	C_6H_6	26	6.10^{-2}M Me-cyclohexane	Butene-2	(178)
0.25	C_6H_6	20	Dil. cyclohexane	Biacetyl	(146)
0.24	C_6H_6	22	Dil. Me-cyclohexane	Octene	(177)
0.23	C_6H_6	25	Dil. cyclohexane	Butene-2	(91)
0.24	C_6H_6	25	Dil. Me-cyclohexane	Butene-2	(91)
0.15	C_6H_6	25	Dil. ethanol	Butene-2	(91)
0.13	C_6H_6	25	Dil. methanol	Butene-2	(91)
0.09	C_6H_6	25	Dil. acetonitrile	Butene-2	(91)
0.07	C_6H_6	25	Dil. water	Butene-2	(91)
0.47	C_6H_6	24	0.IM cyclohexane	Butene-2	(175,178)
0.39	C_6H_6	25	0.3M cyclohexane	Flash photolysis	(179)
0.18	C_6H_6	25	Dil. cyclohexane	Stilbene	(180)
0.23 (5)	C_6D_6	25	Dil. Me-cyclohexane	Butene-2	(178)
0.24	C_6D_6	20	Dil. cyclohexane	Biacetyl	(181)
0.25	C_6D_6	25	Dil. cyclohexane	Butene-2	(127)
0.24	C_6H_6	25	Dil. perfluorohexane	Butene-2	(91)

1. The decrease in quantum yield for $S_1 \rightarrow S_o$ fluorescence with increasing excitation energy has been reported for gas phase (142), liquid phase (141,143,183), and solid phase (144) benzene systems. The decrease in yield shows that the onset of an additional competitive decay process, occurs at an energy above S_1 of the order of 0.3 eV.
2. Low-pressure, narrow band pass excitation studies have shown (98,99,111-113) that the fluorescence yields from single excited vibrational levels of S_1 drop sharply to zero in region of 0.31 eV above the zeroth level of the $^1B_{2u}$ state. The effect is also noted in the perdeuterated molecule (106).
3. Callomon and co-workers (27) report the onset of spectral diffuseness in the S_o - S_1 absorption spectrum of benzene vapor at a level approximately 0.31 eV above S_1.
4. In the medium-pressure gaseous benzene system, Poole (100) has found that the $S_1 \rightarrow S_o$ fluorescence yield drops to zero for excitation of wavelengths of 240 nm or less, despite efficient collisional deactivation.

Such triplet yields, as are available, show that the vibrationally enhanced process is not intersystem crossing.

All of these observations are consistent with the existence of a rapid nonradiative route available for the decay of excited benzene molecules with energy approximately 0.30 eV above S_1. The variation with temperature of the quantum deficit ϕ_{IC} = $1-(\phi_F+\phi_T)$ has been analyzed kinetically in terms of a single process for benzene-h_6 (91,147) and benzene-d_6 (127) in solution. The assumption of a single decay process allows interpretation of the behavior with temperature of the rate constant k_{ICM} (= ϕ_{IC}/τ_F) in terms of the Arrhenius equation which yields an activation energy which is solvent dependent (91), typically of the order of 0.25 to 0.37 eV, and is tentatively interpreted as the mean excess energy required by the thermally equilibrated excited benzene molecules, for decay via this hidden route.

There is also evidence (105) that the quantum deficit increases with temperature in the high-medium-pressure gaseous system. Analysis using the crude model applied to the solution phase data produces an "activation energy" for Channel III decay of 0.14 eV for C_6H_6 and 0.19 eV for C_6D_6. A kinetic scheme involving a nonquenchable radiationless decay of vibrationally excited S_1 states is thought (105) to provide a better fit for the data. Using this approach, the onset of nonradiative deactivation would occur at approximately 0.4 eV above S_1.

The experimental evidence proves, without question, the existence of a Channel III route; and various attempts have been made to provide some explanation. The theories of radiationless transitions applied with varying success to the $T_1 \rightarrow S_o$ step (see section X) have not yet been able to explain satisfactorily the pure singlet decay phenomenon. It has been suggested that photochemical rearrangements could account for the quantum deficit. Evidence (184) (reviewed later in Section XII) has shown that the prismane and Dewar isomers would arise from states of B_{1u} symmetry, while benzvalene and fulvene result from the $^1B_{2u}$ state. Bryce-Smith and Longuet-Higgins (65,185) and Jano (186) have suggested that diradical intermediates consistent with the symmetries of the $^1B_{2u}$ and $^3B_{1u}$ states may be involved in the photoisomerization of benzene and its derivatives. Measured radical yields are very low (180), as are the final yields of isomeric products (187). Though rapid re-aromatization of the diradical, or of benzvalene has been proposed as an explanation of the low isomer yields, it is to be expected that isomerization due to group shifts should occur in the case of benzene derivatives. Examination of the photodecay processes in o-xylene (188) and m-xylene (189) has shown that aromatic isomer products are formed in very small yields, and further, that the activation energy for production of m-xylene from other isomers (0.44 eV) is higher than the activation energy required for the temperature-dependent Channel III process (0.21 eV). Birks (190), adopting the isomer theory of Lamola et al. (191) and Noyes and co-workers (192), suggests a "physical" isomer, the most unusual property of which seems to be that, by definition, it is undetectable. It is suggested that the potential energy surfaces of this "isomer" overlap with those of S_o and S_1 benzene, permitting rapid hidden decay. It is not clear how this differs from what would be conventionally regarded as internal conversion.

Callomon, Leach, and Lopez-Delgado (27) suggested that a σ-π^* state existed in the region between the $^1B_{2u}$ and $^1B_{1u}$ states (see Section II), and selected $^3E_{2u}$ as the most probable state involved. It has been suggested (156) that a singlet state would be more plausible in view of the very high rate constant for the transition from the $^1B_{2u}$ state. The problem of decay of this newly invoked state remains since it is expected that the transition to ground state would be extremely forbidden, leading consequently to a long-lived intermediate. Recent photochemical data (193,194) have shown the existence of a long-lived, non-benzvalene (194) intermediate following 2537 A^o irradiation of benzene. Results obtained by Stein (94,195) in O_2-free solutions of benzene have indicated the formation of benzvalene with an initial quantum yield of 0.18 after excitation with the 2537 A^o line and that this initial yield is independent of temperature in the 9 to 50°C range. This result shows that isomer formation does not account for the thermally activated fluorescence quenching. Stein proposed that benzvalene arises from a higher vibronic state before vibrational relaxation to the S_1 fluorescent state, that is, prior to interaction with the environment. Benzvalene is subsequently quenched by benzene triplet. Results of Lutz and Stein (195) also show an increase in the efficiency of benzvalene formation at higher temperatures, consistent with a decrease of ϕ_T, as determined by Cundall and co-workers (196). Further results of Ilan and Stein (197) on the fluorescence yields of benzene solutions in the presence of xenon still support the assumption.

A recent tunneling effect model for radiationless transitions (198) has been applied to benzene and other aromatic hydrocarbons. The CH stretching vibrations are considered as dominant for the non-radiative process. Rate constants for the radiationless process $S_1 \rightarrow S_0$ calculated by theory are of the same order of magnitude as those obtained by experiment. A further possibility for the interpretation of Channel III is a delay via a configurationally perturbed doubly excited E_{2g} state, though this route seems unlikely following confirmation that the relevant state is of similar energy to $^1B_{1u}$ (see Section II).

IX. THE $T_1 \rightarrow S_0$ PHOSPHORESCENCE TRANSITION

Considering the ease with which phosphorescence emission from solid or glassy phase solutions of benzene is observed, it is perhaps surprising that unambiguous data on the probabilities of radiative and radiationless decay of the triplet state are so difficult to obtain and unreliable. The determination of excited state parameters at a particular temperature requires, at least, evaluation of ϕ_P, ϕ_T and τ_P, and more reliably ϕ_P, ϕ_F, ϕ_T, τ_F, and τ_P. Direct measurement of ϕ_P by comparison of the emission of the given compound with that of a standard sample is one method of approach (148,150). However, since it is necessary that both sample and standard possess identical optical properties (otherwise corrections for effects of refractive index differences, reflectivity losses, and emission polarization on the measurements must be made), this method is of limited applicability. Other methods for measuring ϕ_P have been reported. Kellogg and Bennett (199) estimated phosphorescence yield from measurements of acceptor molecule fluorescence, resulting from triplet energy transfer from the aromatic donor molecule. This method should be reliable, provided solutions are sufficiently dilute to eliminate the effect of triplet-triplet annihilation and provided corrections for anisotropy of emission are made.

A further method involves a comparison of the luminescence from a glassy sample with the light scattered by a MgO surface (149).

Because of the complexity and difficulty in obtaining reliable data, some authors (8,200,201) have made the arbitrary assumption that $k_P = 0.030\ s^{-1}$ and is independent of temperature for all unsubstituted hydrocarbons. From this assumption, and measurements of τ_P, values for k_{GT} have been estimated. The subsequent theoretical analysis of the radiationless transition probability based on the assumption must, therefore, be treated with some caution, since it is arbitrary and probably incorrect.

A method for determining k_P reported by Cundall and Pereira (91) utilizes data from the parallel studies on parameters controlling the spin-allowed emission. In these studies, most exhaustively pursued for benzene, it is found that both k_F and k_{ISC} approach limiting low-temperature values, and estimates of their magnitude in the < 100°K region may be given with some certainty. With this information, the additional data for τ_P and for $\emptyset_P/\emptyset_F$ (χ), k_P can be determined, since

$$k_P = \chi\ k_F\ /\ k_{ISC}\ \tau_F$$

This method does not rely upon any special assumptions, although corrections for photoselection effects (202-205) and for the eigenpolarization of the spectrophotometer must be undertaken. Further studies of fluorescence and phosphorescence at 77°K for simple benzene derivatives using the same spectrophotometer and the same optical arrangement carried out by Cundall and Ogilvie (158) have shown that photoselection and eigenpolarization corrections do not alter the results obtained for k_P (within 2-3%). The above method assumes reliable data for τ_P but measurements are subject to uncertainty. Variations in τ_P for a specific sample can be found which depend upon the rates of cooling of the sample, and also upon the time which elapses between the freezing process and the measurement operation. The cooling rate may affect the decay if two phases of the solid solvent can be formed (206-208). This is the case for cyclohexane matrices in which cubic matrices are produced by rapid freezing, and monoclinic matrices by slow freezing. Variations in τ_P with standing time can be observed if glass relaxation occurs (208-210) and/or thermal equilibration is achieved only slowly (211). Though the effect of this latter factor may be reduced by using very narrow sample tubes, Martin and Kalantar (212) have shown that τ_P varies considerably with time elapsed following freezing, if the solvent utilized undergoes glass transitions in the neighborhood of the temperature used. For the particular case of benzene, τ_P is strongly dependent upon solvent and upon temperature above 60°K. Solvent effects have been reported by

Leubner (213), Hirayama (214), and Hatch et al. (215), among others (Fig. 8). In some instances, apparent solvent effects may be the result of impurity quenching, particularly in solvents of relatively low viscosity. However, this explanation cannot be universally valid and does not assist in understanding the unexpected temperature effect. Leubner (213) could find no correlation between sample viscosity and τ_P, and Ling and Willard (216) reported, from measurements on 3-methyl pentane between 77 and 94°K, that the viscosity had a pseudo-activation energy of 15.9 kcal/mol, a factor of 7 greater than the activation energy measured in much the same temperature range for τ_P. A definitive examination of the problem has been published recently by Kilmer and Spangler (217) (Fig. 9). These authors found that the variation in τ_P for benzene in cyclohexane ($\sim 10^{-3}$M) with temperature could be fitted to the relationship of the type

$$1|\tau_P = 1|\tau_o + k_1 \exp(-\Delta E|KT)$$

For benzene-h_6 it was found that τ_o = 11.44s $\Delta E/hc$ = 554 cm^{-1} and $k_1 = 4.2 \times 10^3 s^{-1}$. The values for C_6D_6 are 14.28s, 705 cm^{-1} and 218 x $10^3 s^{-1}$, τ_P is, of course, a composite function including radiative and nonradiative decay processes and there is still speculation concerning which mode of decay is temperature dependent and which not. The weight of evidence seems to favor the nonradiative route as being that which requires thermal activation. Use of the Kilmer-Spangler equation to extrapolate to room temperature produces phosphorescence lifetimes of 3.8 and 5 ms for C_6H_6 and C_6D_6, respectively. Since phosphorescence is not observed in either liquid or gas phases, it is clear that alternative decay routes operate in fluid media.

Some emission lifetime data are presented in Table 14. Particularly of note are the lifetimes in frozen inert gas solutions, reflecting the enhancement of $T_1 \rightarrow S_o$ transition with increasing atomic number of the solvent; an effect also noted by other authors. The important results of Haaland and Nieman (222) show a marked solvent effect on phosphorescence lifetimes. Perfluorocarbons are the least perturbing solvents on phosphorescence, an effect consistent with results obtained in liquid solution (72,91).

The effect of deuteration whose influence is independent of solvent and temperature, shows the importance of the importance of the C-H/C-D modes in promoting one or both of the triplet decay transitions.

Using the values $\emptyset_F = 0.2^{120}$, $\chi(=\emptyset_P/\emptyset_F)$ = 0.92 (92,219) and τ_P = 5.50 s for benzene in methyl cyclohexane, it is possible to calculate values of 0.0417 and 0.401 s^{-1} for k_P and k_{GT}. The

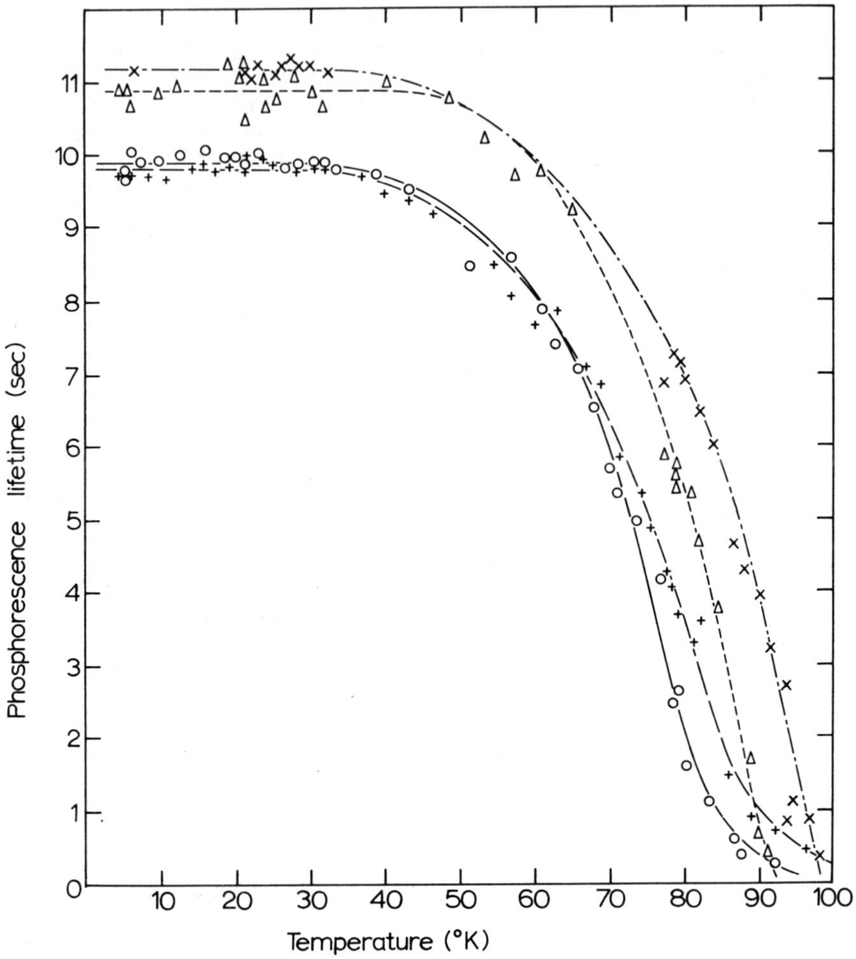

FIGURE 8. Temperature dependence of τ_p for C_6H_6 in ethanol (X), EPA (Δ), 3-methylpentane (O), and MCH-isopentane (+), (Leubner).[213]

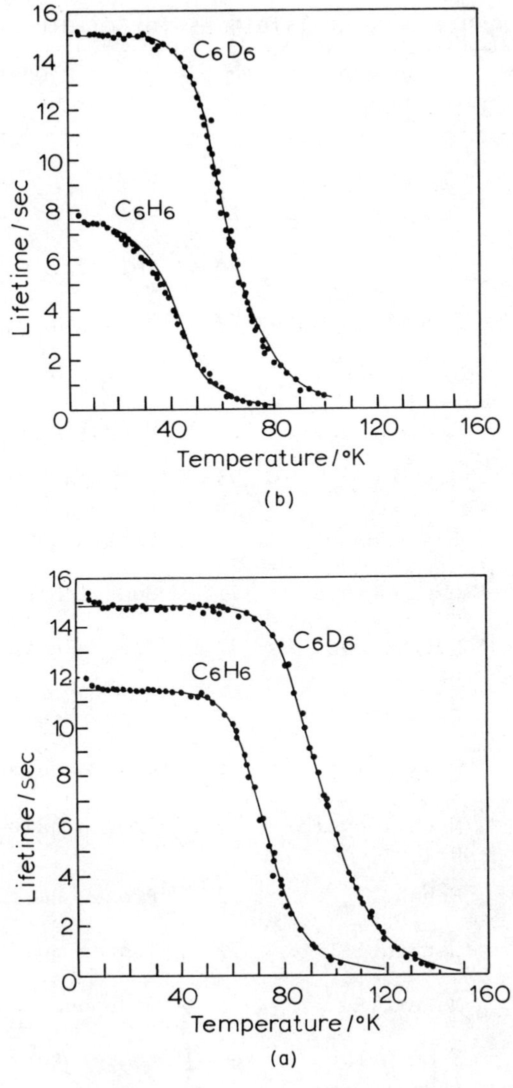

FIGURE 9. Temperature dependence of τ_p for C_6H_6 and C_6D_6 in (a) cyclohexane and (b) cyclopentane, (Kilmer and Spangler).[217]

TABLE 14. Phosphorescence lifetimes for dilute benzene solutions.

τ_P (s)	Isomer	Solvent	Temp. (^oK)	Remarks	Ref.
7.1	C_6H_6	EOA	77		(144)
17.5	C_6H_6	-	20	Two sites for decay	(171)
29.5	C_6D_6	-	20	This for longest	(171)
5.75	C_6H_6	3-Methylpentane	77	Relaxed	(218)
7.48	C_6H_6	Isopropanol	77	Frozen quickly	(218)
4.75	C_6H_6	Cyclohexane	77	Monoclinic	(218)
12.02	C_6D_6	3MP	77	Relaxed	(218)
13.69	C_6D_6	Isopropanol	77	Frozen quickly	(218)
14.10	C_6D_6	Cyclohexane	77	Monoclinic	(218)
5.50	C_6H_6	Methyl-cyclo-hexane	77	Frozen quickly	(219)
6.50	C_6H_6	Ethanol	77	Frozen quickly	(92)
5.50	C_6H_6	Methanol	77	Frozen quickly	(92)
10.50	C_6D_6	Methyl-cyclo-hexane	77	Frozen quickly	(219)
11.20	C_6D_6	Ethanol	77	Frozen quickly	(92)
9.10	C_6D_6	Methanol	77	Frozen quickly	(92)
9.83	C_6H_6	Me-cx/IP	<60	Independent of	(213)
9.92	C_6H_6	3-Methylpen-tane	<60	temperature below	(213)
10.91	C_6H_6	EPA	<60	60^oK but	(213)
11.19	C_6H_6	Ethanol	<60	solvent dependent	(213)
9.0	C_6H_6	Dioxane	77	-	(214)

TABLE 14 (cont.)

5.90	C_6H_6	EPA	77	-	(214)
2.0	C_6H_6	Cyclohexane	77	-	(214)
5.78	C_6D_6	Isopentane	77		(212)
11.79	C_6D_6	3-Methylpentane	77		(212)
12.11	C_6D_6	3-Methylhexane	77		(212)
11.31	C_6D_6	Methyl-cyclo-hexane	77	Constant	(212)
13.70	C_6D_6	Ethanol	77	values after	(212)
13.40	C_6D_6	Iso-propanol	77	solvent	(212)
13.06	C_6D_6	EPA	77	relaxation	(212)
9.46	C_6D_6	Cyclohexane	77		(212)
10.32	C_6D_6	Isooctane	77		(212)
5.75	C_6H_6	3-Methylpentane	77		(212)
8.45	C_6H_6	EPA	77		(212)
3.04	C_6H_6	3-Methylpentane iso-pentane	77		(212)
11.44	C_6H_6	Cyclohexane	-	Limiting	(217)
14.82	C_6D_6	Cyclohexane	-	low-temperature	(217)
7.46	C_6H_6	Cyclopentane	-	values	(217)
14.94	C_6D_6	Cyclopentane	-		(217)
29.5	C_6D_6	Argon	20		(171,220)
14.5	C_6D_6	Argon	20		(171,220)
17.5	C_6H_6	Argon	20		(221)
0.74	C_6H	Krypton	20		(221)

TABLE 14 (cont.)

0.026	C_6H_6	Xenon	20		(221)
29.5	C_6D_6	Argon	20		(221)
0.025	C_6D_6	Xenon	20		(221)
15.4	C_6H_6	Perfluoromethyl-cyclohexane	65		(222)
10.9	C_6H_6	Methyl-cyclo-hexane	65		(222)
23.7	C_6D_6	Perfluoro-cyclo-hexane	65		(222)
14.3	C_6D_6	Methyl-cyclo-hexane	65		(222)
8.0	C_6H_6	EPA	77		(148)
11.8	C_6H_6	Cyclohexane	4.2		(215)
10.2	C_6H_6	Borazine	4.2		(215)
9.5	C_6H_6	C_6D_6	4.2		(215)
15.6	C_6D_6	Cyclohexane	4.2		(215)
12.5	C_6D_6	Borazine	4.2		(215)
6.3	C_6H_6	EPA	77		(154)
9.4	C_6D_6	EPA	77		(154)
4.6	C_6H_6	EPA	77	Monoclinic	(207)

phosphorescence rate constant for benzene is in close agreement with that of Johnson and Ziegler (171) who deduced a value of 0.041 s^{-1} for C_6H_6 and 0.034 s^{-1} for C_6D_6. The latter authors concluded that deuteration affected the vibronic coupling necessary to promote the radiative transition and that the deuteration effect on the radiationless transition is small at variance with early theories of radiationless transitions (201,223).

From the variations of τ_P with temperature, Kilmer and Spangler (217) concluded that the ratio $k_P\ k_{ISC}/k_F$ was independent of temperature and equal to 0.229 ± 0.007 s. Although both k_F (134) and k_{ISC} (147) may increase with temperature above ambient, by analogy with toluene (156) and other methylated derivatives (224) it is expected that both will be nearly constant below room temperature. From a combination of these data it is clear that k_P is independent of temperature.

All of the above data refer to dilute solutions; the effect of solute concentration on the phosphorescence emission yield and lifetime is very marked. At concentrations of 1M and above, a decrease in τ_P and the χ ratio has been reported for EPA (214), EOA (144,152), methylcyclohexane (219), and cyclohexane (207). Phosphorescence is negligible in both pure benzene crystals (225) and polycrystalline powder (144,214). This has been attributed to a rapid triplet-triplet annihilation process, an explanation apparently confirmed by Cundall and Pereira (219), who detected solution some slight rather diffuse phosphorescence together with delayed fluorescence from pure benzene.

X. THE $T_1 \rightarrow S_0$ RADIATIONLESS TRANSITION

The benzene triplet has not been observed to phosphoresce in either the gas or liquid phase, as noted previously but in view of the lifetime measured in glassy media at low temperature this is not surprising, since very small amounts of quenching impurities would reduce emission to undetectable levels. Some detailed investigations have shown that there is an intramolecular process which induces radiationless decay of the benzene triplet state, other than impurity quenching.

By competitive methods Ishikawa and Noyes (226) (sensitized biacetyl phosphorescence) and Cundall and Davies (159) (butene-2 isomerization) both estimated the triplet lifetime to be of the order of 10 μs. In a reexamination of the butene-2 system, Lee (227) estimated a value of 100 μs, a finding confirmed by Cundall and Dunnicliff (105). An examination of the kinetics of the benzene photosensitized composition of cyclopentanone decomposition allowed a value of longer than 3 μs to be deduced. This type of experiment is far from satisfactory since photochemical processes can intervene at low pressures, and impurities and the quenching effects of photoproducts can affect the results. These problems can only be overcome by some form of direct measurement. Parmenter and Ring (228) used a flash method in which 20 torrs of benzene and 0.01 torr of biacetyl were submitted to a 20 J, 3 μs flash of 250 to 260 nm light. From the time dependence of the biacetyl phosphorescence emission, a lifetime of 26 ± 5 μs was deduced. Lee (227) cites a private communication from Parmenter in which the lifetime of the gaseous triplet is given to be 70 ± 20 μs for 2 torrs benzene. Burton and Hunziker (229) have developed a modulated transient absorption technique that uses Hg 6^3P_0 atoms to populate the benzene triplet state. The decay of the triplet is observed by an absorption band at 231.5 nm, assigned to the allowed $^3E_{2g} \rightarrow {}^3B_{1u}$ absorption (36), with f-values of 0.07 and 0.06 for light and heavy benzene, respectively. The decay is pseudo-first-order, resulting from bimolecular triplet-triplet interaction. The results are viewed guardedly by other workers because of effects that may arise from the high light absorption which must be used. First order decay rates of $3 \times 10^3 s^{-1}$ and $2 \times 10^3 s^{-1}$ were calculated for the presence of 300 to 800 torrs of nitrogen, needed to promote the Hg $6^3P_1 \rightarrow 6^3P_0$ transition. The pulse radiolysis of naphthalene/benzene vapor mixtures at 390°K has given data which can be used to calculate $1.3 \times 10^5 s^{-1}$ for the decay constant of the $^3B_{1u}$ state of benzene at a partial pressure of 18 torr (230), from the naphthalene triplet absorption following irradiation.

A new experimental approach to the problem has been used by Hunter and Stock (231) involving measurement of the frequency dependence of the acoustic signal resulting from transformation of electronic excitation from vibrational to translational energy, consequent to excitation of the $^1B_{2u}$ state. This technique was used to measure the decay of the benzene triplet over the range 0 to 6 torrs, and it was found, in agreement with what could be inferred from some of the earlier work, that decay of the triplet depends upon benzene concentration. The pressure-independent decay process was found to have a rate constant of (1.7 ± 0.1) $\times 10^3$ s^{-1} and a collision efficiency of 1.6×10^{-4} deduced for the second order self-quenching process. The authors claim to have discounted the effects of deactivation by traces of oxygen. A strong perturbation of the $^3B_{1u}$ state by ground state $^1A_{1g}$ molecules seems the most likely mechanism for accounting for the observed concentration dependence; and involvement of the triplet excimer is in accord with the following scheme.

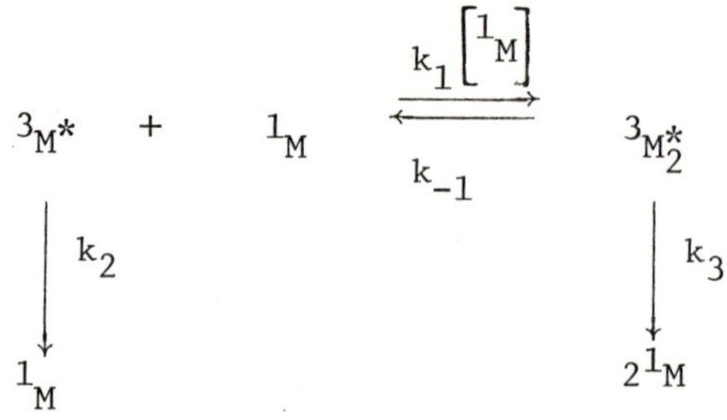

The reciprocal of the triplet lifetime $\tau_T{-1} = k_2 + k_3(k_{-1}+k_3)^{-1} k_1[^1M]$. Analysis of the results shows the $k_3(k_{-1} + k_3)^{-1}$ is 1.6×10^{-4} and thus $k_{-1} >> k_3$. This contrasts with the situation in other aromatic hydrocarbons in solution where triplet-ground state interaction is proposed to explain similar shortening of measured triplet lifetime (232); in these cases, it would seem that $k_3 >> k_{-1}$.

The self-quenching mechanism which has been invoked for the gas phase system applies also to the liquid phase. Order of magnitude agreement between data from the various techniques for measuring τ_T seems to have been achieved. Cundall and Griffiths (233), obtained the first indications from studies on the benzene/butene-2 system, that the triplet state lifetime was very short. The value for $k_{ET}\ \tau_T$ was 21.2M at 25°C where k_{ET} was the overall rate constant for triplet excitation energy transfer from benzene to butene-2, and τ_T the lifetime of the benzene triplet. Assuming a value of about $5 \times 10^9\ M^{-1}s^{-1}$ for k_{ET}, τ_T appears of the order of a few nanoseconds. Although the kinetic analysis may have oversimplified (175), the order of magnitude lifetime is similar to that determined from the work of Kaufman (234) and Hentz and Perkey (176) and also consistent with the results of Dubois (235,236) who failed to detect any evidence for triplet energy transfer to biacetyl in pure benzene. By contrast, Lipsky (237), Sandros (181), and Russian workers (238,239) observed triplet energy transfer from benzene to biacetyl for dilute benzene solutions. This indicates that the triplet lifetime in solution is also concentration dependent as is the case for o-xylene (240) where the lifetime increased from a few nanoseconds to microseconds or more, on dilution. This has been confirmed for benzene (175,189) and further, additional evidence from photochemical (173) and pulse radiolysis (241) systems demonstrate the very short lifetime of the triplet state in pure benzene liquid. It is believed (175) that decay proceeds via the triplet excimer which is only inefficiently quenched by butene-2. Results of electron impact excitation (242) and pulse radiolysis (243) have indicated a triplet lifetime of 500 μs for gaseous benzene at low and high pressures.

Attempts to directly detect the triplet state of benzene under these or similar conditions by transient spectroscopy techniques have so far been unsuccessful. Thomas and Mani (244) found a transitory species with an absorption maximum and a decay time of 112 ns in benzene and benzene cyclohexane mixtures. The results from addition of naphthalene, biacetyl, oxygen, and piperylene suggest that this is a product of reaction between triplet and ground state benzenes. Build-in times for this species, which are 3 ns for pure benzene and 20 ns for 10 percent benzene/cyclohexane mixture are temperature independent. Laser flash photolysis (179) results are consistent with a $T_1 + S_o$ reaction for triplet decay. This is the most that has been achieved to direct observation of the benzene triplet in solution at room temperature.

The lifetime of the triplet state of benzene in fluid solutions (at room temperature and above) is determined by the, seemingly efficient, interaction of that state with ground state molecules. This is not inconsistent with the low temperature behavior from which it has been deduced (51-55), that a thermally activated $T_1 \rightarrow S_o$ transition arises from distortion of the benzene ring which destroys the strict D_{6h} symmetry. Even below 65°K, it is apparent that hydrocarbon solvents are not inert and that perfluorocarbons are the least perturbing of possible solvents (222). The concentration quenching mechanism must involve some, as yet unestablished, form of chemical or charge transfer interaction which reduces the forbidden character of the transition. There is no positive evidence to show that this process occurs at low temperature.

XI. TRANSITIONS FROM HIGHER (n > 1)EXCITED STATES

Several years ago, Braun, Kato, and Lipsky (142) showed that in the vapor solution and bulk liquids, benzene and its methylated derivatives appeared not to undergo efficient internal conversion or intersystem crossing from higher excited singlet states to form either S_1 or T_1 states. This was confirmed by Birks, Conte, and Walker (143) who postulated a scheme involving excimers of higher excited states, to explain their data and noted that benzene proved to be the exception amongst similar compounds in that the effect decreased on dilution.

Some more recent experimental evidence has been presented by Voltz and co-workers (145). Using 2-napthtyl-2-phenyl-5-oxazole-1.3 an excitation energy acceptor, these authors obtained results similar to those of Lipsky (142), but also found that autoionization occurred at wavelengths below 175 nm. There is low efficiency of conversion of B_{1u} and $^1E_{1u}$ to the $^1B_{1u}$ state, and relative efficiencies of fluorescence are constant but different over each of the main bands. Voltz (145) also claims that energy transfer occurs from S_2 and S_3 states, in agreement with Laor and Weinreb (245) but at variance with Lawson, Hirayama, and Lipsky (72,246). Electron scavengers such as O_2 and $CHCl_3$ which affect the S_3 and S_2 states to only a negligible extent, influences the ionization effects at higher excitation energies.

Lawson, Hirayama, and Lipsky (72,246) also measured the variation in quantum efficiencies for internal conversion from S_3 to S_1 in several solvents and found that the values vary very markedly. The efficiencies measured were 0.45 in pure benzene, 0.37 to 0.28 in alcohols and other polar solvents, 0.25 to 0.22 in aliphatic hydrocarbons, 0.04 in n-C_6F_{14}, and about zero in the gas phase. The order of effectiveness in a number of solvents paralleled the enhancement of the 0-0 band in absorption. The implication would seem to be that solvent perturbation produces state mixing and consequent enhancement of internal conversion $S_n \rightarrow S_1$.

Other types of experiment give information about the internal conversion processes of higher excited singlet states. Lumb, Braga, and Pereira (144) measured the fluorescence and phosphorescence excitation spectra of rigid solutions of benzene in ether/isooctane/alcohol glasses at 77°K. The two excitation spectra were identical, showing that ϕ_F and ϕ_P vary in the same manner with λ_{ex}. The efficiencies of conversion from higher levels are not zero, but depend upon concentration; a surface quenching mechanism may be involved.

Direct observation of fluorescence from higher singlet states of benzene and some methyl derivatives has recently been achieved by Hirayama, Gregory, and Lipsky (247). Using apparatus capable of detecting fluorescence yields as low as 10^{-6} they recorded the emission spectra from oxygenated solutions of pure benzene and other aromatics excited at 184.9 nm. Subtraction of the tail of the residual S_1 emission gives a fluorescence spectrum with λ_{max} approximately 235 nm and $\phi_F = 8 \times 10^{-6}$ for benzene. This is assigned to an S_2-S_o emission.

In 0.1M solutions of p-xylene and mesitylene, emissions could also be detected for the S_3-S_o transition. Similar successes have also been achieved with p-xylene, toluene, and mesitylene, in the vapor phase, leading to the hope that data for benzene vapor may soon be forthcoming.

Some information is available on the photochemical effects that occur after the formation of higher excited states, which may help, in part at least, to explain some of the effects. Foote, Mallon, and Pitts (248), and Shindo and Lipsky (249) investigated the photochemistry of benzene vapor, excited into the S_3 state. Both sets of data, when extrapolated to zero pressure, give a quantum yield of unity for chemical reaction. The isomer formed was later (250) shown to be fulvene, with some polymer formation. This route for energy decay will be examined in the following section.

XII. RELEVANT PHOTOCHEMICAL REACTIONS OF EXCITED STATES OF BENZENE

As previously stated, molecular isomerization does not provide an explanation for the quantum deficit following excitation of benzene; the first band does not bear close scrutiny. However, an investigation of the photochemistry of excited benzene is of interest, in view of the variety and complexity of the isomeric products. An extensive review of conditions under which the formation of isomers of benzene and its derivatives can be found has been compiled by Scott and Jones (250). Interconversion of the isomers which would be relevant to photophysical effects are governed by orbital symmetry considerations, and examples of this type of treatment are now appearing (251,252).

In 1966, Bryce-Smith and Longuet-Higgins (65,185) suggested a scheme that seems to provide a generally satisfactory description of the facts. Prefulvene and Dewar-like intermediates consistent (184) with the symmetry of the $^1B_{2u}$, and $^3B_{1u}$ excited states which supposedly give rise to them, have been proposed. The existence of these intermediates has not been proven, and for heavily substituted fluoro-benzenes, Dewar isomeric forms arise from excited singlet states in apparent contravention of the symmetry considerations (253,254). The significance of this is not clear.

No attempt is made here to review all the experimental data, but some investigation relevant to the theme of this article will be cited. Wilzbach, Kaplan, and Harkness (255) showed that benzene-1,3,5,d_3 rearranged to the 1,2,4-isomer when excited at 254 nm in the vapor phase, and deduced that this occurred through a benzvalene intermediate with a yield of about 3 percent. Subsequently, Kaplan and Wilzbach (256) isolated benzvalene, from benzene excited within the first band, with a yield of 0.02. Addition of added gases increases the yield by stabilization of "hot" benzvalene molecules. The data on the effect of exciting wavelength shows that relaxation from higher vibrational levels of the S_1 state is accompanied by a decrease in the yields of <u>both</u> fluorescence and triplet production. The work of Kaplan and Wilzbach provides some evidence for an increase in the yield of benzvalene in going from 254 to 237 nm, but no correlation with the sharp increase in nonradiative decay when the vibrational energy exceeds 2500 cm^{-1} has been achieved. An interesting argument against benzvalene formation providing a complete explanation of the nonradiative processes has been advanced by Spears and Rice (111). The conservation of angular momentum must inhibit conversion of benzene into benzvalene, but no correlation of rotational energies with either radiative or nonradiative rate constants was found by Parmenter and Schuh (257). As pointed out in Section VIII, recent results of Stein (195) have indicated that isomerization cannot explain completely Channel III in benzene.

More work on the photochemistry of benzene cannot fail to be informative. A recent example of the type of experiment that shows promise is that reported by Irina and Kurien (194). Phenol and mucondialdehyde are major products found from the photolysis of benzene vapor in the presence of oxygen, and also that phenol alone is formed in the absence of oxygen on interaction with water. The experimental conditions employed a flow of the vapor through a trap, indicating that the products formed must have resulted from the interaction of a long-lived excited intermediate with water. This intermediate cannot be benzvalene since it has been established (258) that this reacts with water to give a cyclo pentadiene-1-carboaldehyde, via the following mechanism:

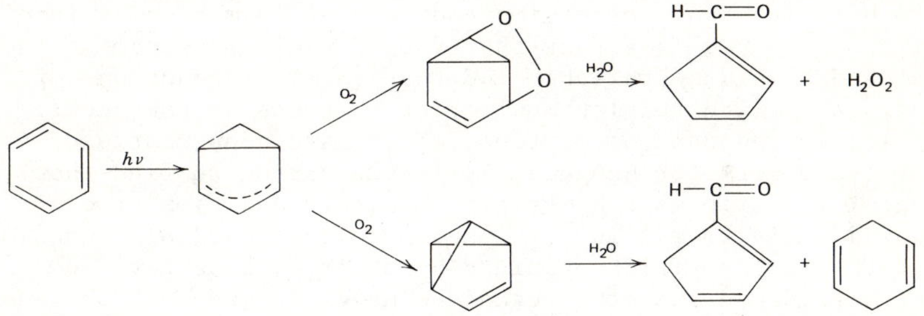

"Internal conversion" in the liquid and solid phases is still obscure. Decomposition and isomerization yields are small, usually of the order of 0.01, and in general are very much less than in the gas phase, presumably due to the rapid collisional stabilization that can occur in condensed phases. Formation of an intermediate of the pre-fulvene type would require several quanta of ring-deforming vibrations, and this could account for the decrease in yields with lowering of temperature.

Since the efficiency of "internal conversion" is greater in solution than in the gas phase, and since the controlling parameters are, to a certain extent, solvent dependent (91,94), the possibility of chemical reactions with solvent by excited state routes cannot be ignored. Experimental difficulties have prevented very clear results. This is illustrated by the discordant data on the reaction of oxygen with benzene under various conditions (130,259-261).

It has been known for some time that hexatrienes can be formed by reaction of excited benzene molecules with solvent in a frozen glass (262-264). Some details of the mechanism of this process have been deduced by Simons and Smith (90). They propose that intermolecular vibronic interaction with solvent is a necessary but not sufficient requirement for production of hexatrienes in viscous media. The reaction is associated with radiationless decay of the triplet state rather than a bimolecular reaction of triplet benzene with neighboring solvent molecules. Since radiationless decay constitutes intersystem crossing, the product of the reaction is also in the ground state. It is stressed by the authors that hexatrienes are not produced in fluid media or by nonphosphorescent benzene derivatives in which there is a rapid competing decay of the triplet state. The formation of the hexatriene only occurs when weak interactions occur between the relaxing triplet and the solvent.

Assignment of nonradiative decay from the S_1 state of benzene to both internal conversion and triplet formation by intersystem crossing requires that the triplet yield measurement techniques used should be reliable and that chemical reactions of excited states with triplet monitoring reagents do not occur. As a rule, in the case of olefins, which have been most used as triplet scavengers for benzene, it is reassuring to find that for the types of olefin used as triplet monitors addition products are usually only formed in very small yields (265,268). Consistency with triplet yields obtained by other methods give some further confidence that the technique used, in general, provide self-consistent results.

Examples of the type of agreement are given for the gas phase in the recent review of Noyes and Al-Ani (269) and comparison of triplet yields in solution obtained by Sandros (146,181) using biacetyl as the scavenger, and Cundall et al. (147,156,157) using butene-2.

XIII. BENZENE EXCIMER

The formation of a benzene dimer from association of a ground state monomer and an excited monomer is a well-established observation of this species was first noted by Dammers de Klerk (270) and Ivanova and co-workers (271) in a study of the effects of concentration on the fluorescence spectrum of benzene. Broad structureless emission on the long wavelength side of monomer emission at room temperature is clearly observed at concentrations of about 1M, and a further increasing concentration becomes the dominant component in the spectrum.

The existence of such a dimeric species, stable in the excited state, and experiencing dissociative repulsive interaction in the ground state, is explained (272) in terms of configurational interaction between excitation and charge resonance states of the dimer. Its structure is generally believed to be of an eclipsed parallel plate form, though Birks and co-workers (273-275) favor a slightly displaced structure with one molecule overlapping the other. The study of Birks and Christophorou (276) has shown that the net excimer binding energy is dependent upon intermolecular separation, and it seems well established that the particular ring separation in benzene is 0.31-0.33 nm.

The excimer so described is singlet in character, and many authors (272,277,279) have suggested that emission to the repulsive ground state is dipole forbidden. It is possible (272) that a small rotation from the strict parallel plate configuration is sufficient to remove this restriction, and Chesnut et al. (280) have shown that a rotation of 30 degrees decreases the excimer binding energy by 0.01 eV. This would suggest that excimer emission probability may be temperature dependent; and such has been found to be the case by Hirayama and Lipsky (125, 281) and Cundall and Robinson (134), but not by Birks and co-workers (282).

The parameters that describe the properties of the excimer state may be extracted from kinetic analysis of transient and steady state luminescence data. The scheme due to Birks, Braga, and Lumb (131,274), although criticized for its simplicity (125,283), accounts well for experimental decay of the singlet excimer. The observed quantum yield of monomer fluorescence $\emptyset_{FM}$ is reduced in the presence of excimer from the intrinsic yield $\emptyset^{o}_{FM}$ according to:

$$\emptyset_{FM} = \emptyset^{o}_{FM}c(1 + c|c_h)$$

where $1|c_h$ is the Stern Volmer quenching parameter such that $\emptyset_{FM}$ at $c = c_h$. A similar expression for excimer fluorescence is

$$\emptyset_{FD} = \emptyset^{o}_{FD}\ (1 + c_h|c)$$

Two evaluations of c_h are thus possible for all temperatures where excimer band emission can be observed. Cundall and Robinson (134) showed that the temperature of c_h led to the deduction that $(E_b + E_{ICM} - E_{ICSD}) = 0.38$ eV. (here subscripts refer to processes described in the introductory section). Excimer fluorescence lifetimes, τ_D can be extracted from knowledge of the measured emission lifetime of more concentrated solutions, and the intrinsic monomer lifetime τ_M, according to:

$$\tau_{obs} = \frac{\tau_M + \tau_D(c/c_h)}{1 + c/c_h}$$

Combination of this data with that of the intrinsic excimer emission yield ϕ^o_{FD} allows computation of k_{FD}, the excimer fluorescence probability. Cundall and Robinson (134) found that the rate constant so calculated had an associated activation energy of 0.15 eV, assigned to thermal distortion of the dimer to overcome dipole restrictions on the emission. Helman (136,139) has found a concentration dependence for k_{FD}, and he speculates that this may be due to the formation of trimeric excited complexes.

Ludwig and Amata (135) also found it necessary to postulate the involvement of multimolecule complexes to explain some anomalies in the CCl_4 quenching of emission data. There is no direct evidence for such species, though they may exist, provided the exciplex is of lower potential than the singlet monomer molecule. The binding energy for such a species would be very small. This, together with the probable highly forbidden emission, might lead to difficulties in observation. Proof of the involvement of the types of species in the decay of the excimer remains unavailable.

The excimer binding energy deduced by Gregory and Helman (136) from transient luminescence decay data at low temperature (0.26 eV) is at variance with the conclusions of other authors (125,134) ($\geq$0.36 eV) using steady state luminescence data. The problem lies in determining the temperature dependence, if any, of the ratio k_{FD}/k_{FM}, since all authors agree that the ratio k_{DM}/k_{MD} has a complete activation energy of approximately 0.25 ($\pm$0.03) eV. Theoretical estimates of Vala, Hillier, Rice and Jortner (284) and Chestnut et al. (280) place the expected binding energy in the range 0.3 to 0.5 eV depending upon ring separation. The various parameters describing the singlet benzene excimer are summarized in Table 15, compiled with reference to Fig. 10.

The experimental difficulties involved in measurement of fluorescence yields and lifetimes for benzene, and other systems also, as a function of concentration make it certain that the various thermodynamic quantities have not yet been firmly established.

It is reasonable to presume that the routes for decay of the excimer are similar to those available for the excited monomer. There is no direct evidence for the existence of the triplet excimer although its existence has been invoked to explain anomalous triplet yield data (175). Hillier, Glass and

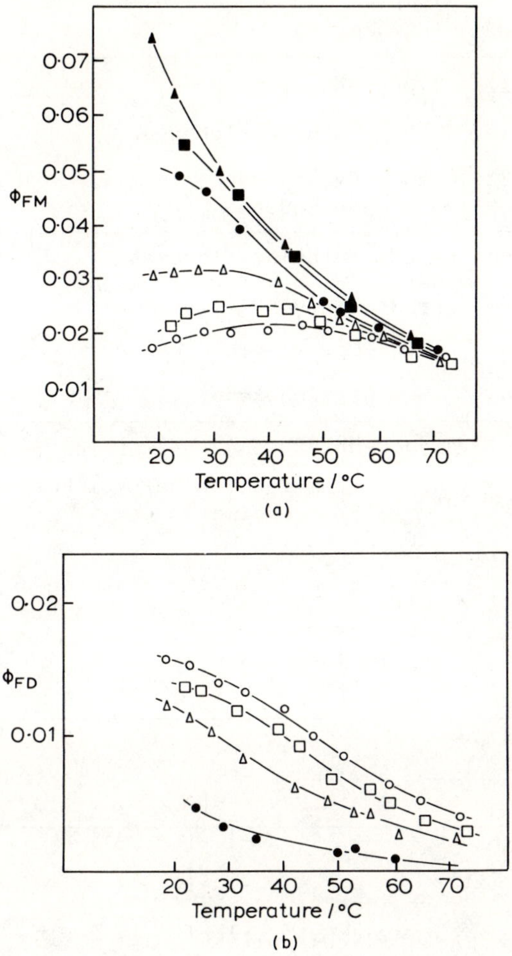

FIGURE 10. (a) Variation of monomer fluorescence yield for benzene with concentration and temperature. ▲, 0.12 M; □, 0.51 M; ●, 1.12 M; △, 4.5 M; □, 7.9 M; O, 11.2 M. (b) Variation of excimer fluorescence yield with concentration and temperature. O, 11.2 M; □, 7.9 M; △, 4.5 M; ●, 1.12 M.

TABLE 15. Decay parameters of the singlet excimer of benzene.

Parameter	Value	Temp. (°C)	Comments	Ref.
ϕ^{o}_{FD}	0.012	20	Hexane	(124)
	0.016	25	Methylcyclohexane	(125)
	0.022	20	Cyclohexane, T dependence	(134)
E_b	0.26	-30°C	Methyl-cyclohexane	(136)
	0.22	-	Cyclohexane	(282)
	≥0.36	-	Methyl-cyclohexane	(125)
	0.34	-	Cyclohexane	(134)
c_h	4.8	24	Hexane	(285)
	2.1	25	From monomer Me-cyclohexane	(125)
	5.0		From excimer	
	5.0	20	From monomer	
	16.1		From excimer cyclohexane	(283)
	5.0	25	From monomer	(134)
	4.65		From excimer (very T dependent)	
R_m	0.46 eV			(134)
	0.48 eV			(282)
ΔS^o	-29.6		(cal $mol^{-1}K^{-1}$)	(134)
	-28		(cal $mol^{-1}K^{-1}$)	(125)
	-20.3		(cal $mol^{-1}K^{-1}$)	(282)
E_{FD}	0.145		(eV)	(134)
τ_D	12.3	25	From ionization studies	(286)
	12.	20	Hexane	(282)
	30	20	Cyclohexane	(134)

Rice (287) inferred its existence from elegant experiments in which they examined the phosphorescence emission of the para-cyclophane compounds. With the two benzene rings held together by methylene bridges in a sandwich structure, they observed shifted phosphorescence spectra with associated loss of structure; the most significant effects being at a ring separation of approximately 0.3 nm. Phillips and Schug (288) identified emission at about 4.30 nm in 1 MeV electron irradiated benzene at $138^{o}K$ with the triplet excimer.

Conclusive evidence of this elusive species awaits systematic studies using transient absorption techniques. Further, the characteristic excimer emission has not been observed in the solid phase, where intermolecular distances are such as to minimize the interaction and because achievement of the necessary orientation is inhibited in the frozen samples.

XIV. CONCLUSION

It is clear that the photophysical behavior of benzene is complex and that the nonradiative $S_1 \rightarrow S_o$(?) process vaguely described at Channel III is especially important. It is not possible to conclude whether benzene is a "typical" aromatic molecule until comparably detailed studies of other systems have been made.

ACKNOWLEDGMENTS

Part of the work presented here was completed during July of 1974 at the University of Lourenco Marques, Mozambique. We wish to express our gratitude to ULM for the hospitality and for financial support for one of us (R.B.C.) and also to NATO for a research grant. One of us (D.A.R.) gratefully acknowledges additional support from the Research and Publications Committee, ULM.

REFERENCES

(1) J. I. Steinfeld, "Molecules and Radiation", Harper and Row, New York, 1974.

(2) For example: (a) M. Goeppart-Mayer and A. L. Sklar, J. Chem. Phys., 6, 645 (1938); (b) R. C. Parr, D. P. Craig, and I. C. Ross, J. Chem. Phys., 18, 1561 (1950).

(3) H. Sponer, G. Nordheim, A. L. Sklar, and E. Teller, J. Chem. Phys., 7, 207 (1939).

(4) J. R. Platt and H. B. Klevens, Chem. Rev., 41, 301 (1949).

(5) S. Leach, R. Lopez-Delgado, and F. Delmas, J. Mol. Spectrosc., 7, 304 (1962).

(6) R. B. Merrithew and C. V. Marusak, J. Mol. Spectrosc. 25, 269 (1968).

(7) H. C. Wolf, "Solid State Physics" (F. Seitz and D. Turnbull, Eds.), Vol. 9, Academic Press, New York, 1959, p. 1.

(8) J. B. Birks, "Photophysics of Aromatic Molecules", Wiley, New York, 1970.

(9) J. Koutecky, J. Cizek, J. Dubsky, and K. Hlavaky, Theoret. Chim. Acta, 2, 462 (1964).

(10) J. Koutecky, K. Hlavaky, and P. Hochman, Theor. Chim. Acta, 3, 341 (1965).

(11) J. E. Bloor, J. Lee, and S. Gartside, Proc. Chem. Soc., 413, (1960).

(12) R. J. Buenker, J. L. Whitten, and J. D. Pekke, J. Chem. Phys., 49, 261 (1968).

(13) T. Dunn and C. K. Ingold, Nature, 176, 65 (1965).

(14) S. D. Peyerinholf and R. J. Buenker, Theor. Chim. Acta, 19, 1 (1970).

(15) J. Karwowski, Chem. Phys. Lett., 18, 47 (1973).

(16) P. J. Hay and I. Shavitt, Chem. Phys. Lett., 22, 33 (1973).

(17) J. Karwowski, Acta Phys. Polon., A., 42, 655 (1972).

(18) J. N. Murrell and T. A. Pople, Proc. Phys. Soc. (London), 69A, 245 (1956).

(19) (a) For benzene: B. Katz, M. Brith, B. Sharf and J. Jortner, J. Chem. Phys., 52, 88 (1970); (b) Confirmed with toluene and xylene: B. Katz, M. Brith, B. Sharf and J. Jortner, J. Chem. Phys., 54, 3924 (1971); (c) E. Pantos and T. D. S. Hamilton, Chem. Phys. Lett., 17, 588 (1972).

(20) M. Brith, R. Lubert and I. T. Steinberger, J. Chem. Phys., 54, 5104 (1971).

(21) E. N. Lassettre, A. Skerbele, M. Dilton and K. Ross, J. Chem. Phys., 48, 5066 (1968).

(22) A. M. Taleb, I. H. Munro, and J. B. Birks, Chem. Phys. Lett., 21, 454 (1973).

(23) G. C. Morris and J. G. Angus, J. Mol. Spectrosc., 45, 271 (1973).

(24) B. Katz, M. Brith, B. Sharf, and J. Jortner, J. Chem. Phys., 52, 88 (1970).

(25) P. M. Johnson, J. Chem. Phys., 62, 4562 (1975).

(26) J. B. Birks, Chem. Phys. Lett., 3, 567 (1969).

(27) J. H. Callomon, J. E. Parkin, and R. Lopez-Delgado, Chem. Phys. Lett., 13, 125 (1972).

(28) G. Diesken and J. W. Moskowitz, J. Chem. Phys., 47, 3491 (1967).

(29) J. Momigny, C. Goffert and L. K'Or, Int. J. Mass. Spectrom. Ion. Phys., 1, 53 (1968).

(30) A. D. Baker, C. R. Brundle, and D. W. Turner, Int. J. Mass. Spectrom. Ion. Phys., 1, 443 (1968).

(31) E. Lindholm and B. O. Johnson, Chem. Phys. Lett., 1, 501 (1967).

(32) B. F. Minaev, Isv. Vyssh. Vchet. Zaved. Fiz., 14, 118 (1971).

(33) L. Burnelle and M. J. Kraneppol, J. Mol. Spectrosc., 37, 383 (1971).

(34) D. Kearns, J. Chem. Phys., 36, 1608 (1962).

(35) For summary see: S. P. McGlynn, T. Azumi and M. Kinoshita, "Molecular Spectroscopy of the Triplet State", Prentice-Hall, N.J., 1969.

(36) J. Leclerq, and J. M. Leclerq, Chem. Phys. Lett., 18, 411 (1973).

(37) R. Astier and Y. H. Meyer, Chem. Phys. Lett., 3, 399 (1968).

(38) J. B. Doering, J. Chem. Phys., 51, 2866 (1969).

(39) S. D. Colson and E. R. Bernstein, J. Chem. Phys., 43, 2661 (1965).

(40) R. M. Hochstrasser and J. E. Wessel, Chem. Phys. Lett., 24 1 (1974).

(41) R. M. Hochstrasser, H. N. Sung and J. E. Wessel, Chem. Phys. Lett., 24, 7 (1974).

(42) L. Wunsch, H. J. Neusser and E. W. Schlag, Chem. Phys. Lett., 31, 433 (1975).

(43) J. Metcalfe, M. G. Rockley and D. Phillips, J. Chem. Soc. Faraday II, 70, 1660 (1974).

(44) M. Kasha, Disc. Faraday Soc., 9, 14 (1950).

(45) T. A. Gregory, F. Hirayama and S. Lipsky, J. Chem. Phys., 58, 4697 (1973).

(46) See also: R. B. Cundall and T. F. Palmer, Annual Reports A, Chemical Society, London, 1973.

(47) H. Schull, J. Chem. Phys., 17, 295 (1949).

(48) G. S. Kembrovskii, V. P. Bobrovich and A. N. Sercherichо, Opt. Spectrosc., 24, 213 (1968).

(49) G. N. Lewis and M. Kasha, J. Am. Chem. Soc., 66, 2100 (1944).

(50) G. W. King and E. H. Pinnington, J. Mol. Spectrosc., 15, 394 (1965).

(51) M. S. de Groot and J. H. van der Waals, Mol. Phys., 6, 545 (1963).

(52) J. van Egmond and J. H. van der Waals, Mol. Phys., 26, 1147 (1973).

(53) J. van Egmond and J. H. van der Waals, Mol. Phys., 28, 457 (1974).

(54) J. van der Waals, A. M. D. Berghuis and M. S. de Groot, Mol Phys., 21, 497 (1971).

(55) J. van der Waals, A. M. D. Berghuis, and M. S. de Groot, Mol. Phys., 13, 301 (1967).

(56) A. A. Gwaiz, M. A. El-Sayed and D. S. Tinti, Chem. Phys. Lett., 9, 454 (1971).

(57) D. M. Haaland and G. C. Nieman, J. Chem. Phys., 59, 4435 (1973).

(58) C. C. Nieman and D. S. Tinti, J. Chem. Phys., 46, 1432 (1967).

(59) D. M. Burland, G. Castro and G. W. Robinson, J. Chem. Phys., 52, 4100 (1970).

(60) A. Devaquet, J. Am. Chem. Soc., 94, 9013 (1972).

(61) Y. Fujimura, H. Yanaguchi and T. Nakajima, Bull. Chem. Soc. Japan, 45, 304 (1972).

(62) A. D. Liehr, J. Phys. Chem., 67, 389 (1963).

(63) D. Bryce-Smith, A. Gilbert and D. A. Robinson, Angew. Chem., 83, 803 (1971).

(64) E. E. van Tamlen and S. P. Pappas, J. Am. Chem. Soc., 85, 3297 (1963).

(65) D. Bryce-Smith and H. C. Longuet-Higgins, Chem. Commun., 593 (1966).

(66) G. V. Klimusheva, Teor. Khim. Akad. Nauk. SSSR., 2, 37-43 (1966).

(67) For brief reviews see: (a) B. Linder and S. Abdulnur, J. Chem. Phys., 54, 1807 (1971). (b) J. B. Birks, ref. 8. pp. 109-119.

(68) For brief review of observable effects of Davydoy splitting, see: H. Suzuki, "Electronic Absorption Spectra and Geometry of Organic Molecules", Academic Press, New York, 1967, pp. 112-113.

(69) Three of the components are allowed and are easily observed (see ref. 68); the fourth factor group is forbidden but has also been observed: S. D. Colson, J. Chem. Phys., 44, 3324 (1968).

(70) J. W. Eastman and S. J. Rehfeld, J. Phys. Chem., 74, 1438 (1970).

(71) For a description of the simple theory and an account of deviations noted for benzene, see ref. 68, pp. 100-101.

(72) C. W. Lawson, F. Hirayama and S. Lipsky, J. Chem. Phys., 51, 1590 (1969).

(73) J. S. Ham, J. Chem. Phys., 21, 756 (1953).

(74) N. S. Bayliss and L. Hulme, Austral. J. Chem., 6, 257 (1953).

(75) Y. Kanda, Y. Gondo and R. Shimada, Spectrochim. Acta, 17, 424 (1961).

(76) N. S. Bayliss and N. W. Cant, Spectrochim. Acta, 18, 1287 (1962).

(77) J. R. Platt, J. Mol. Spectrosc. 9, 288 (1962.

(78) R. Pariser, J. Chem. Phys., 24, 250 (1956).

(79) N. S. Ham and K. Ruedenberg, J. Chem. Phys., 25, 1 (1956).

(80) G. W. Robinson, J. Chem. Phys., 46, 572 (1967).

(81) M. Koyanagi and Y. Kanda, Spectrochim. Acta, 20, 993 (1964).

(82) G. Durocher and C. Sandorfy, J. Mol. Spectrosc. 14, 400 (1964).

(83) G. Durocher and C. Sandorfy, J. Mol. Spectrosc., 20, 410 (1966).

(84) M. Koyanagi, J. Mol. Spectrosc., 25, 273 (1968).

(85) N. S. Bayliss, J. Mol. Spectrosc., 31, 406 (1969).

(86) A. Trombeti and C. Zauli, J. Chem. Soc., (B), 1968, 1574-1577.

(87) Y. Munakami and Y. Kanda, Bull. Chem. Soc. Japan 41, 2599 (1968).

(88) N. C. Perrins, J. P. Simons and A. L. Smith, Trans. Faraday Soc., 67, 3415 (1971).

(89) J. P. Simons and A. L. Smith, J. Chem. Soc. Faraday II., 70, 53 (1974).

(90) J. P. Simons and A. L. Smith, Chem. Phys. Lett., 16, 536 (1972); ibid., 25, 97 (1974).

(91) R. B. Cundall and L. C. Pereira, Chem. Phys. Lett., 18, 371 (1973).

(92) R. B. Cundall, and L. C. Pereira, submitted for publication.

(93) L. C. Pereira, Ph.D. Thesis, University of Nottingham (1972).

(94) M. Luria, M. Ofran and G. Stein, J. Phys. Chem., 78, 1904 (1974).

(95) D. A. Robinson, Unpublished results.

(96) R. B. Cundall and A. J. Harrison, To be published.

(97) C. S. Parmenter, Adv. Chem. Phys., 22 (1972).

(98) C. S. Parmenter and M. W. Schuyler, J. Chem. Phys., 52, 5366 (1970).

(99) C. S. Parmenter and M. W. Schuyler, Chem. Phys. Lett., 6, 339 (1970).

(100) J. A. Poole, J. Phys. Chem., 69, 1343 (1965).

(101) A. E. Douglas and C. W. Matthews, J. Chem. Phys., 48, 4788 (1968).

(102) E. M. Anderson and G. B. Kistiakowsky, J. Chem. Phys., 48, 4787 (1968).

(103) W. A. Noyes, Jr., W. A. Mulac and D. A. Harter, J. Chem. Phys., 44, 2100 (1966).

(104) E. M. Anderson and G. B. Kistiakowsky, J. Chem. Phys., 51, 182 (1969).

(105) K. Dunnicliff, Ph.D. Thesis, University of Nottingham (1971): R. B. Cundall and K. Dunnicliff, To be published.

(106) A. S. Abramson, K. G. Spears and S. A. Rice, J. Chem. Phys., 56, 2291 (1972).

(107) (a) G. M. Breuer and E. K. C. Lee, J. Chem. Phys., 51, 3130 (1969); (b) G. M. Breuer and E. K. C. Lee, J. Chem. Phys., 51, 3615 (1969).

(108) T. Chen and E. W. Schlag, "Molecular Luminescence" (E. C. Lim, Ed.), W. A. Benjamin, New York, 1969, p. 381.

(109) M. Nishikawa and P. K. Ludwig, J. Chem. Phys., 52, 107 (1970).

(110) G. S. Burton and H. E. Hunziker, J. Chem. Phys., 52, 3302 (1970).

(111) K. G. Spears and S. A. Rice, J. Chem. Phys., 55, 5561 (1971).

(112) B. K. Selinger and W. R. Ware, J. Chem. Phys., 53, 3160 (1970).

(113) W. R. Ware, B. K. Selinger, C. S. Parmenter and M. W. Schuyler, Chem. Phys. Lett., 6, 342 (1970).

(114) J. R. Lockwood, Ph.D. Thesis, Nottingham (1974).

(115) B. K. Selinger and W. R. Ware, J. Chem. Phys., 52, 5482 (1970).

(116) R. B. Cundall, K. Dunnicliff and A. J. R. Voss., Joint Ann. Mtg. Chem. Soc. and R.I.C., Nottingham, April 1969.

(117) S. J. Strickler and R. J. Watts, J. Chem. Phys., 44, 5366 (1970).

(118) G. B. Kistiakowsky and C. S. Parmenter, J. Chem. Phys., 42, 2942 (1965).

(119) R. B. Cundall and L. C. Pereira, J. Chem. Soc. Faraday II, 68, 1152 (1972).

(120) J. W. Eastman, J. Chem. Phys., 49, 4617 (1968).

(121) M. D. Lumb and M. I. Barradas, L. C. Pereira and C. L. Braga, Rev. Fiz. Quin. Eng. U.L.M., 1, 1 (1969).

(122) G. B. Evans, Ph.D. Thesis, Nottingham (1968).

(123) W. R. Dawson and M. W. Windsor, J. Phys. Chem., 72, 325 (1968).

(124) M. D. Lumb and D. A. Weyl, J. Mol. Spectrosc., 23, 365 (1967).

(125) F. Hirayama and S. Lipsky, J. Chem. Phys., 51, 1939 (1969).

(126) I. B. Berlman, "Handbook of Fluorescence Spectra of Aromatic Molecules", Academic Press, New York, 1971.

(127) R. B. Cundall, S. McD. Ogilvie, and D. A. Robinson, J. Photochem., 1, 417 (1973/1974).

(128) C. A. Parker, Anal. Chem., 34, 502 (1962).

(129) M. Luria and G. Stein, J. Phys. Chem., 76, 165 (1972).

(130) K. Kaplan, L. A. Wendling and K. E. Wilzbach, J. Am. Chem. Soc., 93, 3819 (1971).

(131) J. B. Birks, C. L. Braga and M. D. Lumb, Proc. Roy. Soc., A 283, 83 (1965).

(132) T. V. Ivanova, G. A. Mokeeva and B. Ya. Sveshnikov, Opt. Spectrosc., 12, 325 (1962); T. V. Ivanova, P. I. Kudryashev, and B. Ya. Sveshnkov, Sov. Phys. Dokl., 6, 407 (1961).

(133) M. D. Lumb, Private communication, 1971.

(134) R. B. Cundall and D. A. Robinson, J. Chem. Faraday II, 68, 1133 (1972).

(135) P. K. Ludwig and C. D. Amata, J. Chem. Phys., 49, 326 (1968).

(136) T. A. Gregory and W. P. Helman, J. Chem. Phys., 56, 377 (1972).

(137) J. R. Greenleaf and T. A. King, Proc. Int. Conf. Luminescence, Budapest, 212 (1968).

(138) P. M. Froehlich and H. A. Morrison, J. Phys. Chem., 76, 3566 (1972).

(139) W. P. Helman, J. Chem. Phys., 51, 354 (1969).

(140) See, for example, ref. 86, 119.

(141) S. Kato, S. Lipsky, and C. L. Braun, J. Chem. Physics, 37, 190 (1962).

(142) C. L. Braun, S. Kato and S. Lipsky, J. Chem. Phys., 39, 1645 (1963).

(143) J. B. Birks, J. C. Conte and G. Walker, J. Phys. B, 1, 934 (1968).

(144) M.D. Lumb, C. L. Braga and L. C. Pereira, Trans. Faraday Soc., 65, 1992 (1969).

(145) C. Fuchs, F. Heissel and R. Voltz, J. Phys. Chem., 76, 3867 (1972).

(146) K. Sandros, Acta Chem. Scand., 25, 3651 (1971).

(147) R. B. Cundall and D. A. Robinson, J. Chem. Soc. Faraday II 68, 1145 (1972).

(148) C. A. Parker and C. C. Hatchard, Analyst, 87, 664 (1962).

(149) E. H. Gilmore, G. E. Gibson and D. S. McClure, J. Chem. Phys., 20, 829 (1952); ibid, 23, 399 (1955).

(150) E. C. Lim, J. Chem. Phys., 36, 3497 (1962).

(151) M. D. Lumb and A. O. Al Chalabi, Chem. Phys. Lett., 11, 365 (1971).

(152) M. D. Lumb and L. C. Pereira, "Organic Scintillators and Liquid Scintillation Counting" (D. L. Horrocks and C. Peng, Eds.), Academic Press, New York, 1971, p. 239.

(153) E. C. Lim, Private Communication, cited in D. M. Haaland and G. C. Nieman, J. Chem. Phys., 59, 1013 (1973).

(154) R. Li and E. C. Lim, J. Chem. Phys., 57, 605 (1972).

(155) S. F. Fischer and E. C. Lim, Chem. Phys. Lett., 14, 40 (1972).

(156) R. B. Cundall, L. C. Pereira and D. A. Robinson, J. Chem. Soc. Faraday II, 69, 701 (1973).

(157) R. B. Cundall and L. C. Pereira, Chem. Phys. Lett., 16, 371 (1972).

(158) R. B. Cundall and S. Mc. Ogilvie, To be published, 1976.

(159) R. B. Cundall and A. S. Davies, Trans. Faraday Soc., 62, 1151 (1966).

(160) W. A. Noyes, Jr., and D. A. Harter, J. Chem. Phys., 46, 674 (1967).

(161) S. Y. Ho and W. A. Noyes, Jr., J. Am. Chem. Soc., 89, 5091 (1967).

(162) H. O. Denschlag and E. K. C. Lee, J. Am. Chem. Soc., 90, 3628 (1968).

(163) K. Nakamura, J. Am. Chem. Soc., 93, 3138 (1971).

(164) G. C. Nieman and G. W. Robinson, J. Chem. Phys., 39, 1298 (1963).

(165) S. D. Colson and G. W. Robinson, J. Chem. Phys., 48, 2550 (1968).

(166) F. M. Garforth, G. K. Ingold and H. G. Poole, J. Chem. Soc., 491 (1948).

(167) E. R. Bernstein, S. D. Colson, D. S. Tinti and G. W. Robinson, J. Chem. Phys., 48 (1968); G. Nonnemacher and R. Mecke, Spectrochim. Acta, 17, 1049 (1961).

(168) F. Langenbucher and R. Mecke, Spectrochim. Acta, 21, 1287 (1965).

(169) A. R. Gee and G. W. Robinson, J. Chem. Phys., 46, 4847 (1967).

(170) J. W. Rabalais, H. J. Maria and S. P. McGlynn, J. Chem. Phys., 51, 2259 (1969).

(171) P. M. Johnson and L. Ziegler, J. Chem. Phys., 56, 2169 (1972).

(172) G. S. Hammond and A. A. Lamola, Unpublished, cited in "Photochemistry", J. G. Calvert and J. N. Pitts, Jr., Wiley, New York, 1966.

(173) R. B. Cundall and W. Tippett, Trans. Faraday Soc., 66, 350 (1970).

(174) R. R. Hentz and H. G. Altmiller, J. Phys. Chem., 74, 2646 (1970).

(175) R. B. Cundall and D. A. Robinson, J. Chem. Soc. Faraday II, 68, 1691 (1972).

(176) R. R. Hentz and L. M. Perkey, J. Phys. Chem., 74, 3047 (1970).

(177) R. R. Hentz and R. M. Thibault, J. Phys. Chem., 77, 1105 (1973).

(178) R. B. Cundall, L. C. Pereira, and D. A. Robinson, Chem. Phys. Lett., 13, 253 (1972).

(179) R. V. Bensasson, J. T. Richards, and J. K. Thomas, Chem. Phys. Lett., 9, 13 (1971).

(180) V. M. Berenfel'd and V. A. Krongauz, Izn. Akad. Nauk. SSSR Sen. Fiz., 32 (9), 1575 (1968).

(181) K. Sandros, Acta Chem. Scand., 23, 2815 (1969).

(182) D. A. Robinson, Ph.D. Thesis, University of Nottingham (1971).

(183) See also J. B. Birks, ref. 8, p. 172.

(184) See, for example: I. Jano and Y. Mori, Chem. Phys. Lett., 2, 185 (1968).

(185) D. Bryce-Smith, Pure Appl. Chem., 16, 47 (1968).

(186) I. Jano, Chem. Phys. Lett., 2, 205 (1968).

(187) See, for example: (a) K. E. Wilzbach, L. Kaplan and A. L. Harkness, J. Am. Chem. Soc., 90, 1116 (1968); L. Kaplan and K. E. Wilzbach, J. Am. Chem. Soc., 90, 3291 (1968).

(188) R. B. Cundall, D. A. Robinson and A. J. R. Voss, J. Photochem., 2, 239 (1973/1974).

(189) A. J. R. Voss, Ph.D. Thesis, Nottingham (1969).

(190) J. B. Birks, Organic Molecular Photophysics, Vol. I, Wiley, New York (1974).

(191) A. A. Lamola, G. S. Hammond and F. B. Mallory, Photochem. Photobiol., 4, 259 (1965).

(192) D. Phillips, J. Lemaire, C. Burton and W. A. Noyes, Jr., Adv. Photochem., 5, 329 (1968).

(193) G. Stein, Private communication.

(194) J. Irina and K. C. Kurien, Chem. Commun., 1973, 739.

(195) H. Lutz and G. Stein, J. Phys. Chem., 78, 1909 (1974) and references therein.

(196) See, for example: R. B. Cundall and S. McD. Ogilvie, "Organic Molecular Photophysics:, Vol. II, Wiley, New York (1974).

(197) Y. Ilan and G. Stein, Chem. Phys. Lett., 31, 441 (1975).

(198) S. J. Formosinho, J. Chem. Soc. Faraday II, 70, 605 (1974); S. J. Formosinho and J. Dias da Silva, Mol. Photochem., 6 (4), 409 (1974).

(199) R. E. Kellogg and R. G. Bennett, J. Chem. Phys., 41, 3942 (1964).

(200) W. Siebrand and D. F. Williams, J. Chem. Phys., 46, 403 (1967).

(201) W. Siebrand, "The Triplet State" (A. B. Zahlan, Ed.), Cambridge University Press, 1967.

(202) M. J. Shinitsky, J. Chem. Phys., 56, 5979 (1972).

(203) A. H. Kalantar, J. Chem. Phys., 48, 4992 (1968).

(204) M. Almgren, Photochem. Photobiol., 8, 231 (1968).

(205) E. D. Cehelnik, K. D. Mielenz and R. A. Velapoidi, J. Res. NBS, 79A, 1 (1975).

(206) T. E. Martin and A. H. Kalantar, J. Chem. Phys., 50, 1486 (1969).

(207) J. D. Spangler and N.G. Kilmer, J. Chem. Phys., 48, 698 (1968).

(208) T. E. Martin and A. H. Kalantar, J. Chem. Phys., 49, 244 (1968).

(209) T. E. Martin and A. H. Kalantar, J. Phys. Chem., 74, 2030 (1970).

(210) A. Bernas and T. B. Truong, Photochem. Photobiol., 15, 311 (1972).

(211) J. H. Leubner, J. Phys. Chem., 73, 2088 (1969).

(212) T. E. Martin and A. H. Kalantar, J. Phys. Chem., 72, 2265 (1968).

(213) J. H. Leubner, J. Phys. Chem., 74, 77 (1970); J. H. Leubner and J. E. Hodgkins, J. Phys. Chem., 73, 2545 (1969).

(214) F. Hirayama, J. Chem. Phys., 42, 3726 (1965).

(215) G. F. Hatch, M. D. Erlitz and G. C. Nieman, J. Chem. Phys., 49, 3723 (1968); E. C. Lim, Ed., "Molecular Luminescence", W. A. Benjamin, New York, 1969.

(216) A. C. Ling and J. E. Willard, J. Phys. Chem., 72, 1918 (1968).

(217) N. G. Kilmer and J. D. Spangler, J. Chem. Phys., 54, 604 (1971).

(218) T. E. Martin and A. H. Kalantar, J. Chem. Phys., 48, 4996 (1968).

(219) R. B. Cundall and L. C. Pereira, Chem. Phys. Lett., 29, 561 (1974).

(220) P. M. Johnson and M. C. Studez, Chem. Phys. Lett., 18, 371 (1973).

(221) Yei-Ping Hsu and P. M. Johnson, J. Chem. Phys., 59, 136 (1973).

(222) D. M. Haaland and G. C. Nieman, J. Chem. Phys., 59, 1013 (1973).

(223) G. W. Robinson, J. Chem. Phys., 47, 1967 (1967), and references therein.

(224) S. McD. Ogilvie, Ph.D. Thesis, University of Nottingham, (1974).

(225) H. Sternlicht, G. C. Nieman and G. W. Robinson, J. Chem. Phys., 38, 1326 (1963).

(226) H. Ishikawa and W. A. Noyes Jr., J. Chem. Phys., 37, 583 (1962).

(227) E. K. C. Lee, H. O. Denschlag, and G. A. Hanniger, Jr., J. Chem. Phys., 48, 4547 (1968).

(228) C. S. Parmenter and B. L. Ring, J. Chem. Phys., 46, 1198 (1967).

(229) C. S. Burton and H. E. Hunziker, Chem. Phys. Lett., 6, 352 (1970).

(230) M. Nishikawa and M. C. Sauer, J. Chem. Phys., 51, 1 (1969).

(231) T. F. Hunter and M. G. Stock, Chem. Phys. Lett., 22, 368 (1973); T. F. Hunter and M. G. Stock, J. Chem. Soc. Faraday II (1974).

(232) J. Langelaar, G. Jansen, R. P. H. Rettschnick, and G. J. Hoytink, Chem. Phys. Lett., 12, 86 (1971).

(233) R. B. Cundall and P. A. Griffiths, Disc. Faraday Soc., 36, 111 (1963); Trans. Faraday Soc., 61, 1968 (1965).

(234) R. G. Kaufman, Disc. Faraday Soc., 36, 262 (1963).

(235) J. T. Dubois and F. Wilkinson, J. Chem. Phys., 38, 2541 (1963); ibid., 39, 377 (1963).

(236) J. W. van Loben Sels and J. T. Dubois, J. Chem. Phys., 45, 1522 (1966).

(237) S. Lipsky, J. Chem. Phys., 38, 2786 (1963).

(238) I. A. Lisovska, M. D. Sakharova and I. C. Kaplan, Optics Spectrosc., 23, 348 (1967).

(239) V. A. Krongauz, Dokl. Akad. Nauk SSSR, 155, 658 (1964).

(240) R. B. Cundall and A. J. R. Voss, Chem. Commun., 1969, 116 (1969); R. B. Cundall, D. A. Robinson and A. J. R. Voss, J. Photochem., 2, 234 (1973/1974).

(241) R. B. Cundall, G. B. Evans and J. P. Keene, J. Phys. Chem. 72, 3871 (1968).

(242) K. C. Smyth, J. A. Schiavone and R. S. Freund, J. Chem. Phys., 61, 1789 (1974).

(243) Y. Hatano, S. Takao and T. Veno, Chem. Phys. Lett., 30, 429 (1975).

(244) J. K. Thomas and J. Mani, J. Chem. Phys., 51, 1834 (1969).

(245) V. Laor and A. Weinreb, J. Chem. Phys., 43, 1565 (1965).

(246) C. W. Lawson, F. Hirayama and S. Lipsky, in: "Molecular Luminescence", (E. C. Lim, Ed.), W. A. Benjamin, New York, 1969, p. 837.

(247) F. Hirayama, T. A. Gregory and S. Lipsky, J. Chem. Phys., 58, 4886 (1973).

(248) J. K. Foote, M. H. Mallon and J. N. Pitts, Jr., J. Am. Chem. Soc., 88, 3698 (1966).

(249) K. Shindo and S. Lipsky, J. Chem. Phys., 45, 2292 (1966).

(250) L. F. Scott and M. Jones, Jr., Chem. Rev., 72, 181 (1972).

(251) P. Lechtken, R. Breslaw, A. H. Schmidt, and N. J. Turro, J. Am. Chem. Soc., 95, 3025 (1973).

(252) C. A. Renner, T. J. Katz, J. Pouliquen, N. J. Turro and W. H. Waddell, J. Am. Chem. Soc., 97, 2568 (1975).

(253) G. Camoggi, F. Gozzo and C. Cevidalli, Chem. Commun., 1966, 313.

(254) G. P. Semeluk and R. S. Stevens, Chem. Commun., 1970, 1720.

(255) K.E. Wilzbach, L. Kaplan and A. L. Harkness, J. Am. Chem. Soc., 90, 1116 (1968).

(256) L. Kaplan and K. E. Wilzbach, J. Am. Chem. Soc., 90, 3291 (1968).

(257) C. S. Parmenter and M. D. Schuh, Chem. Phys. Lett., 13, 120 (1972).

(258) J. Irina and K. C. Kurien, Chem. Ind., 1972, 763.

(259) K. Wei, J. C. Mari and J. N. Pitts, Jr., J. Am. Chem. Soc., 89, 4225 (1967).

(260) C. Farenhorst, Tetrahedron Lett., 4835 (1968).

(261) M. Luria and G. Stein, Chem. Commun., 1970, 1650.

(262) E. Migirdicyan, J. Chem. Phys., 63, 520 (1966); ibid., 63, 535 (1966).

(263) S. Leach, J. Chem. Phys., 55, 719 (1958).

(264) E. J. Anderson, H. I. J. Chilton and G. Porter, Proc. Chem. Soc., 1960, 352.

(265) K. E. Wilzbach and L. Kaplan, J. Am. Chem. Soc., 88, 2066 (1966).

(266) J. Cornelisse, V. Y. Merritt and R. Srinivasan, J. Amer. Chem. Soc., 95, 6197 (1973).

(267) W. Tippett, Ph.D. Thesis, University of Nottingham (1968).

(268) D. Bryce-Smith, A. Gilbert, B. Orger and H. Tyrell, Chem. Commun., 1974, 334.

(269) W. A. Noyes, Jr., and K. Al-Ani, Chem. Rev., 74, 29 (1974).

(270) A. Dammers-de-Klerk, Mol. Phys., 1, 141 (1958).

(271) T. V. Ivanova, G. A. Mokeeva and B. Ya. Sveshukov, Opt. Spectros., 12, 325 (1962).

(272) E. Konijnenberg, Ph.D. Thesis, University of Amsterdam (1963).

(273) J. B. Birks, A. A. Kazzaz, and T. A. King, Proc. Roy. Soc., A291, 556 (1966).

(274) J. B. Birks, "Progress in Reaction Kinetics", (G. Porter, Ed.), Vol. 5, Pergammon, New York, 1970, p. 263.

(275) B. L. Barnes and J. B. Birks, Proc. Roy. Soc., A291, 570 (1966).

(276) J. B. Birks and L. G. Christophorou, Proc. Roy Soc., A277, 571 (1964).

(277) J. Amuzumi and H. Azumi, Bull. Chem. Soc. Japan, 39, 2317 (1966).

(278) A. K. Chandra and E. C. Lim, J. Chem. Phys., 48, 2589 (1968); ibid, 49, 5066 (1968).

(279) A. K. Chandra, Chem. Phys. Lett., 5, 229 (1970).

(280) D. B. Chestnut, C. J. Fritchie and H. E. Simmons, J. Chem. Phys., 42, 1127 (1965).

(281) F. Hirayama and S. Lipsky, in: "Molecular Luminescence", (E. C. Lim, Ed.), W. A. Benjamin, New York, 1969.

(282) J. R. Greenleaf, M. D. Lumb and J. B. Birks, J. Phys. B., 1, 1157 (1968).

(283) Y. Yguerabide, J. Chem. Phys., 49, 1018 (1968).

(284) M. T. Vala, I. H. Hillier, S. A. Rice and J. Jortner, J. Chem. Phys., 49, 326 (1966).

(285) M. D. Lumb, Private communication, 1971.

(286) L. G. Christophorou, J. G. Carter and M. E. M. Abu-Zeid, J. Chem. Phys., 49, 3775 (1968).

(287) I. H. Hillier, L. Glass and S. A. Rice, J. Chem. Phys., 45, 3015 (1966).

(288) D. H. Phillips and J. C. Schug, J. Chem. Phys., 50, 3297 (1969).

Primary Photoprocesses of Organo-Transition Metal Compounds

C. RANDOLPH BOCK* AND E. A. KOERNER VON GUSTORF+
Institut für Strahlenchemie
im Max-Planck-Institut für Kohlenforschung
D-4330 Mülheim a.d. Ruhr
West Germany

CONTENTS

---

*Present address: Department of Chemistry, Massachusetts Institute of Technology, Cambridge, Massachusetts 02139.

+Deceased.

I. INTRODUCTION

The photosensitivity of organo-transition metal (OTM) compounds has been exploited for synthetic purposes for many years (1). The primary photoprocesses of OTM compounds, however, have only recently been very extensively investigated. In this chapter an appraisal is made of recent developments (1969 to mid-1975) in the determination and interpretation of the primary photobehavior of OTM complexes.

"OTM compounds" refers here to species which contain transition metals and organic molecules as ligands. In this definition no distinction is made regarding the mode of metal ligand bonding. Primarily, the investigations included in this article deal with complexes of low oxidation state transition metals with organic ligands, especially with π-acceptors. In a few instances, however, results concerning nontransition metal compounds have been used to illustrate principles and phenomena of potential importance in OTM complex photochemistry.

"Primary process" has been used in accordance with this recently suggested definition (2): "Any continuous sequence of one or more primary steps which starts with the light absorption step." In this sense a primary step is "any one of the elementary transformations of an excited state molecule of the species which absorbs light. The absorption step itself is also a primary step" (2). Important primary processes of OTM compounds which are described here include: (i) absorption, (ii) dissociative reactions, (iii) intramolecular "twisting" isomerizations, (iv) intermolecular energy transfer, (v) intermolecular electron transfer, (vi) luminescence. Reactions involving OTM compounds as quenchers have also been included.

Information concerning the photoreactivity of OTM compounds, that is, types and quantum yields of primary photoreactions, is in general not available from analyses of the net products of photoinduced reactions. The primary photochemical products of OTM compounds, particularly coordinatively unsaturated species, are often highly reactive. Secondary thermal reactions involving them are, therefore, common and in fact produce a tremendous variety of net chemical products (1,3,4). The net products, however, may show little resemblance to the primary photoproducts, as illustrated by the recently studied (5) photoinduced chemistry of [η-(1,2-dihaloethylene)]tetracarbonyliron (L-$Fe(CO)_4$). The primary photoprocesses that result from ultraviolet photolysis of these compounds are loss of L and CO:

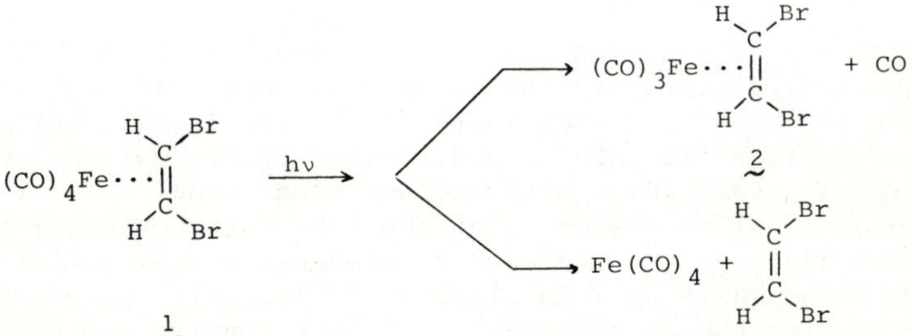

1

The $Fe(CO)_3L$ product then rearranges and recombines with CO:

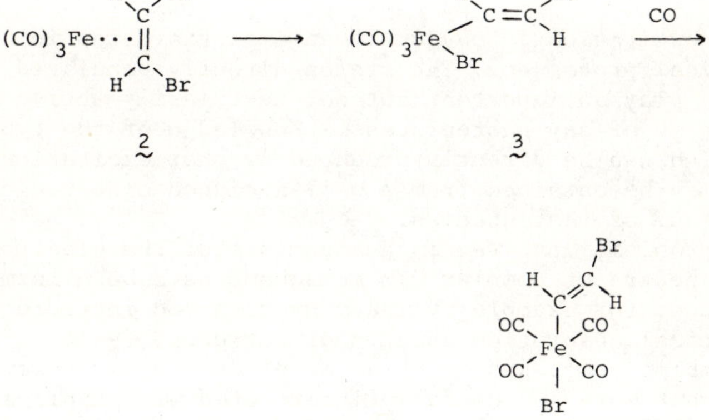

4

The net product, however, results from reaction of [4] with the other primary product, Fe(CO)4:

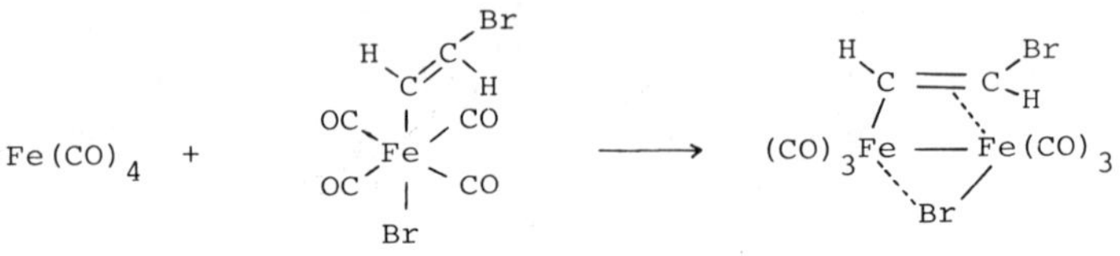

4 5

Several reviews dealing with the photochemistry of OTM compounds are accessible. Synethetic aspects have been emphasized in the article by Koerner von Gustorf and Grevels (1), which covers the earlier literature; and in the more recent review by Leenders et al. (3), which also treats recent studies relevant to photobiological processes. The photochemistry of metal carbonyl compounds has been reviewed comprehensively by Wrighton (4) and with emphasis on mechanisms by Vogler (6). Excellent books concerning the photochemistry of coordination compounds (7) and inorganic photochemistry (8) are also available. Other review articles of related topics are mentioned in the text.

II. FORMATION OF EXCITED STATE OTM COMPLEXES: ELECTRONIC TRANSITIONS AND SPECTRAL PROPERTIES

Photoexcitation is obviously an important step in all primary photochemical processes. The states directly populated by light absorption play an important but not necessarily active role in the mechanism of any photoprocess. Knowledge of the types of states which can be directly produced by photoexcitation of an OTM compound may be obtained from a well-assigned electronic absorption spectrum of that species.

The problems involved in assignments of the electronic absorption spectra of complex OTM compounds have been formidable, as indicated, for example by the many reported interpretations of the optical absorption spectra of ferrocene (9-14). The spectra of simpler complexes, such as metal carbonyls, however, have been somewhat more successfully interpreted with applications of MO theory (1,3-5,15). In this section very simple considerations are employed to illustrate the various types of electronic transitions in OTM complexes. While qualitative aspects of the spectral properties associated with the transitions are briefly discussed here, additional information may be found elsewhere (3,8,16).

As a model to facilitate discussion we choose the complex M-CO, a species which could be formed, for instance, from a transition metal atom and a CO molecule in a rare gas matrix (17,18). The bonding interaction between the metal and CO may be described with simple MO theory in which only the frontier orbitals (19) of M and CO are considered (Fig. 1). A σ bond results from interaction of an empty metal orbital of σ symmetry along the MCO axis with the carbon lone pair of CO. The π bonding, also called back donation, results from interaction of a filled metal orbital of π symmetry (e.g., the d_{xz} or d_{yz} where the z-axis is taken as the molecular axis) with a π^* orbital of CO. These effects lead to the frontier molecular orbitals shown in Fig. 1 and described energetically by the MO scheme (Fig. 2).

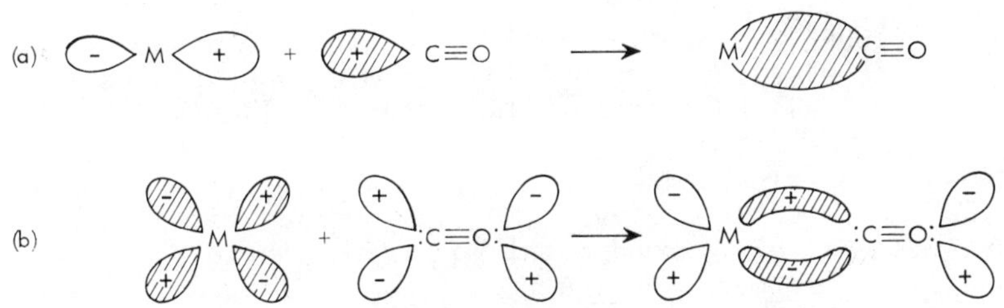

FIGURE 1. Orbital diagram of the bonding interactions in MCO (20). a) Formation of the M-CO σ bond. b) Formation of the M-CO π-bond.

Electronic transitions have been somewhat arbitrarily categorized according to the "characters" of the molecular orbitals involved. Some of the molecular orbitals of MCO are composed mainly of metal orbtials (π_b,σ^*) and thus have predominantly metal character. Others, however, have major contributions from CO orbitals (σ_b,π^*) and therefore have mainly ligand (CO) character. Three important categories of electronic transitions may be identified. These are numbered according to their representations in Fig. 2:

1. Transitions between MO's predominantly of d-orbital character. These are called metal localized, ligand field (LF), and d,d transitions.

2. Transitions between MO's having major contributions from ligand orbitals. These are termed intraligand, internal ligand (IL), or ligand localized transitions.

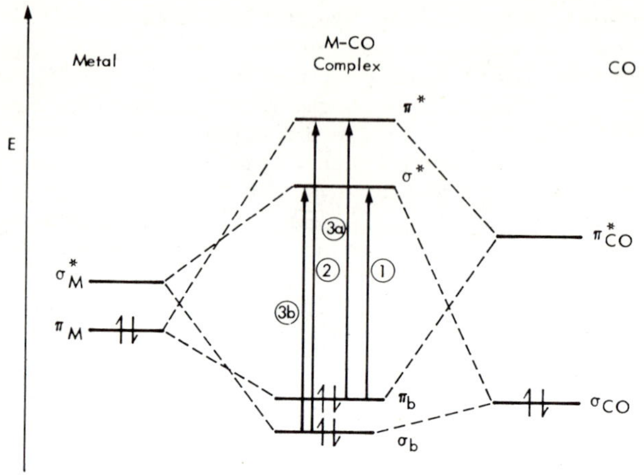

FIGURE 2. Schematic MO energy level diagram and electronic transitions of MCO. Nonbonding orbitals have been excluded for simplicity.

3. Cross transitions, that is, those occurring from a "metal type" MO to a "ligand type" MO (3a) or vice versa (3b). These are respectively termed metal to ligand charge transfer (MLCT) and ligand to metal charge transfer (LMCT) transitions.

Other electronic transitions that occur in OTM complexes include:

(a) Charge transfer to solvent (CTTS), which is exhibited for example by ferrocene in haloalkane solvents (Section III-B-5).

(b) The recently discussed interligand transmetallic charge transfer (ITCT) transitions (21), which occur between molecular orbitals which are localized on different ligands in the same OTM molecule. The term transmetallic refers to the function of the metal orbitals which in this case is to provide an overlap allowed pathway for the transition.

(c) In metal-metal bonded compounds transitions may occur between the orbitals involved in the metal-metal bond (Section III-B-3).

(d) Intervalence transfer (IT) transitions may occur in bi- or polynuclear complexes between metallic centers that differ strongly in their redox properties (22-25). Such transitions may involve bridging ligand orbitals (23):

$$[(NH_3)_5Ru^{III}(pyz)Ru^{II}Cl(bipy)_2]^{4+} \xrightarrow{h\nu} [(NH_3)_5Ru^{II}(pyz)Ru^{III}Cl(bipy)_2]^{4+*}$$

The chemical consequences of IT transitions have only recently been investigated (26).

From the above discussion it should be evident that a single OTM compound may possess a rich variety of orbitally different excited states. In many cases this array of levels is seemingly compressed into a narrow energy range. In fact, the number, variety, and close proximity of the excited states in OTM compounds often result in electronic absorption spectra that are ill-resolved. Detailed assignments of the spectral bands may then require extensive MO calculations (15).

Fortunately, the spectra of all OTM complexes are not so difficult to interpret. In more favorable cases, qualitative assignments of the various bands are easily obtained and are often as useful to the photochemist as the quantitative determinations. Since the different types of absorption bands in general have distinguishing characteristics (ε_{max}, half-width, etc.) which are affected differently by alterations in the molecule or in the environment, qualitative assignments usually can be determined with a few experimental data. Supporting evidence for such assignments can often be obtained from more detailed assignment determinations made for similar compounds. Presented below are some of the more important characteristics of the various types of electronic absorption bands in OTM complexes.

Ligand Field Spectra. Ligand field (d,d) transitions are overlap forbidden and in centrosymmetric complexes also Laporte forbidden. For these reasons even the spin-allowed LF absorption bands are generally weak ($10^{-2} < \varepsilon_{max} < 10^2$)(16). There are many exceptional complexes, however, in which the intensities of such bands are considerably greater ($\varepsilon_{max} \approx 10^3$) (27). This intensity enhancement may arise from the mixing of ligand and metal orbitals or of d- and p-metal orbitals. The degree of mixing depends on both bonding and symmetry properties (5,27). Mixing of the weakly allowed LF excited states with energetically close lying electronically allowed states may also produce an intensity enhancement in the LF transitions (intensity borrowing) (16,27).

Spin-allowed LF transitions are always interconfigurational. In complexes of π-acceptor ligands these transitions also involve promotion of an electron from orbitals which are π-bonding to orbitals which are σ-antibonding with respect to the M-L bonds. The transitions effect a net decrease in the M-L bond strength and thus the equilibrium M-L bond lengths are expected to be greater in the excited state than in the ground state of such molecules. The excited state potential energy surface can be thought of as being displaced from that of the ground state as shown schematically in Fig. 3.

Vibrations in the ground state molecules thus allow vertical transitions to occur over a wide range of energy to closely spaced upper vibrational levels of the excited state. The spectral bands which correspond to these transitions, therefore, have large half-widths and appear as smooth gaussian shaped curves (Fig. 3)(28-30). Other factors that may contribute to the broadening of ligand field bands include spin-orbit coupling and Jahn-Teller effects (16).

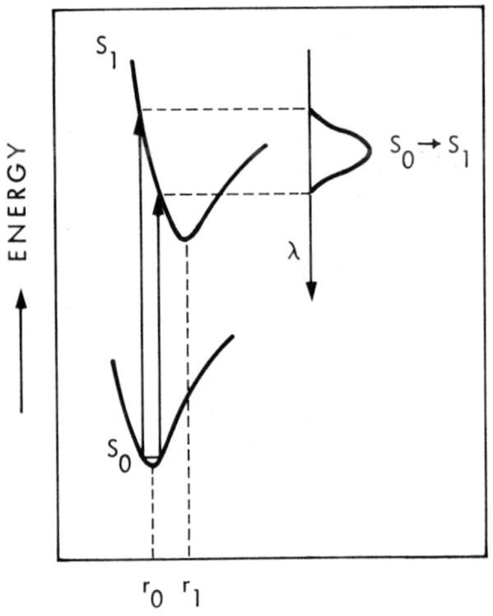

FIGURE 3. Schematic representation of an interconfigurational ligand field transition and the resulting spectral absorption band (7).

Intraligand Spectra. Although IL absorption bands are rarely observed in the UV-VIS spectra of complexes containing only "inorganic" ligands, they play an important role in the spectra of OTM compounds. Frequently, the π,π^* transitions of free ligands are only slightly perturbed by coordination of the ligand to a transition metal. Thus the spectral bands associated with IL transitions can usually be identified by comparing the spectrum of the complex with the spectrum of the uncoordinated ligand. For complexes having positively charged metal centers the spectrum of the protonated ligand may provide a better comparison (31,32).

IL absorption bands are probably least distinguishable from CT bands. Since both IL and CT transitions may be strongly allowed, the associated absorption bands have maximum extinction coefficients in the same order of magnitude (10^3 - 10^6 M^{-1} cm^{-1}). Compared with CT bands, however, the position of an IL band is relatively insensitive to the oxidizing or reducing strength of the central metal. Thus by comparing the spectra of several complexes that differ only in the identity or oxidation state of the metal, a clear distinction between IL and CT bands may be facilitated (7).

CT Spectra. The apparent features of spectral absorption bands associated with CT transitions are so diverse that relatively few general statements concerning them may be made. However, since CT bands commonly show a much stronger dependence on solvent polarity than either LF or IL bands, they are usually identifiable on this basis (33,34).

The degree of solvent dependence displayed by CT bands varies greatly (35-37). With increasing solvent polarity, a CT band may shift to either higher or lower energies and in unusual cases not at all. An example of a strong dependence is that shown by the lowest energy MLCT band of [6], which is illustrated in Fig. 4. In this case the position, half-width and maximum extinction coefficient of the band vary greatly. However, the oscillator strength is essentially invariant (37).

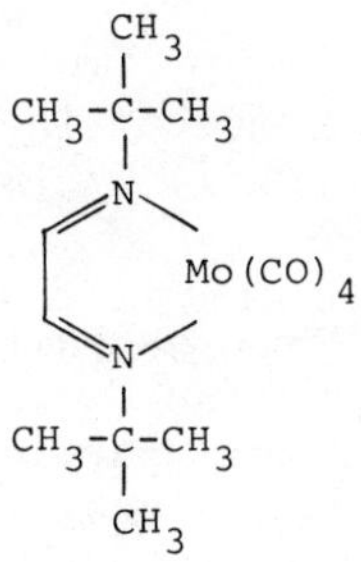

6

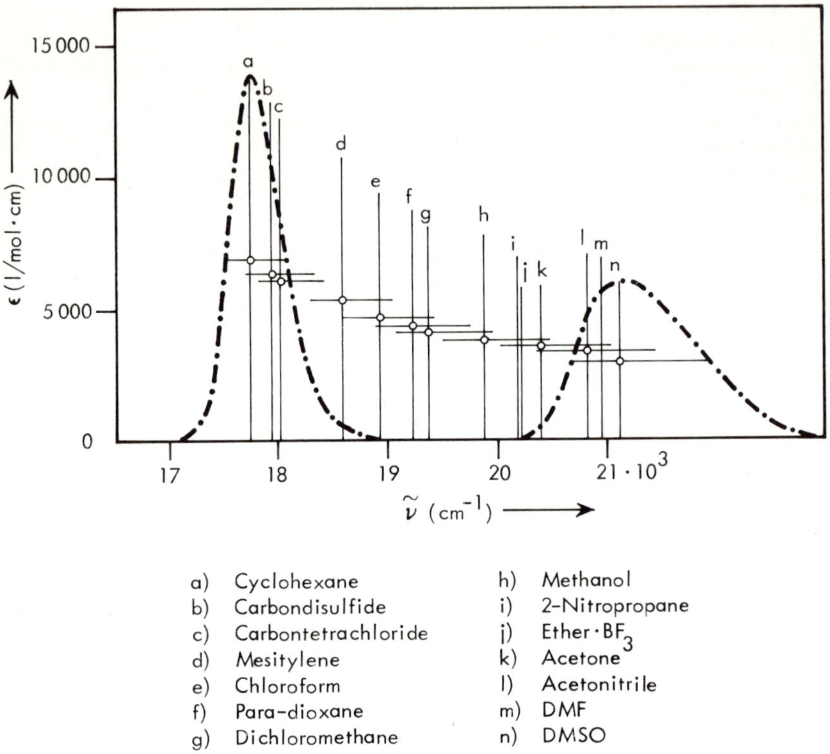

FIGURE 4. Dependence of the long wavelength absorption band (MLCT) of [6] on the solvent (38).

These effects arise from the different degrees of solvent stabilization of the ground and excited state molecules. The ground state molecules of this particular compound have a much greater dipole moment than the excited state species and are thus stabilized to a greater extent in the more polar media (37,38). More detailed discussions of solvent effects on CT spectra may be found elsewhere (35,36,39).

Ligand substituents may also have a pronounced effect on the position of CT absorption bands. A common and not unexpected result is that the energies of MLCT bands are lowered by electron withdrawing ligand substituents and increased by electron donors

(40,41). In the case of $Ru(NH_3)_5(py\text{-}x)^{2+}$, where py-x represents a para or meta substituted pyridine, the para substituents have a greater effect due to their possible π-interactions with the ring system (40).

The maximum extinction coefficients of CT bands are typically in the range 10^3 to 10^6, but are obviously sensitive to environmental factors. One feature of CT spectra which is not greatly affected by solvent or temperature variations is the oscillator strength, f, which is given approximately by

$$f = 4.6 \times 10^{-9} \varepsilon_{max} \Delta\nu_{1/2}$$

where ε_{max} is the maximum extinction coefficient of the band and $\Delta\nu_{1/2}$ is the band width in cm^{-1} at half height (16,38). Since CT transitions are in general strongly allowed, the corresponding oscillator strengths are large ($f \geq 0.1$).

III. PRIMARY DECAY PROCESSES

A. Introduction

1. Photoreactivity Models. The diverse photoreactivity exhibited by transition metal complexes is attributed in part to the variety of orbitally different excited states possessed by these systems. From simple considerations certain reactivity tendencies are associated with each particular type of excited state. Thus, for example, ligand substitution is expected to occur mainly from ligand field excited state compounds since LF transitions should in general produce the greatest net decrease in metal-ligand bond strengths (1,7). Similarly, intramolecular redox processes are expected to occur from CT excited states (34) and ligand rearrangements from IL transitions (1,7,8).

Recently, attempts have been made to formulate theoretical models upon which the primary chemical decay processes of electronically excited transition metal complexes may be qualitatively interpreted and predicted. The attempts have focused primarily on the substitutional behavior of ligand field excited state compounds (42-48) and the redox processes associated with LMCT excited states (34).

The two models that treat the photosubstitutional reactivity are based upon somewhat different assumptions. In this section the basic aspects of the models are discussed. Application of the simpler approach of Wrighton et al. (42,49) is illustrated in the next section.

The basic assumptions made in the Zink model are (43,44,47):

1. When an electronic transition takes place in a transition metal complex the bonding properties of the system change in a way which can be predicted with ligand field theory.
2. The ligands whose bonds to the metal are labilized to the greatest extent are preferentially lost from the excited complex.
3. The lowest excited state of a given spin manifold will be the most photoactive state of that manifold.

Three steps are involved in predicting the photoreactions of a transition metal complex with this model. First, the molecular axes that are labilized in the excited state are determined. If there are two different ligands on the most labilized axis, the model predicts the one whose bond to the metal is weakened more. Finally, the relative quantum yields of substitution (involving displacement of the two ligands on the labilized axis) are also predicted.

The treatment developed by Wrighton et al. (42,49) is based upon two of the same assumptions (1 and 2 above) used by Zink. However, in this case the lowest lying excited states are not assumed to be the most photoactive. Accordingly, the Wrighton et al. model may be used to interpret wavelength-dependent photoactivity which apparently involves more than one photoactive ligand field excited state (vide infra).

Although the models have proved to be useful tools for rationalizing some aspects of the photosubstitutional behavior of simple transition metal complexes, they are not without deficiencies. For example, the predictions of reactivity made with the models are only qualitative. Thus a reaction that is predicted for a particular complex may not occur at all. Another important deficiency of the models was recently discussed by Ford (50). In the series of analogous rhodium(III) complexes, $Rh(NH_3)_5X^{n+}$ ($X = NH_3$, H_2O, OH^-, Cl^-, Br^-, and I^-), relative quantum yields of ligand substitution are strongly dependent on the rates of physical radiationless decay of the excited complexes to the ground state species. According to the Zink model, however, relative quantum yields within such a series should reflect "reactivities" of the excited state complexes (44,47).

A perhaps more fundamental oversimplification made in the models is the implicit assumption of ground state molecular symmetries for the corresponding reactive excited state species. Adamson et al. (51,52) have questioned the general validity of this assumption. They suggest that thermally relaxed excited state complexes may show large geometrical distortions and may actually have symmetries other than those of the ground state molecules. Thus calculation of excited state bonding properties based on this assumption may be inappropriate. Highly distorted excited states are in fact predicted for many metal complexes (53-59).

At present it seems that before such theoretical considerations may be used in a truly predictive manner, further development and refinement of the methods are necessary. This development may be derived from consideration of: (i) the reactivity of other types of states, for example, IL and CT excited state compounds (34,48); (ii) theoretical models for radiationless transitions of reactive excited state compounds (60-66); (iii) the symmetry requirements of excited state reactions (67); and (iv) correlations of photoreactivity with detailed calculations of excited state bonding properties, geometry, and so on (56, 58,68,69). The accumulation of data regarding primary photoprocesses should provide a solid basis for this maturation and also a means of evaluating more highly developed models.

2. Wavelength-dependent Phenomena. "Wavelength-dependent phenomena" or "wavelength effects" refer to photoprocesses which have experimental quantum yields (2) that depend on the wavelength of exciting light.

From many studies of the photochemical and photophysical properties of organic compounds, it has been established that in these systems luminescence and chemical reactions having significant efficiency ($\Phi \gtrsim 0.01$) almost always occur from the lowest excited electronic levels of the possible spin manifolds (70). (Recent studies with more sensitive instrumentation have revealed that in a number of cases luminescence of small efficiency occurs from upper excited states (71,72).)

This behavior is well understood: photoexcitation of a typical aromatic hydrocarbon molecule with a singlet ground state (S_o) produces electronic transitions almost exclusively within the singlet manifold ($S_o \rightarrow S_n$, n = 1,2,3...). The vanishingly small oscillator strengths of the absorption bands for the singlet-triplet transitions ($S_o \rightarrow T_n$, n = 1,2,3...) preclude direct production of triplet excited states. Fast vibrational relaxation and internal conversion between the excited singlet states result in the eventual population of the lowest excited singlet (S_1). Since internal conversion from S_1 to S_o is relatively slow, other deactivating processes may occur. In many systems a partitioning results between a variety of decay pathways such as fluorescence, intersystem crossing, chemical reaction and internal conversion. Thus, in general, the only excited states which show appreciable behavior other than internal conversion are the lowest singlet and triplet (S_1 and T_1). Therefore wavelength effects are very rare in the field of organic photochemistry.

In contrast to the typical behavior of organic compounds discussed above, many photoreactions of transition metal complexes have wavelength-dependent quantum yields (7). Generally, these wavelength effects have been interpreted in terms of more than one reactive excited state of the photolyzed species. The photoreactivity of $W(CO)_5L$ (L = amine), for example, has been interpreted in this manner with the previously mentioned model of substitutional photoreactivity proposed by Wrighton et al. (42, 49,73). Assuming ligand dissociation to be the only primary photochemical process (Section III-B-1), photolysis of $W(CO)_5L$ could produce three primary products:

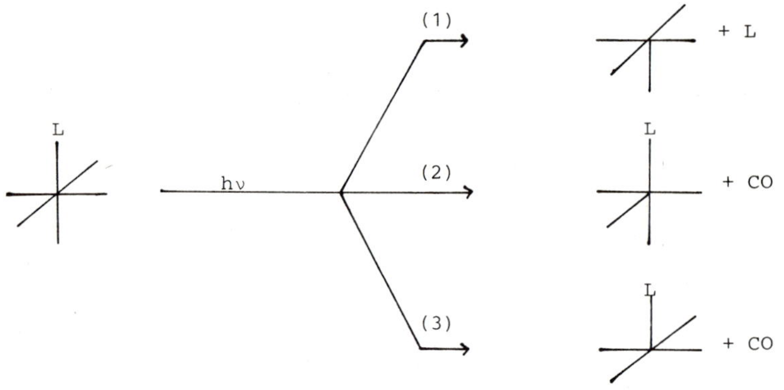

Although not conclusive, the data in Table 1 suggest that with L = py, only processes 1 and 2 occur with substantial efficiency. Although process 1 is always dominant, as the wavelength of the exciting light is shortened, the quantum yield of 1 decreases while the quantum yield of 2 increases. Similar trends are apparent in the photobehavior of other $M(CO)_5L$ compounds (M = Mo, W; L = saturated amine) (73).

The splitting of the d-orbitals in $M(CO)_5L$ is given approximately by the simplified MO scheme in Fig. 5. For d^6 compounds, such as $M(CO)_5L$ (M = Mo,W), the lowest energy ligand field excited states result from the one electron transition $e \rightarrow a_1$.

According to the photoreactivity model (42,49), this transition should labilize the ligands on the Z-axis (OC-M-L). Since the d_{xz} and d_{yz} metal orbitals are involved in π-bonding with these π-acceptor ligands, the decrease in electron population of these orbitals results in weaker metal-ligand π-bonds along

TABLE 1. Quantum yields for photoreactions of $W(CO)_5(py)$ (49)

$$1: \quad W(CO)_5(py) \xrightarrow{h\nu} W(CO)_5(\text{1-pentene})$$

$$2: \quad W(CO)_5(py) \xrightarrow{h\nu} \underline{cis}\text{-}W(CO)_4(py)_2$$

Reaction	Φ 436 nm	Φ 366 nm	φ 313 nm	Φ 254 nm
1[a]	0.63	0.50	0.38	0.3_4
2[b]	0.00_2	0.01_3	0.03_9	0.04
3	-	-	-	-

[a]Quantum yields measured at room temperature in 3.66M 1-pentene in isooctane.
[b]Quantum yields for formation of cis-$W(CO)_4(py)_2$ in room temperature isooctane containing 0.25M py.

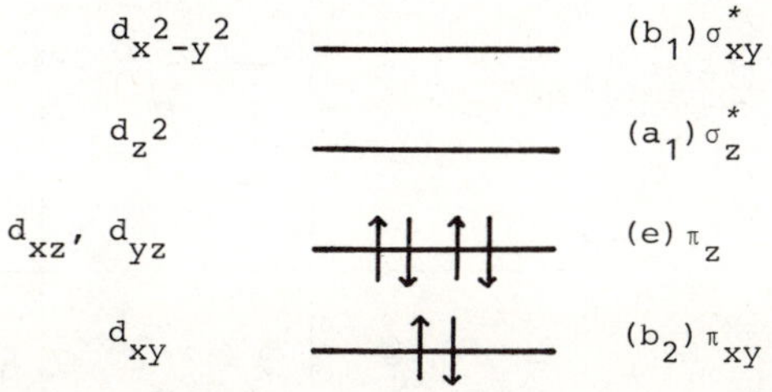

FIGURE 5. Arrangement of the metal d orbitals in $W(CO)_5L$ (42).

the Z-axis. The metal-ligand σ-bonds on this same axis are also weakened by population of the d_{z^2} (σ_z*) orbital in the excited state configuration. Thus the model "predicts" that the lowest LF excited states in $M(CO)_5L$ should decay by processes 1 and 3. Higher energy excitation is expected to produce LF excited states derived from the $b_2 \rightarrow b_1$ one electron transitions. The photosubstitution model of Wrighton et al. similarly predicts the reactivity of these excited states should involve process 2.

TABLE 2. Wavelength dependent quantum yields for the substitution of CO (Φ^{CO}) and dienes (Φ^{D}) in diene carbonyl complexes (74,75).

	[complex] (mol/l)	[P(OCH$_3$)$_3$] (mol/l)	254 nm		405 nm	
			Φ^{CO}	Φ^{D}	Φ^{CO}	Φ^{D}
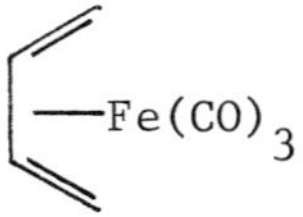	$1.1\cdot10^{-2}$	0.1	0.28	0.08	0.05	0.01
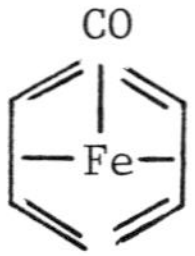	$5.7\cdot10^{-3}$	0.2	0.14	0.03	0.01	0.01
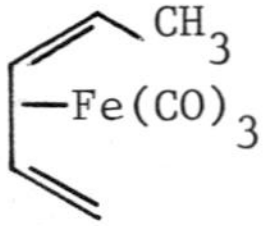	$9.6\cdot10^{-3}$	0.1	0.46	0.64		
	$9.5\cdot10^{-3}$	0.1			0.05	0.088
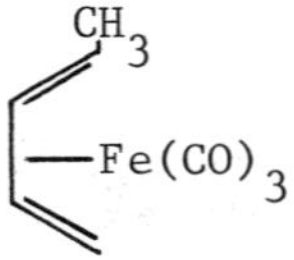	$10.4\cdot10^{-3}$	0.1	0.23	0.016		
	$10.2\cdot10^{-3}$	0.1			0.03	0.009

The experimental results (Table 1) are in fair agreement with the predictions of the model. Exceptional is the apparent lack of efficient axial CO substitution (3) from the low energy excited states. This disagreement represents the inadequacy of the model to quantitatively predict efficiencies of decay processes. It was suggested that although the axial CO may be strongly labilized, the rate of CO loss may not be great enough to compete with the other deactivating processes.

Some other examples of substituted metal carbonyl compounds which show wavelength effects in their photosubstitution reactions are given in Table 2.

Processes other than ligand substitution may also show wavelength dependent quantum yields. Photolysis of the series of ruthenium(II) compounds, $Ru(bipy)_2ClL^+$ and cis-$Ru(bipy)_2L_2^{2+}$ (L = cis- and trans-4-stilbazole) produces cis-trans isomerization of coordinated stilbazole as the only detectable photoreaction (76,77). Low energy excitation of these complexes, which directly produces MLCT excited states, results in large quantum yields of cis → trans isomerization. The trans → cis isomerization, however, is more efficient with photolysis at shorter wavelengths, in which case IL excited states are produced (Table 3). To account for these results, both IL and MLCT (L = stilbazole) excited states were proposed to be reactive. A fast rate of reaction from the upper IL state rather than a slow rate of radiationless decay was suggested to produce the wavelength effects in these systems.

Excitation wavelength dependent emission quantum yields have been observed in a few cases also (64,77). The emission

TABLE 3. Quantum yields of isomerization by photoexcitation of ruthenium(II)-4-stilbazole (L) complexes (77).

Complex		Excitation Wavelength (nm) 313	366	436
$Ru(bipy)_2(\text{trans-}L)_2^{2+}$	$(\Phi_{t\to c})$	0.15	0.05	0.05
$Ru(bipy)_2(\text{cis-}L)_2^{2+}$	$(\Phi_{c\to t})$	0.15_6	0.15	0.51
$Ru(bipy)_2CL(\text{trans-}L)^+$	$(\Phi_{t\to c})$	0.09_2		0.03_5
$Ru(bipy)_2Cl(\text{cis-}L)^+$	$(\Phi_{c\to t})$	0.12_3		0.66_5

quantum yields of cis-$Rh(bipy)_2Cl_2^+$ and cis-$Rh(phen)_2Cl_2^+$ are larger with excitation into LF absorption bands than with excitation into higher energy IL bands (60,64). Interestingly, other workers found that in these compounds the upper ^1IL state decays more efficiently via intersystem crossing to the ^3IL state than via internal conversion to the lower ^1LF excited state (78).

Observation of multiple decay pathways having wavelength-dependent quantum yields are relatively common among transition metal complexes. It is fairly certain that these effects result from the competition of chemical decay with physical decay in the higher excited states of compounds. At this point, however, due to a general lack of knowledge concerning the rates of primary steps from upper excited states, it is impossible to draw general conclusions regarding the origin of these effects. The relative rates of chemical reaction and physical decay will determine the efficiencies of such processes.

It has been suggested (79) that slow interconversion of relatively isolated states of different orbital nature (IL, LF, CT) may occur due to the different degrees to which these states are affected by spin-orbit coupling. Relatively small rate constants ($\sim 10^6 s^{-1}$) for interconversion between ligand localized (^3IL) and metal localized (^3LF) excited states have in fact been measured in a few cases (78). In other studies, however, the observed wavelength effects have been attributed to very fast reactions of upper excited states (49,76,77). Thus either factor (slow internal conversion or fast reaction rate) or a combination thereof may produce wavelength effects.

B. Photochemical Primary Processes

1. Ligand Photosubstitution in Solutions. In 1961 Strohmeier and Gerlach (80) proposed that the primary process involved in photosubstitution reactions of the Group VIb hexacarbonyls is dissociative loss of CO. From quantitative studies of photosubstitution reactions involving a series of donor ligands, it was concluded that the quantum yield of this primary process is unity in all cases (80,81):

$$M(CO)_6 + h\nu \xrightarrow[M=Cr,Mo,W]{\Phi = 1.0} M(CO)_5 + CO$$

This conclusion was further supported by the determination of a unit efficiency for the benzophenone sensitized reaction of $Cr(CO)_6$ with pyridine (82) (Section III-D-1).

Recent flash photolysis investigations have provided direct evidence that the primary photochemical process of these compounds is indeed CO dissociation. The reactivity of the primary product from photolysis of $Cr(CO)_6$, $Cr(CO)_5$, has also been determined through these studies.

The results of early flash studies were irreproducible and led to various conclusions concerning the nature of the primary photoproduct. For example, flash photolysis of $Cr(CO)_6$ in room temperature degassed cyclohexane solutions and in polystyrene film was reported to produce a transient having a maximum absorbance in the visible at 450 nm and which decayed with second-order (equal concentration) kinetics (83). The following scheme was proposed for the overall mechanism:

$$Cr(CO)_6 + h\nu \longrightarrow Cr(CO)_6^*$$

$$Cr(CO)_6^* \longrightarrow Cr(CO)_5 + CO$$

$$Cr(CO)_5 + CO \xrightarrow{k} Cr(CO)_6$$

Values of k/ε, where ε is the extinction coefficient of the transient at 450 nm, of 0.317 and 0.00025 cm/s were obtained for cyclohexane and polystyrene as solvent, respectively.

Under supposedly the same conditions (i.e., degassed cyclohexane at room temperature) flash photolysis of $Cr(CO)_6$ was reported by Nasielski et al. (84) to give two successive transients. The first showed maximum visible absorption at 483 nm and decayed in a first-order process with a lifetime of $\tau = 6$ ms. The second transient decayed much more slowly with approximately second-order kinetics. Added CO enhanced the rate of decay of the second transient but did not affect that of the first. It was suggested that the transients were the same species supposedly found by Stoltz et al. (85,86) in low temperature photolyses of $Cr(CO)_6$: C_{4v} and D_{3h} $Cr(CO)_5$. The following mechanism was proposed:

$$Cr(CO)_6 + h\nu \longrightarrow Cr(CO)_5(C_{4v}) + CO$$

$$Cr(CO)_5(C_{4v}) \longrightarrow Cr(CO)_5(D_{3h})$$

$$Cr(CO)_5(D_{3h}) + CO \longrightarrow Cr(CO)_6$$

Similar results were obtained by Kelly and Morris (87) for the flash photolysis of $Cr(CO)_6$ and $Mo(CO)_6$ in room temperature degassed cyclohexane solutions. With $Cr(CO)_6$ the decay of the first transient followed first-order kinetics with a half-life of $t_{1/2}$ = 5 ms. This species showed a maximum absorbance near 470 nm. The decay of the second transient followed second-order kinetics for two half-lives but the second-order rate constant varied by an order of magnitude and the wavelength of maximum absorption varied by 30 nm. A mechanism involving the reaction of $Cr(CO)_5$ with impurities was suggested as a possible explanation of the results.

More extensive studies of $Cr(CO)_6$ in highly purified cyclohexane have demonstrated that the earlier contradictory and irreproducible results were almost certainly due to reactions of $Cr(CO)_5$ with impurities in the solutions (88). Flash photolysis of $Cr(CO)_6$ produced a transient species within 50 ns of excitation and having λ_{max} = 503 ± 5 nm. In degassed cyclohexane this transient decayed (τ > 200 μs) to form further intermediates which themselves decayed (τ < 1s) to reform $Cr(CO)_6$.

When the experiment was carried out under 1 atm of CO, the initial transient decayed with a half-life of $t_{1/2}$ = 25 μs to regenerate $Cr(CO)_6$. This species was thus assigned as $Cr(CO)_5$. In addition, a small quantity (≤ 10%) of other products was formed. The efficiency of production of the side products from $Cr(CO)_5$ was found to be independent of flash intensity and of the concentration of $Cr(CO)_6$. No decomposition of cyclohexane could be detected in irradiated solutions of $Cr(CO)_6$, and it was concluded that the side products arose from reactions of $Cr(CO)_5$ with small amounts of impurities in solution. The overall scheme for the photolysis of $Cr(CO)_6$ in cyclohexane was proposed to be:

$$Cr(CO)_6 + h\nu \longrightarrow Cr(CO)_5 + CO$$

$$Cr(CO)_5 + L \underset{k_2}{\overset{k_1}{\rightleftharpoons}} Cr(CO)_5L \qquad (L = \text{impurity})$$

$$Cr(CO)_5 + CO \xrightarrow{k_3} Cr(CO)_6 \qquad k_3 = 3 \pm 1 \times 10^6\ \underline{M}^{-1}\ s^{-1}$$

Further evidence for this scheme was obtained from a laser flash spectroscopic study of the interaction of $Cr(CO)_5$ with various solvents (L) capable of coordination (89). In these experiments $Cr(CO)_6$ was flash photolyzed in CO saturated (1 atm) cyclohexane solutions which also contained small quantities of L. Rate constants for formation (k_1) and dissociation (k_2) of the solvent complexes, $Cr(CO)_5L$, were determined. These data, along with the calculated stability constants of $Cr(CO)_5L$, are given in Table 4. It is apparent from these data that solvents having a polar π-electron system stabilize the $Cr(CO)_5$ fragment more than simple σ-donors.

TABLE 4. Rate constants and stability constants for the reaction of $Cr(CO)_5$ with L:

$$Cr(CO)_5 + L \underset{k_2}{\overset{k_1}{\rightleftharpoons}} Cr(CO)_5L \quad (K = \frac{k_1}{k_2}) \quad (89).$$

L	k_1 ($\underline{M}^{-1}s^{-1}$)	k_2 (s^{-1})	K ($\underline{M}^{-1}$)	λ_{max}(nm) for $Cr(CO)_5L$
Benzene	7.0×10^6	2.1×10^5	3.4×10^1	475
Diethyl ether	1.1×10^7	1.4×10^2	7.6×10^4	455
Methanol	3.1×10^6			460
Ethyl acetate	9.2×10^7	8.1×10^1	1.1×10^6	445
Acetone	1.3×10^8	1.4×10^1	9.3×10^6	438
Acetonitrile	1.6×10^8	$< 1 \times 10^{-2}$	$> 1 \times 10^{10}$	391

The flash studies have confirmed the qualitative nature of the primary photochemical decay process in $Cr(CO)_6$ and related compounds. The results of recent quantum yield studies, however, are not consistent with previous conclusions concerning the quantitative aspects of the decay process. For example, photolysis of the hexacarbonyls in cyclohexane containing pyridine gave wavelength independent quantum yields of $\Phi = 0.67$ (Cr) and $\Phi = 0.75$ (W) for formation of $M(CO)_5py$ (90). These values do <u>not</u> increase with increasing concentrations of pyridine as would be expected from an associative mechanism. Furthermore, the insensitivity of the quantum yields to solvent viscosity indicates that cage recombination of the primary fragments is unimportant. Although the quantum yields from benzophenone sensitized reactions are equal to those obtained with direct photolysis, peculiar results were obtained with other donors (Section III-D-1). Further interpretation of these sensitization results would, therefore, be speculative.

Photosubstitution reactions of OTM complexes may also involve associative primary processes:

$$ML_n + h\nu \rightarrow ML_n^*$$

$$ML_n^* + L' \rightarrow ML_nL'$$

$$ML_nL' \rightarrow ML_{n-1}L' + L$$

These reactions may be identified by the dependence of the quantum yield for the reaction on the concentration of entering ligand. An exemplary case, in which the photosubstitution of CO occurs partially by the associative mechanism outlined above, was studied by Keeton and Basolo (59):

$$Mn(CO)_4NO + PPh_3 \xrightarrow{h\nu} Mn(CO)_3(NO)(PPh_3) + CO$$

This reaction was indicated to occur partially via the associative mechanism by the dependence of the quantum yield on the concentration of PPh_3 (0.85-19.9 mM) and by the observation (UV) of a long-lived intermediate formed in the photolysis. The overall mechanism, involving both the dissociative pathway and the associative step is shown in Fig. 6. The latter is thought to be a nucleophilic attack on a relaxed excited state which has square pyramidal rather than the ground state trigonal bipyramidal structure. With the weaker nucleophile, $AsPh_3$, the quantum yield is independent of [$AsPPh_3$] (1.14-20.1 mM) and no intermediate is observed. In this case only the dissociative pathway occurs.

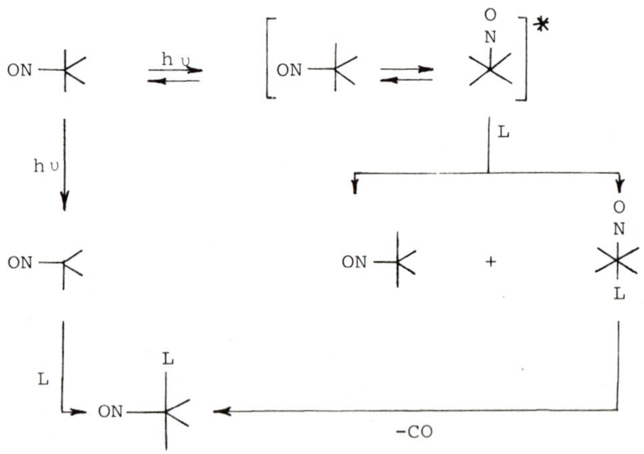

FIGURE 6. Proposed mechanisms for the photosubstitution of $Mn(CO)_4(NO)$ (59).

2. Geometrical Photoisomerizations. In this section geometrical photoisomerization reactions of OTM compounds are considered. Recent accounts of photoinduced rearrangements of coordinated ligands and OTM complex photoassisted isomerization of olefins, dienes, and so on are available elsewhere (3,4,91).

By analogy with thermal processes (92), photoinduced isomerizations may occur either by intramolecular or intermolecular mechanisms. Intramolecular isomerizations are further classified either as "twisting" mechanisms, which involve no metal-ligand bond breaking, or as bond-rupture mechanisms. The latter term applies to reactions of chelate compounds in which one of the ligands is for a short time partially dissociated. Intermolecular mechanisms involve species other than those that make up the reactant complex.

It is worthwhile to consider which primary photoprocesses might be involved in these mechanisms. The intermolecular photoinduced isomerizations, like photosubstitutions, may involve either dissociation or association of a ligand as a primary process (7):

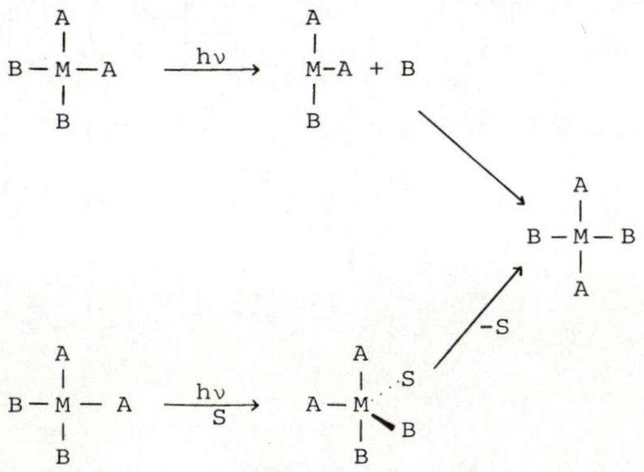

(S = Solvent)

Intramolecular "twisting" mechanisms may involve the decay of a distorted excited state complex. Thermally equilibrated ligand field excited state complexes are expected in many cases to show geometrical distortions (Section III-A-1), and may therefore be suspected as precursors in these reactions.

A particularly well studied system which undergoes photochemical cis → trans isomerization is Pt(gly)$_2$. The work on these compounds through 1969 has already been discussed in detail elsewhere (7), and thus only the major conclusions thereof will be mentioned here. A definitive study by F. Scandola et al. (93) showed that the photochemical cis → trans isomerization of this species occurs with an intramolecular "twisting" mechanism (Fig. 7). The thermally relaxed reactive excited state was proposed to have a pseudo-tetrahedral geometry as shown in Fig. 7.

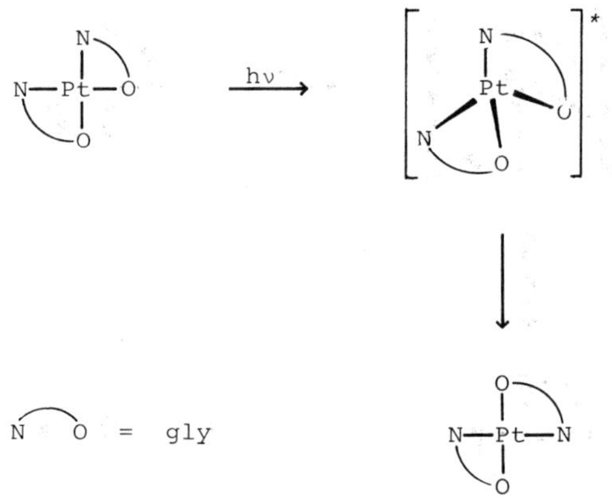

FIGURE 7. Twisting mechanism proposed for the cis → trans photoisomerization of Pt(gly)$_2$ (7,93).

Large geometrical distortion in the lowest excited states of Pt(gly)$_2$ is also suggested by spectroscopic studies, which have indicated that similar square planar complexes have pseudotetrahedral excited states (93a), by semi-empirical molecular orbital calculations (56), and also by recent quenching and sensitization studies of the cis → trans reaction (Section III-D-1).

In a more recent study, the effects of certain environmental factors on this cis → trans photoisomerization have been ascertained (94). Although unaffected by temperature, oxygen, or ethanol, the quantum yield of this process is strongly dependent on solution pH. In neutral and slightly acidic (pH 6) solutions, the only detectable photoproduct from cis-$Pt(gly)_2$ is the trans isomer. As the pH of the photolyzed solution is lowered, however, the quantum yield of isomerization decreases while the quantum yield of a substitution reaction increases. Finally, at pH $\leq$ 2 the substitution reaction accounts for essentially all photochemistry. At all pH values the sum of the quantum yields for isomerization and substitution is fairly constant ($\Phi \approx 0.11$). This competition between the two reaction pathways might have been misinterpreted in terms of a unique dissociative primary process. Recombination could then produce isomerization in neutral media while ligand protonation would trap the primary product at low pH resulting in net substitution. Actually, the dissociative pathway is acid catalyzed, involving an interaction between the excited state complex and hydronium ions (94).

Cis-$Pt(py)_2Cl_2$ also undergoes photoisomerization to the trans isomer in chloroform solutions (95). The isomerization is the predominant process (Φ~0.04), but is accompanied by a dissociative photoreaction (Φ~0.025). Quantitative studies showed that most of the isomerization occurs via an intramolecular mechanism; however, the details of the process are not yet certain.

Although other platinum(II) complexes undergo cis → trans photoisomerization, mechanistic information concerning these reactions is scarce. At present it is thus impossible to make reasonable structure/reactivity correlations concerning excited state decay in these compounds.

The synthetic utility of several cis → trans photoisomerizations has been emphasized. In addition to cis-$Pt(py)_2Cl_2$, the cis isomers of the following compounds can be photochemically isomerized to their respective trans isomers with high chemical yields: $Pt(PPh_3)_2X_2$ (X = Cl,Br) and $Pt(isoquinoline)_2Cl_2$ (96,97), $Pt(PEt_3)_2Cl_2$ (97).

An intermolecular mechanism involving a primary ligand photodissociation step has been proposed for the cis → trans isomerization of the dicarbene complexes, $M(CO)_4L_2$, (M = Cr,Mo, W; L = 1,3-dimethyl-4-imidazoline-2-ylidene). The thermodynamically less stable trans isomers reconvert thermally to the cis-complexes (98,99) (Fig. 8).

Many six-coordinate complexes of ruthenium(II), osmium(II), iridium(III), and rhodium(III) also undergo photoinduced

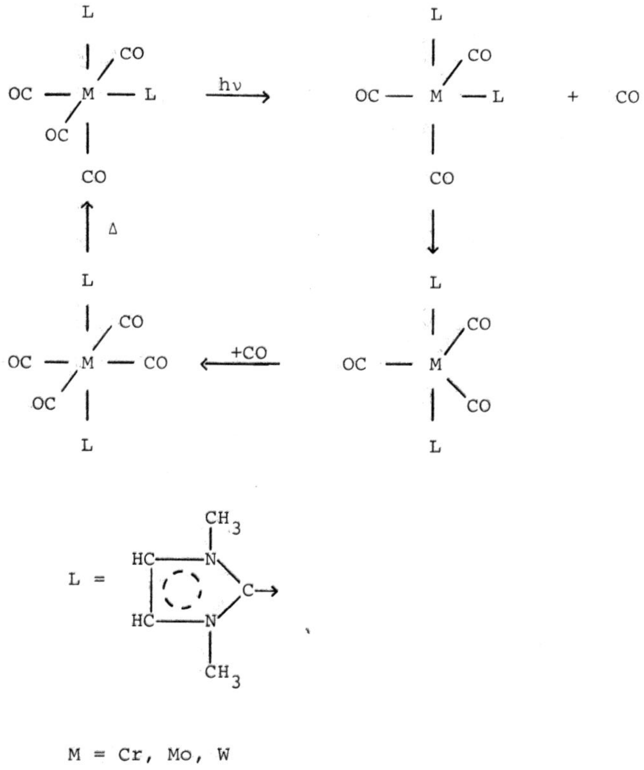

FIGURE 8. Proposed mechanism for the thermally reversible photo-induced _cis_ → _trans_ isomerization of some dicarbene complexes (98,99).

geometrical isomerizations (100-104). Thermal reversibility of these reactions has also been noted in some cases (105):

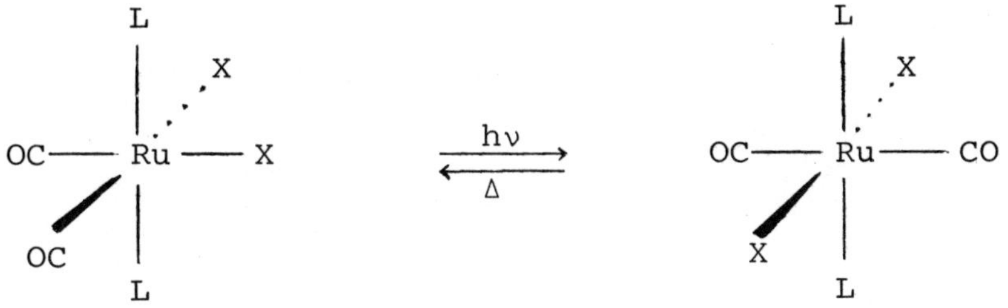

L = phosphorus(III) or arsenic(III) ligand

X = Cl, Br, I

The intimate mechanistic details of these processes, however, are not yet available.

Photoisomerization has also been observed in matrix-isolated OTM compounds (Section III-B-4).

3. Compounds Containing Metal-Metal Bonds. There is growing evidence that homolytic metal-metal bond cleavage is in general an important decay process of photoexcited compounds containing metal-metal bonds. This evidence is derived from a series of experiments involving quantitative aspects of the photochemical reactions as well as from flash photolysis studies, in which the behaviors of the photolytically generated mononuclear fragments are observed.

The pertinent features of the electronic absorption spectra of the compounds of interest are shown in Table 5. Most of these compounds exhibit an intense ($\varepsilon_{max} \approx 20{,}000$) band in the near ultraviolet and a weaker maximum at somewhat longer wavelengths. The more intense bands have been assigned to the $d\sigma(MM) \rightarrow d\sigma^*(MM)$ transitions (106-113). The lower-energy, less intense bands are also thought to arise from d,d transitions, such as $d\pi(ML) \rightarrow d\sigma^*(MM)$ (106-113).

The results of recent quantitative photochemical studies of these compounds suggest that irradiation into either of the two main absorption bands results in efficient homolysis of the metal-metal bonds. For example, the photoreactions of these complexes in CCl_4, which result in mononuclear chloride substituted products, occur with high quantum yields and clean stoichiometry (Table 6) (106-108,111). Other chemical results that indicate that homolytic cleavage is an important consequence of photoexcitation are:

TABLE 5. Electronic absorption spectral data of metal-metal bonded compounds at 25° (106-108).[a]

Compound	Band maxima (nm); (ε, $l.mol.^{-1}cm^{-1}$)	
$Mn_2(CO)_{10}$	~390 (sh)	342(21,400)
$Re_2(CO)_{10}$	~313(15,500)	275(12,800)
$Mn_2(CO)_9(PPh_3)$	430(sh)	360(24,200)
$Mn_2(CO)_8(PPh_3)_2$[b]	450(sh)	373(23,700)
$ReMn(CO)_{10}$	323(13,700)	~263(sh)
$[CpMo(CO)_3]_2$	512(1720)	388(20,400)
$[CpW(CO)_3]_2$[c]	493(2450)	362(20,200)
$[CpW(CO)_3Mo(CO)_3Cp]$[c]	475(389)	320(sh)
$[CpMo(CO)_3Mn(CO)_5]$	450(1,400)	373(14,000)
$[CpMo(CO)_3Re(CO)_5]$	454(642)	365(6,500)
$[CpW(CO)_3Mn(CO)_5]$	450(1,460)	362(14,900)
$[CpW(CO)_3Re(CO)_5]$	411(1,400)	331(9,000)

[a]Isooctane solution unless otherwise indicated.
[b]CH_2Cl_2 solution.
[c]CCl_4 solution.

1. Irradiation (350 nm) of $Mn_2(CO)_{10}$ in degassed THF produced a paramagnetic species which slowly reverted back to $Mn_2(CO)_{10}$; this species reacted with I_2 in THF under nitrogen to produce 30 to 35 percent $Mn(CO)_5I$. Although this species is almost certainly not $Mn(CO)_5\cdot$ (vide infra), infrared spectral data suggest that is may be some derivative thereof (114).
2. Photoreactions in the presence of small concentrations of iodine donors occur with high quantum yields (106):

TABLE 6. Quantum yields for reactions of metal-metal bonded complexes in degassed CCl_4.

Starting Compound	Product	Excitation Wavelength	Φ_{dis}[a]	Φ_{prod}[b]	Ref.
$Mn_2(CO)_{10}$	$Mn(CO)_5Cl$	366	0.41	0.72	(106)
		313	0.48	1.02	
$Re_2(CO)_{10}$	$Re(CO)_5Cl$	313	0.60	1.20	(106)
$Mn_2(CO)_9(PPh_3)$	$Mn(CO)_5Cl$	366	0.45	0.36	(106)
	$Mn(CO)_4(PPh_3)Cl$	366		0.40	
$Mn_2(CO)_8(PPh_3)_2$	$Mn(CO)_4(PPh_3)Cl$	366	0.70	0.48	(106)
$[CpMo(CO)_3]_2$	$CpMo(CO)_3Cl$	550	0.35	0.77	(107)
		405	0.42	0.94	
		366	0.45	1.04	
$[CpW(CO)_3]_2$	$CpW(CO)_3Cl$	550	0.12	0.25	(107)
		366	0.21	0.41	
$CpMo(CO)_3Mn(CO)_5$	$CpMo(CO)_3Cl$				
	$Mn(CO)_5Cl$	366	0.51		(108)
$CpMo(CO)_3Re(CO)_5$	$CpMo(CO)_3Cl$ $Re(CO)_5Cl$	366	0.56		(108)
$CpW(CO)_3Mn(CO)_5$	$CpW(CO)_3Cl$	366	0.35		(108)
	$Mn(CO)_5Cl$	436	0.05_6		
$CpW(CO)_3Re(CO)_5$	$CpW(CO)_3Cl$	366	0.23		(108)
	$Re(CO)_5Cl$	436	0.05		

[a]Quantum yields for disappearance of starting materials.
[b]Quantum yields of production formation.

$$Re_2(CO)_{10} \xrightarrow[10^{-3}-10^{-4}M\ I_2]{366\ nm,\ isooctane} 2Re(CO)_5I \quad (\Phi_{formation} = 1.14)$$

3. Photolysis of mixed solutions of homonuclear compounds produces heteronuclear species; photolysis of heteronuclear compounds in inert solvents produces homonuclear species (106-108,115):

$$Mn_2(CO)_{10} + Re_2(CO)_{10} \xrightarrow[hexane]{h\nu} MnRe(CO)_{10}$$

$$[CpM(CO)_3]_2 + M'_2(CO)_{10} \xrightarrow[N_2/benzene]{h\nu} CpM(CO)_3\text{-}M'(CO)_5$$

$$M = W, Mo; \quad M' = Mn, Re$$

$$2Cp(CO)_2Fe\text{-}Mn(CO)_5 \underset{366\ nm}{\overset{546\ nm}{\rightleftarrows}} [CpFe(CO)_2]_2 + Mn_2(CO)_{10}$$

4. Photolyses of compounds in the presence of Ph_3CCl and $PhCH_2Cl$ produce, in addition to mononuclear chloride substituted metal complexes, ESR detectable $Ph_3C\cdot$ radicals and nearly quantitative yields of $PhCH_2CH_2Ph$, respectively (106,107,111):

$$[CpM(CO)_3]_2 \xrightarrow[isooctane]{h\nu/0.1\underline{M}\ PhCh_2Cl} \underset{(quantitative)}{CpM(CO)_3Cl} + \underset{(90\%)}{PhCH_2CH_2Ph}$$

$$M = Mo, W$$

Taken together, these results provide strong evidence that homolytic metal-metal bond breaking is the principal primary chemical decay process in these complexes.

The results of recent kinetic flash photolysis studies are consistent with this interpretation, but indicate that other processes may also be important. For example, flash photolysis of $(CpMo(CO)_3)_2$ in cyclohexane, THF and acetonitrile, with either visible ($\lambda > 460$ nm) or ultraviolet ($340 < \lambda < 405$ nm) light results in the formation of <u>two</u> transient products, both of which were detected at the earliest experimental observation times ($t \geq 50$ μs) (110,115). Two subsequent thermal processes result in quantitative regeneration of the dimer complex.

The intermediate involved in the first process is indicated to be the monomer fragment, $CpMo(CO)_3^{\cdot}$, by the following: (i) It is relatively nonabsorbing in the spectral region where $(CpMo(CO)_3)_2$ strongly absorbs due to the σ(Mo-Mo) $\rightarrow$ σ^*(Mo-Mo) transition. (ii) The intermediate's decay follows second-order kinetics at essentially diffusion controlled rates. The second order rate constants at 20° are k = (2 ± 1) x 10^9 (THF), (3 ± 1) x 10^9 (acetonitrile) and (5 ± 1) x $10^9 \underline{M}^{-1}s^{-1}$ (cyclohexane). (iii) The rate of decay of this species is independent of added inert electrolyte and of added CO (110,115).

The slower reaction is suggested to be the recombination with CO of a coordinatively unsaturated species which has retained the Mo-Mo bond. However, it is uncertain whether or not this species is a primary photoproduct (110,115).

Qualitatively similar results have been obtained with $Mn_2(CO)_{10}$ (112). Ultraviolet (310-380 nm) flash photolysis of this compound in both THF and cyclohexane also produces two intermediates. The first, shorter-lived transient is virtually transparent in the spectral region associated with the $\sigma \rightarrow \sigma^*$ transition of the Mn-Mn bond in the parent dimer complex. For this reason and because this species decays with second-order kinetics at near diffusion limited rates (k = 3.3 x $10^9 \underline{M}^{-1}s^{-1}$ in cyclohexane at 25°), it is suggested to be $Mn(CO)_5^{\cdot}$.

In cyclohexane the second intermediate has λ_{max} = 500 nm and decays in a comparatively slow, first-order process ($t_{1/2}$ = 26 s) to give $Mn_2(CO)_{10}$. In THF, similar spectral characteristics are displayed (λ_{max} = 492 nm); but in this solvent the transient is considerably longer lived. The nature and origin of the second transient are not certain. It is suggested to be the paramagnetic species produced by conventional photolysis of $Mn_2(CO)_{10}$ in solutions of various ethers and originally observed by Hallock and Wojcicki (114). However, the identity of this product is still a matter of discussion (116,117). Further investigation appears to be necessary to clearly determine the nature and origin of these transient products.

From the flash photolysis and quantitative photochemical studies discussed above, it has been shown that homolytic metal-metal bond fission is an important primary photoprocess in these compounds. The net chemical reactions which follow the primary processes are somewhat more varied, however. In addition to the net photochemistry discussed above, these compounds also undergo: (i) photosubstitution reactions in which the metal-metal bond is retained in the products (118-121), (ii) disproportionations (120,122), and (iii) elimination reactions such as (123,124):

$$[(\eta^5-C_5(CH_3)_5Mo(CO)_3]_2 \xrightarrow{h\nu} [(\eta^5-C_5(CH_3)_5Mo(CO)_2]_2 + 2CO$$

Much of this varied net photochemistry can be accounted for in terms of initial metal-metal bond homolysis followed by thermal reactions of the primary fragments. Such a mechanism, involving initial homolysis of the Mn-Mn bond followed by associative substitution of a $Mn(CO)_4PPh_3$ fragment, has been proposed for this net thermal process (125):

$$[Mn(CO)_4PPh_3]_2 + P(OPh)_3 \xrightarrow[\Delta]{} Mn_2(CO)_8(PPh_3)P(OPh)_3 + PPh_3$$

The quantitative experiments by Wrighton et al. (106-108,111) have provided strong evidence for this general interpretation of the net photochemistry. More definitive, however, would be direct information, perhaps obtainable from laser flash photolysis studies, regarding the origins of the photolytically produced transients.

Photoinduced cleavage of the quadruply metal-metal bonded dimer, $Re_2(Cl_8)^{2-}$, also occurs in acetonitrile solutions under ultraviolet irradiation (λ_{ex} = 313 nm, 366 nm, 254 nm) but not upon photolysis into the longest wavelength absorption band (λ_{ex} = 632.8 nm) (126). With λ_{ex} = 313 nm, the quantum yield for disappearance of the dimer is $\Phi = 0.017 \pm 0.005$. Under conventional photolysis conditions the first observable product is $ReCl_4(CH_3CN)_2^-$. However, the mechanistic details of the photoreaction are not yet certain.

4. Matrix Studies. In the past few years the photochemistry of organometallic compounds (especially carbonyl complexes) isolated in rigid matrices at very low temperatures has been investigated. The matrix isolation technique involves cocondensation of substrate molecules with a suitable matrix material, usually argon, on an optical window at temperatures commonly between 4 and 20°K. Photochemical transformations of the substrate are usually monitored by infrared and UV/VIS absorption spectroscopy. Reactions requiring significant thermal activation or diffusion are strongly inhibited in the matrix environment. Thus primary photoproducts, which in room temperature fluid media are reaction intermediates and can be observed only with great difficulty if at all, may be casually investigated with the matrix technique. Since a current and detailed review of this topic is appearing elsewhere (3), only a brief summary of the more important results is presented here.

The matrix photochemistry of binary metal carbonyls has been extensively investigated. In all cases the first observable consequence of UV/VIS irradiation is expulsion of a CO group. In many cases interaction of the coordinatively unsaturated product with the matrix molecules (even with such "inert" matrices as the rare gases) has been demonstrated. Prolonged UV/VIS irradiation usually results in the loss of additional CO groups. The binary CO complexes that have been investigated include the following: $M(CO)_6$, M = Cr, Mo, and W, (127-133); $Fe(CO)_5$ (134-138); $Fe_2(CO)_9$ (139); $Ni(CO)_4$ (140,141); $Os(CO)_5$ (142).

Vacuum ultraviolet irradiation of matrix-isolated complexes may result in ionic cleavage of M-L bonds. $Ni(CO)_3^-$, $Cr(CO)_5^-$, and $Fe(CO)_4^-$ have been produced in this manner from $Ni(CO)_4$, $Cr(CO)_6$, and $Fe(CO)_5$, respectively (143).

Although the solution photochemistry of substituted metal carbonyls is indeed very rich, the behavior of these compounds in low-temperature matrices is rather limited. As explained by Black et al. (144,145), cage effects in matrices probably prevent the occurrence of many reactions which readily take place in solution. This explanation appears reasonable for the behavior of carbonyl complexes which also contain bulky ligands, for example, bis(triisopropylphosphine)tetracarbonyltungsten (144). This complex, which undergoes photosubstitution of a phosphine ligand by CO in a CO saturated solution, does not undergo metal-phosphine bond cleavage in low-temperature matrices.

The photochemical behavior of tricarbonylcyclobutadiene-iron (TCI) has been investigated in perfluoropropane and neopentane matrices at 77°K (146) and in rare gas matrices at 8°K (147). In the fluoro- and hydrocarbon matrices, irradiation of TCI ($\lambda > 220$ nm) was reported to result in the loss of cyclobutadiene. An intense ESR signal ($g \approx 2.0$) observed from the photolyzed samples was attributed to $Fe(CO)_3$ (146).

Irradiation ($\lambda > 280$ nm) of TCI in rare gas matrices produced different results (147). In this case IR spectral studies showed that the primary photoproducts are dicarbonylcyclobutadieneiron and CO, which recombine upon warming the matrix:

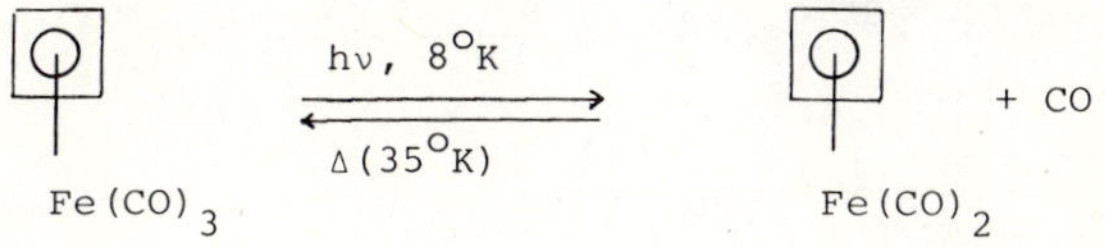

Photoisomerization of five-coordinate compounds has also been observed in low-temperature media. Photolysis of $Mo(CO)_5(PCx_3)$, (Cx = cyclohexyl) in a hydrocarbon glass at 77°K produces two isomers of $Mo(CO)_4(PCx_3)$ (145):

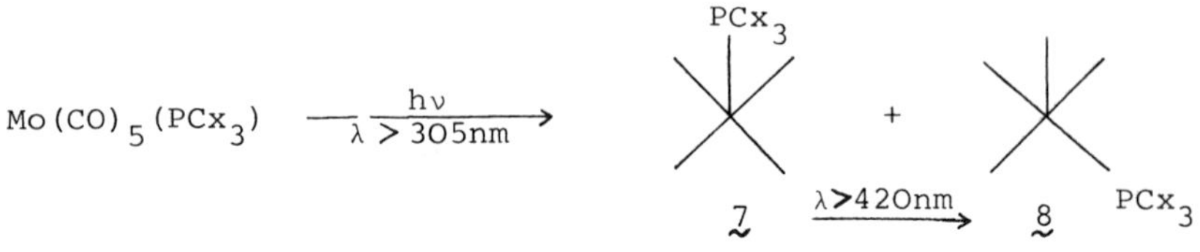

Isomer [7] is converted to isomer [8] upon photolysis (λ > 420 nm), but not when the glass is softened by warming.

Similarly structured isomers of $M(CO)_4(CS)$, (M = Cr, W) are formed by the photolysis of $M(CO)_5CS$ in argon and methane matrices at 20°K (148):

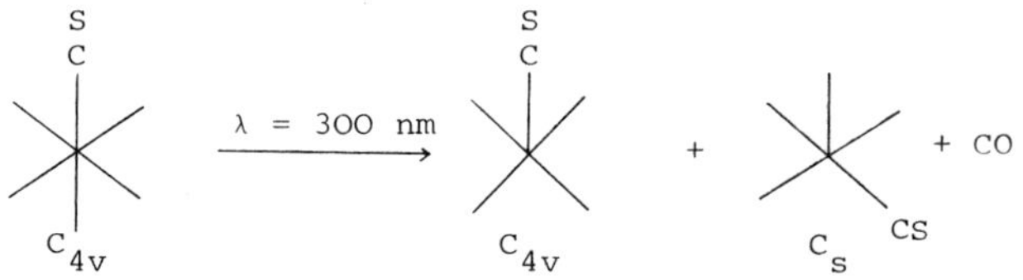

Both products show long-wavelength absorption maxima. Photolysis of either isomer results in isomerization as well as partial regeneration of $M(CO)_5CS$:

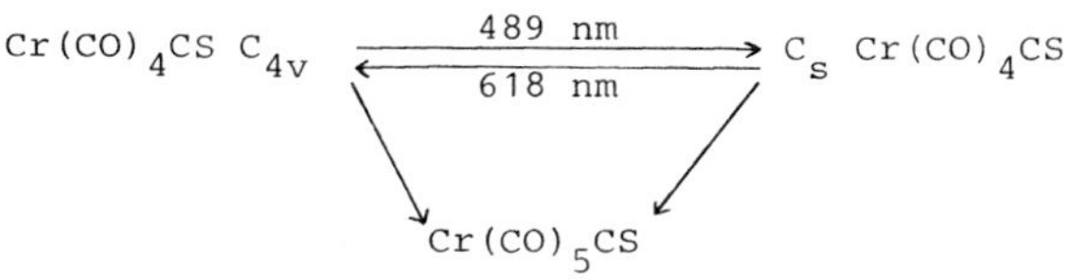

The photoisomerization of the five-coordinate species probably occurs via decay of D_{3h} (trigonal bipyramid) excited states (57,130).

Reversible ionic dissociation has also been observed in an argon matrix at 20°K (149,150):

$$(\pi\text{-}C_5H_5)NiNO \underset{(290<\lambda<350\ nm)}{\overset{(230<\lambda<280\ nm)}{\rightleftarrows}} [(\pi\text{-}C_5H_5)Ni^+,NO^-]$$

Photolyses of complexes isolated in matrices of ligating molecules have also produced some interesting results. For example, irradiation of $Fe(CO)_2(NO)_2$ in a CO matrix produces $Fe(CO)_5$, while in argon or methane the only observable process is loss of CO (142,149). Dinitrogen complexes have been generated by photolysis of both carbonyl and substituted carbonyl complexes in N_2 matrices. $Ni(CO)_3N_2$, for example, has been formed in this manner (140). Nitrogen complexes of iron are also formed by the photolysis of $Fe(CO)_5$ in a nitrogen matrix (137).

Photochemical studies of OTM compounds in low-temperature matrix environments have provided useful information concerning excited state decay properties. Many unstable primary products and intermediates have been characterized with this method.

Unfortunately, however, conclusions based upon matrix studies may not be directly extrapolated to the photochemical behavior of compounds in room temperature fluid media. While allowing the observation of certain photoproducts, the very low temperatures and rigidity of matrices may concurrently prevent the formation of others. The dissociation of bulky ligands discussed above is an illustrative case. Primary photoprocesses that require much thermal activation are also obviously impeded in matrix media.

The photobehavior of molecules having excited states strongly perturbed by spin-orbit coupling may show large variations at low temperatures due to changes in the Boltzman populations of the various component levels in the excited states of such species. The temperature dependence of the luminescent properties of several heavy metal complexes have been interpreted in this manner (Section III-C).

5. Photodissociation of CT Complexes. Solutions of either ferrocene (Cp_2Fe) or ruthenocene (Cp_2Ru) in halocarbon-containing solvents show ultraviolet absorption bands which have been attributed to charge-transer-to-solvent transitions (151-153). These are perhaps more realistically viewed as transitions of metallocene-solvent CT complexes:

$$Cp_2M\text{-}RX \xrightarrow{h\nu} (Cp_2M\text{-}RX)^*$$

Such complexes may either be thermodynamically stable species or "contact" complexes, which exist only during short-lived encounters of the components.

The primary photochemical reactions effected by irradiation into the CTTS bands of (i) ferrocene in ethanol/CCl_4 or ethanol/

$CHCl_3$ and (ii) ruthenocene in CCl_4 or 1,2-dichloroethane are in general the same (152,153):

$$Cp_2M\text{-}RX \xrightarrow[\text{CTTS}]{h\nu} Cp_2M^+ + X^- + R\cdot$$

Quantum yields for the primary reactions are in all cases $\Phi \approx 1$ with CTTS photoexcitation. Irradiation producing intramolecular metallocene excited states, however, is ineffective. Sensitization of the ferrocene reaction has also been investigated (Section III-D-1).

Photolyses of ferrocene and ruthenocene in noncomplexing solvents generally leads to no net photochemistry. Other metallocenes, however, particularly d^{17} and d^{19} species, are more photosensitive than the d^{18} analogs (154).

Finally, mechanisms of secondary thermal reactions involving the ferricenium produced in the primary photoreaction of ferrocene in ethanol/halocarbon solvents have also been investigated (155-157).

C. Luminescence

Our understanding of the nature and decay properties of excited state OTM complexes has been tremendously advanced by the discovery and investigation of a number of luminescent compounds. Luminescence allows spectroscopic characterization of compounds in electronically excited states and also, in favorable cases, provides a means of monitoring the behavior of excited state complexes under photochemically important conditions.

Two comprehensive reviews on the luminescence of transition metal complexes cover the literature through 1968 (32,158). The solution luminescence of metal complexes has also been reviewed (159). Finally, two recent articles in Accounts Chem. Res. (60,160) should provide the interested reader with detailed summaries of important contributions in this area.

1. Ruthenium(II) and Osmium(II) Compounds.

The first published report on the emission from a ruthenium(II) complex was in 1959 (161). In this paper the emission of $Ru(bipy)_3^{2+}$ at 77°K was reported to be "line-like" and to have a lifetime of less than 10^{-4}s. The radiative transition was suggested to occur from a singlet metal-to-ligand charge-transfer (MLCT) excited state. The emission from $Ru(bipy)_3^{2+}$, which is not "line-like" at all but spans about 6000 cm^{-1} and has a maximum intensity at $\lambda \approx 600$ nm (162), has been reassigned several times since the initial report by Paris and Brandt (161). Besides a charge transfer fluorescence (161,163), the emission was suggested to be a ligand field (d-d) fluorescence (162), and a ligand field phosphorescence (158).

G.A. Crosby and co-workers finally found that the energies of the emissions of several related complexes correlated with energies of the MLCT absorption bands and not with energies predicted for the LF transitions. They concluded that the emissions from all of these complexes (Table 7) occur from MLCT excited states (164-166). Recently, however, Houten and Watts found that the emission lifetime and quantum yield of $Ru(bipy)_3^{2+}$ are strongly affected by deuteration of the solvent while only slightly increased by deuteration of the bound bipyridines (167). They have thus proposed that the emitting excited state of $Ru(bipy)_3^{2+}$ has some CTTS character, which may be important in electron transfer quenching reactions of this complex (Section III-D-2).

The emission lifetimes of the bipy and phen complexes of ruthenium(II) at 77°K are generally in the range τ = 0.5-10 μs. (Table 7). Since these values are intermediate to those generally observed for the fluorescence and phosphorescence of organic compounds, the radiative transition in the ruthenium complexes was suggested to be a heavy-atom perturbed spin-forbidden process (168,169). From a determination of the absolute quantum yields as well as lifetimes of a series of ruthenium(II) and osmium(II) complexes, the associated radiative lifetimes were calculated (170). The variations in these inherent lifetimes within the series could be rationalized with a semi-emipirical spin-orbit coupling model thus affording further evidence that the radiative transitions are formally spin forbidden in these systems. It was also demonstrated that population of the emitting states from upper excited states generally occurs with unit efficiency in these compounds.

The influence of ligand substituents on the nature and decay properties of excited bipy and phen complexes was also studied (171,172). $Ru(diphenylbipy)_3^{2+}$ and $Ru(diphenylphen)_3^{2+}$ have optical properties differing somewhat from those of the parent unsubstituted complexes. The MLCT absorption bands of the substituted systems are more intense (ε_{max} values are about twice as large as those of the parent complexes), the emission quantum yields are considerably larger and the emission lifetimes at 77°K are shorter (Table 7). While the radiative rate constant is larger for the substituted complexes, the nonradiative decay constant is smaller.

The photophysical differences between the parent and substituted complexes were interpreted with a semi-empirical model for intensity borrowing via spin-orbit coupling (i.e., mixing of the

TABLE 7. Quantum yields and experimental lifetimes of emissions from ruthenium(II) and osmium(II) complexes.[a]

Compound	Φ	τ (μs)	Ref.
$[Rh(bipy)_3]Cl_2$	0.376	5.21	(170)
$[Rh(phen)_3]I_2$	0.584	9.79	(170)
$[Ru(tripy)_2]I_2$	0.479	10.66	(170)
$[Ru(bipy)_2(en)]I_2$	0.0222	0.96	(170)
cis-$[Ru(CN)_2(bipy)_2]$	0.269	3.96	(170)
$[Ru(bipy)_2(ox)]$	0.0124	0.61	(170)
cis-$[Ru(bipy)_2(py)_2]I_2$	-	5.3	(168)
cis-$[Ru(bipy)_2Cl_2]$	-	0.69	(168)
$[Ru(diphenylbipy)_3]Cl_2$	0.573	4.68	(171)
$[Ru(diphenylphen)_3]Cl_2$	0.682	9.58	(171)
$[Os(bipy)_3]I_2$	0.0348	0.89	(170)
$[Os(phen)_3]I_2$	0.126	2.43	(170)
$[Os(tripy)_2]I_2$	0.124	3.88	(170)

[a]All data measured from complexes in ethanol-methanol (4:1, v/v) at 77°K.

spin-allowed and spin-forbidden MLCT states) (171). According to this model, spin-orbit coupling effects are greater in the parent complexes presumably because their excited states are more localized on the metal center. The nonradiative decay constants correlate well with the magnitude of spin-orbit coupling. The radiative rate constants, however, are more dependent upon the intensity of the spin allowed MLCT transition.

A more detailed description of the nature and decay properties of the luminescent excited states of these compounds has come from investigations of the effects of temperature (2-77°K) and applied magnetic fields on the emission efficiencies, lifetimes and band shapes (173-176). As the temperature is lowered from 77°K to 2°K, the emission lifetime of $Ru(bipy)_3^{2+}$ in PMM increases monotonically from 10 μs to 190 μs, while the emission quantum yield decreases by about 50 percent (174). Such behavior is inconsistent with a simple one-level emitting state model having a temperature-independent radiative rate constant and a temperature dependent nonradiative rate constant. The results were interpreted instead with a group theoretical model for electronic coupling in charge transfer excited states. According to this model, the MLCT state of $Ru(bipy)_3^{2+}(D_3)$ is split into three components: A_2, E and A. Furthermore, the model assumes that (i) each component level has its own temperature-independent probabilities for radiative and nonradiative decay to the ground state; (ii) a Boltzman equilibrium always exists between the various levels; and (iii) the luminescing manifold is populated with 100 percent efficiency from upper excited states regardless of the temperature and the excitation wavelength. The component properties of the emitting level of $Ru(bipy)_3^{2+}$ which were derived from an analysis of the low temperature emission data are shown in Fig. 9 (174).

The temperature dependence (2-10°K) of the $Ru(bipy)_3^{2+}$ emission band shape was also investigated (176). The dramatic changes that occur between 5 and 1.65°K were attributed to the changes in the relative populations of the lowest two components of the emitting level (Fig. 9). Analysis of this effect with the Boltzman law yielded an energy gap between the two levels of 10.2 cm^{-1}. This is in excellent agreement with the value obtained from the temperature dependence of the emission decay properties.

Finally, magnetically induced circularly polarized emissions ("MICE") have been reported for several charge transfer emitters containing ruthenium(II) and analyzed in terms of Crosby's proposed model for MLCT excited states (177).

$Os(bipy)_3^{2+}$, $Os(phen)_3^{2+}$, and $Os(tripy)_2^{2+}$ display absorption and emission spectra which are only grossly similar to the spectra of analogous ruthenium(II) complexes. Besides intense spin-allowed MLCT absorption bands, whose maxima occur between 400 and 500 nm for both the ruthenium(II) and osmium(II) complexes, less intense ($\varepsilon \approx 4 \times 10^3$) but fairly structured (at 77°K) bands are displayed by these osmium(II) analogs at somewhat longer wavelengths (500-700 nm). While an unambiguous assignment of these bands is not readily available, they have been suggested to arise from the spin-orbit perturbed spin-forbidden MLCT transitions (170).

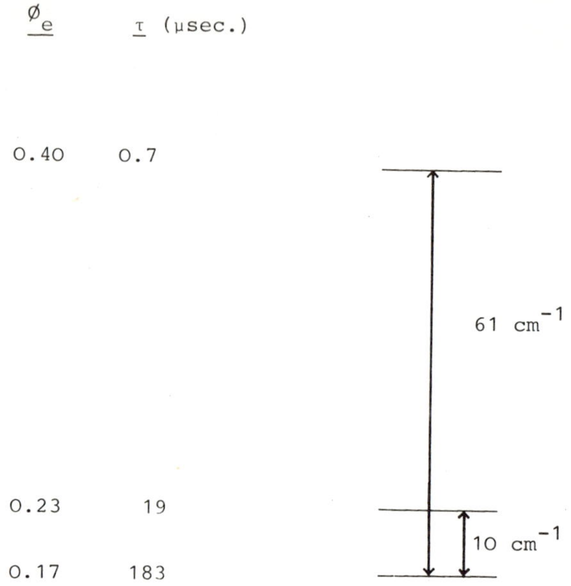

FIGURE 9. Emission quantum efficiencies (ϕ_e) and decay constants (ε) for the components of the emitting level of $Ru(bipy)_3^{2+}$ in PMM (174).

The emissions from the osmium(II) complexes of bipy, and so on, have in general received assignments analogous to those of ruthenium(II), that is, to formally spin-forbidden charge transfer ($\pi^* \rightarrow d$) transitions (79,163,170). A notable difference in the behavior of the two sets of compounds is that the lifetimes of the osmium(II) emitters are generally shorter than those of ruthenium(II) (Table 7). This difference is due to the greater affect of spin-orbital coupling on the radiative decay rates of the excited state osmium(II) species (170).

From recent studies of the temperature dependence (77-4°K) of the emission spectra, quantum yields and decay times for a series of osmium(II) complexes containing π-conjugated ligands, both major and minor splittings in the emitting levels have been inferred (178). The major splittings proposedly arise from spin-orbital coupling while the minor splittings (~100 cm^{-1}) are determined by coulombic interactions.

2. Rhodium(III) and Iridium(III) Complexes. In contrast to the bipy and phen containing complexes of ruthenium(II), which invariably show a CT ($\pi_L^* \rightarrow d$) type emission, the rhodium(III) analogs display either a metal localized ($d^* \rightarrow d$) or ligand localized ($\pi_L^* \rightarrow \pi_L$) luminescence (64,179-182). The _tris_ complexes, for example, $Rh(bipy)_3^{3+}$ and $Rh(phen)_3^{3+}$, have emissions at 77°K in glassy solvents which are very similar to the phosphorescent emissions of the respective free ligands under the same conditions (64,179-181). The complex emissions have a maximum intensity at $\lambda \approx 480$ nm, are highly structured and have wavelength independent quantum yields at 77°K (Table 8) (64,179). Since the long-wavelength absorption spectra as well as the emission spectra of the complexes are so similar to those of the respective free ligands, the emissions were assigned to a ligand localized ($\pi_L^* \rightarrow \pi_L$) transition (64,179-181). The lifetimes of the emissions are much shorter than those of the ligand phosphorescences ($\tau_L \sim 1$ s). This difference is a manifestation of enhanced spin-orbit coupling in the IL excited states brought about by the presence of the heavy metal center (179,180). Thus the complex emissions are considered to be metal perturbed but ligand localized phosphorescences (79,179,180). The nonexponential decay of the emissions from $Rh(bipy)_3^{3+}$ and $Rh(phen)_3^{3+}$ at low temperatures (T = 7-9°K) is also a result of the significant splitting in the luminescent excited level brought about by spin-orbit coupling (183).

The mixed ligand complexes $Rh(bipy)_2(phen)^{3+}$ and $Rh(phen)_2(bipy)^{3+}$ exhibit nonexponential emission decay at 77°K (184) The emissions of both compounds are composed of two components having lifetimes similar to those of the parent _tris_ compounds. The emission quantum yields are strongly wavelength dependent but are always within the range of the quantum yields of $Rh(bipy)_3^{3+}$ and $Rh(phen)_3^{3+}$ (Table 8). The multiple emission properties of these complexes have been interpreted in terms of exciton states which are localized on the individual ligand molecules.

Other rhodium(III) complexes which exhibit IL phosphorescence include some β-diketone complexes, for example, RhL_3 (L = aca, tfaca, and hfaca) (185).

Another series of luminescent rhodium(III) complexes which have been investigated are _cis_-$Rh(bipy)_2X_2^+$ and _cis_-$Rh(phen)_2X_2^+$ where X = Cl^-, Br^- and I^-. Instead of the highly structured, relatively long-lived "ligand type" emission shown by the parent _tris_ compounds, these systems have a broad, structureless, and short-lived emission (Table 8) which is located in the near IR($\lambda_{max} \sim 700$-800 nm) (64,179-181).

TABLE 8. Emission properties of some rhodium(III) complexes at 77°K.

Complex	λ_{ex}(nm)[g]	Φ_e	τ_e	Conditions	Ref.
$[Rh(phen)_3]Cl_3$	365	0.80		a	(184)
	313	0.90	47.2 ms	b	(64)
	265	~1.0		b	
$[Rh(bipy)_3]Cl_3$	313	0.29	2.29 ms	b	(64)
	265	~0.33		b	
cis-$[Rh(phen)_2Cl_2]Cl$	405	0.030	46.9 μs	b	(64)
	313	0.015		b	
	313	0.032		a	
	366	0.0417	33.9 μs	c	(182)
cis-$[Rh(bipy)_2Cl_2]Cl$	405	0.018	47.3 μs	b	(64)
	313	0.013		b	
	313	0.020		a	
	366	0.0369	31.8 μs	c	(182)
cis-$[Rh(bipy)_2Br_2]NO_3$	366	0.177	18.0 μs	c	(182)
cis-$[Rh(phen)_2Br_2]NO_3$	366	0.194	16.0 μs	c	(182)
cis-$[Rh(bipy)_2I_2]I$	366	~0.55		c	(182)
	-	-	9.7 μs	e	(179)
cis$[Rh(phen)_2I_2]I$	366	~0.46		c	(182)
	-	-	10.8 μs	e	(179)

TABLE 8. (continued)

Complex	λ_{ex}(nm)[g]	Φ_e	τ_e	Conditions	Ref.
trans-$[Rh(py)_4Br_2]Br$	436	0.0806	499 μs	c	(182)
	-	-	~500 μs	d	(179)
trans-$[Rh(py)_4Cl_2]Cl$	405	0.052	51.9 μs	b	(64)
	313	0.025			
	-	-	~704 μs	d	(179)
$[Rh(bipy)_2(phen)]Cl_3$	365	0.57		a	(184)
	313	0.32	37 ms	f	
	297	0.30			
	254	0.56	2.3 ms	f	
$[Rh(bipy)(phen)_2]Cl_3$	365	0.72		a	(184)
	313	0.66	36 ms	f	
	297	0.75			
	254	0.60	2.3 ms	f	

[a] ETOH-MeOH (5:1).

[b] Ethyleneglycol-water (1:1).

[c] ETOH-MeOH (4:1, v/v).

[d] Solid.

[e] Water-methanol (1:4, v/v).

[f] Ethanol.

[g] Wavelength of exciting light.

The emissions in these cases have been assigned to spin-forbidden ligand field transitions ($d^* \rightarrow d$) (179-181). The short lifetimes (relative to the ligand localized emitters) are due to the stronger perturbing effect of the metal in such transitions where metal orbitals are directly involved (180, 182).

Changing the halide substituents from chloro to bromo to iodo causes a red shift in the emission band maximum, an increase in the emission quantum yield and a decrease in the lifetime (Table 8) (179,181,182). The spectral shifts were explained on the basis of changes in the average ligand field strength (179). The quantum yield and lifetime effects are due to changes in the radiative rate constant, which increases with the weight of the coordinated halogen. No effect was shown by the nonradiative rate constants, which for all six complexes are similar in magnitude (182).

The mechanism and efficiency of relaxation from upper excited states to the emitting levels in the $Rh(phen)_2X_2^+$ and $Rh(bipy)_2X_2^+$ complexes have also been investigated (Section III-A-2).

The D_{4h} complexes, trans-$Rh(py)_4X_2^+$ ($X = Cl^-$, Br^-), also display ligand field luminescence in glassy solvents at 77°K (64,179,182). The emission spectra of these systems are similar to those of the C_{2v} complexes cis-$Rh(bipy)_2X_2^+$ (broad and structureless), and their quantum efficiencies are low (Table 8) (64,179,180,182). While the emission quantum yield of the bromo compound is independent of excitation wavelength ($350 < \lambda < 500$ nm) (182), that of the chloro complex is evidently very sensitive (64). The longer lifetimes of these compounds compared to the C_{2v} systems are due to the orbital forbiddeness of the $d^* \rightarrow d$ transition in these cases (179).

Many other rhodium(III) complexes exhibit ligand field phosphorescence, including a number of amine complexes (186-188); $RhCl_4L^{1-}$ (L = phen and substituted phen) (189); $Rh(dtp)_3$ and $Rh(dmtc)_3$ (190); $Rh(dppe)_2Cl_2^+$ and $Rh(dpae)_2Cl_2^+$ (191); and $Rh(py)_2Cl_4^-$ (163).

The following types of iridium(III) complexes have been found to luminesce (L,L' = phen, bipy, and related ligands):

1. IrL_3^{3+} (192,193)
2. $IrCl_2L_2^+$ (64,170-172,180,194,195)
3. $IrCl_4L^{1-}$ (189,195)
4. $Ir(dtp)_3$, $Ir(dmtc)_3$ (190)
5. $Ir(dppe)_2Cl_2^+$, $Ir(dpae)_2Cl_2^+$ (191)

6. $IrCl_2LL'^+$ (196)

7. $IrCl_4(py)_2^-$ (163)

Several early reports concerned the emissions from $Ir(phen)_3^{3+}$ and $Ir(bipy)_3^{3+}$ (166,181,197). However, it appears that the actual identities of the complexes studied in these cases were $IrCl_2(phen)_2^+$ and $IrCl_2(bipy)_2^+$ (32,171). More recently, the group 1 compounds have been synthesized and their emission properties and energy donor abilities have been examined (192,193).

The emission characteristics of $Ir(bipy)_3^{3+}$ are very similar to those displayed by $Rh(bipy)_3^{3+}$ (192). The band shapes and energies are almost identical. The highest energy emission band component of the iridium complex is at 22.3 kK, only 0.1 kK less than that of the rhodium compound. The lifetime of $Ir(bipy)_3^{3+}$ in an ethanol-methanol glass at 77°K (ca. 80 μs) is, however, considerably shorter than that of $Rh(bipy)_3^{3+}$ under the same conditions (τ = 2.2 ms). The emitting excited state of $Ir(bipy)_3^{3+}$ is proposed to have mainly intraligand character. The shorter lifetime compared to that of $Rh(bipy)_3^{3+}$ is consistent with the greater spin-orbit coupling expected for complexes of the heavier metal. The greater spin-orbit coupling is also manifested by the weak (ε < 100) absorption bands of $Ir(bipy)_3^{3+}$ at low energies which are probably associated with transitions to the triplet luminescent state. The luminescence and energy donor properties of $Ir(bipy)_3^{3+}$ and $Ir(phen)_3^{3+}$ in methanol solutions at 21° have also been determined (Section III-D-1).

Emission data for the group 2 complexes, cis-$IrCl_2L_2^+$, are given in Table 9. Assignments of the luminescence from the parent compounds (L = bipy and phen) have been controversial. Crosby and co-workers (170,171,195) proposed that the radiative transitions in these cases are of the CT($\pi_L^* \rightarrow$ d) type. DeArmond et al. (60,64,180) have argued that the metal and ligand orbitals are strongly mixed in these compounds; accordingly, terms such as MLCT, which imply that the metal and ligand orbitals are localized, are inappropriate for these transitions. They suggest instead that the transitions are labeled "delocalized molecular" $(d - \pi^*) \rightarrow (d - \pi)$.

Recent investigations of ligand substituent effects on the luminescence properties of these complexes have resulted in rather detailed assignments of the emitting excited states (Table 9). According to Crosby et al., the changes in the excited state decay properties are effected by two mechanisms: (i) changes in the degree of intensity borrowing via spin-orbit coupling (i.e., mixing of the spin-allowed and spin-forbidden MLCT states) (171); (ii) changes in the extent of configuration interaction between the MLCT and IL excited states (172,194).

TABLE 9. Emission data for $IrCl_2L_2^+$ in ethanol-methanol (4:1, v/v) at 77°K (171,194).

L	E_e(kK)[a]	Φ_e	τ_e(μs)	k_r(μs^{-1})	k_{nr}(μs^{-1})
bipy	21.1	0.516	5.94	0.0869	0.0815
4,4'-dimethylbipy	21.1	0.534	5.91	0.090	0.0789
4,4'-diphenylbipy	20.4	0.881	5.03	0.175	0.0237
phen	21.1	0.496	6.92	0.0714	0.0731
4,7-dimethylphen	21.1	0.617	21.7	0.0284	0.0176
5,6-dimethylphen	20.5	0.770	66.3	0.0116	0.0034
4,7-diphenylphen	19.5	0.829	15.0	0.0556	0.0114

[a] Energy of the shortest wavelength maximum in the vibrational progression of the emission band

The emitting states in the bipy complexes are estimated to be energetically well separated from the corresponding triplet IL excited states. Thus essentially no contributions from IL states are expected in the emitting levels of these systems. The emissions in all cases are assigned to CT ($\pi_L^* \rightarrow d$) transitions. The variations in the decay constants among the bipy compounds have been interpreted with the semi-empirical model for intensity borrowing via spin-orbit coupling (170-172,194).

The changes in the physical decay properties of the phen complex which are brought about by ligand substituents are rather different (Table 9), and are inconsistent with predictions based on the spin-orbit coupling mechanism (171). The proximity of the emission bands of the substituted phen complexes to the respective free ligand phosphorescence bands, and the relatively long emission lifetimes of these complexes led Crosby et al. to propose that the emitting levels have contributions from both IL and MLCT excited states (171,172).

The parent complex, cis-$IrCl_2(phen)_2^+$, has an essentially pure MLCT emitting level. The decay constants of this state are similar to those of the MLCT emitting states of the bipy series. The luminescent state of $IrCl_2(5,6\text{-dimethylphen})_2^+$, however, has almost complete IL character. As expected for ligand localized transitions, the radiative and nonradiative rate constants of this state are much smaller than those of the MLCT emitting states in the bipy system (Table 9). The emitting levels of the other substituted phen complexes are proposed to have fairly equal contributions from the zero-order IL and MLCT states (194).

Strong evidence for these interpretations was obtained from a study of the solvent dependence of the emission properties (194,198). Here it was found that the emission lifetimes and, except in the case of $IrCl_2(5,6\text{-dimethylphen})_2^+$, energies are very sensitive to solvent polarity. Increases in the energy of an emission band were in all cases accompanied by increases in the lifetime of the emission. To account for these results it was suggested that the shorter-lived zero-order MLCT state energy is solvent dependent whereas that of the longer-lived zero-order IL state is unaffected. As the solvent is made more polar, the energy gap between the zero-order states decreases and the degree of configuration interaction between them increases.

Although Crosby et al. (194) did not consider contributions to the emitting levels of these complexes from LF excited states, recent studies (199) of the photochemical and photophysical decay properties of $IrCl_2(phen)_2^+$ and $IrCl_2(5,6\text{-dimethylphen})_2^+$ in fluid media indicate that photochemically active LF excited states can be thermally populated from the emitting levels in these systems.

In addition to the problem of assigning unique configurations to the emitting states of these compounds, it has been suggested that assignment of distinct spin labels is also inappropriate in some cases (79). MLCT emitting states such as those of $IrCl_2L_2^+$ (L = bipy and phen) are proposed to be so strongly spin-orbit coupled that no spin label may reasonably be applied to them. The IL emitting state of the 5,6-dimethylphen derivative, however, may be described as a triplet. Finally, the intermediate cases, $IrCl_2L_2^+$ (L = 4,7-diphenylphen and 4,7-dimethylphen), have intermediate multiplicity character as well as intermediate orbital nature.

Dependence of the emission lifetimes of $IrCl_2L_2^+$ (L = 4,7-diphenylphen and 4,4'-diphenylbipy) in PMM on temperature (4.2-77°K) and on external magnetic field strength (0 and 13 kG) have also been investigated (173). In the absence of the external field, the experimental emission lifetimes at 4.2°K exceed the radiative lifetimes of these species at 77°K. This behavior was interpreted in terms of multicomponent emitting levels analogous to what has been proposed for the excited states of several ruthenium(II) complexes (vide supra). The emission lifetimes are somewhat diminished in the presence of the magnetic field probably because the energy of one or more of the component levels in the excited states are altered through a Zeeman interaction.

The analogous mixed ligand complex of group (6), $IrCl_2$(phen)(5,6-dimethylphen)$^+$, exhibits nonexponential emission decay at 77°K in an EtOH/MeOH glass (196). The emission could be separated into two components: the longer-lived, which was monitored at 485 nm, has a lifetime of $\tau \approx 70$ μs; the shorter-lived portion has a lifetime of $\tau \approx 10$μs measured at 560 nm. These lifetimes are similar in magnitude to those of the parent complexes, $IrCl_2$(5,6-dimethylphen)$_2^+$ and $IrCl_2$(phen)$_2^+$ (Table 9). Two non-thermally-equilibrated emitting states (IL and MLCT) are proposed to account for this behavior. The results are quite similar to what has been observed with heterobischelated rhodium(III) compounds (vide supra).

The emissions from the iridium(III) complexes in groups 3, 4, 5, and 7 have all been assigned to LF ($d^* \rightarrow d$) transitions. Although in some cases (163,190,191) the emitting levels have also been assigned a triplet multiplicity, the appropriateness of spin labels for LF excited states of such heavy metal complexes has been questioned (79). In all of these complexes the emission bands are broad, smooth, gaussian shaped and considerably red-shifted from the lowest energy absorption bands as expected for LF emissions (195). In general, the lifetimes are ca. 1 μs at 77°K. The emission lifetimes of $IrCl_4L^-$ vary from 0.094 to 2.7 μs at 77°K depending on substituents in L. However, the mechanism of this dependency is not certain (189).

3. Compounds Containing CO as a Ligand. Until recently, few CO-containing, luminescent transition metal compounds were known (1,32). The low emission yields in these compounds have been generally attributed to fast excited state ligand dissociation reactions as well as to radiationless decay. In the past few years, however, a large number of luminescent transition metal carbonyl compounds have been reported. These relatively highly luminescent systems contain a third transition series metal and/or have emitting states assigned to CT configurations.

a. $W(CO)_5X$ (X = n-donor). Several tungsten complexes of the general formula $W(CO)_5X$ (X = n-donors such as amines, alcohols and ethers) are reported to emit in methylcyclohexane at 77°K (200,201) (Table 10). The luminescence spectra in all cases appears as a broad structureless band with maximum intensity at approximately 530 nm. The lack of an observable emission in the analogous chromium and molybdenum compounds was correlated with their lack of an absorption band which is exhibited by the tungsten compounds. This low-intensity, longest wavelength band was assigned to the lowest spin-forbidden ligand field transition in the tungsten complexes. Overlap of the luminescence with this band suggested assignment of a spin-forbidden ligand field transition to the emission. This assignment was further substantiated by the emission lifetimes, which are considered typical for spin-forbidden emission of heavy metal complexes (32).

Since the radiative rate constants are calculated to be relatively constant, the data for the aliphatic amine complexes suggest that the nonradiative decay rates decrease as the number of hydrogen atoms bound to the nitrogen atom is decreased. This result is consistent with the theory of Robinson and Frosch (202), which predicts that internal conversion processes are facilitated by high-energy vibrational modes. However, the very similar lifetimes for all of the compounds of oxygen donors suggest that other factors may be involved. Information regarding excited state decay processes in $W(CO)_5X$ has also been obtained from studies of quenching reactions (Section III-D-1) and of the wavelength dependence of ligand substitution reactions (Section III-A).

b. $M(CO)_4X$ (M = Cr, Mo, W; X = bipy and related ligands. The luminescence of $M(CO)_4X$, where M = Cr, Mo and W, and X = bipy and related ligands has also been discussed. In one article,

TABLE 10. Emission lifetimes of $W(CO)_5X$ in methylcyclohexane at 77^oK (201).

X	τ(μs)
Ethylamine	0.92
t-Butylamine	1.2
Isopropylamine	0.65
Cyclohexylamine	1.1
Diethylamine	5.1
Dimethylamine	2.6
Piperidine	3.1
Diethylmethylamine	25.5
Triethylamine	9.7
Tri(η-butyl)amine	6.9
Dimethyl(1-phenyl)ethylamine	15.5
Ethanol	11.7
2-Propanol	6.6
t-Butylalcohol	6.3
Diethylether	7.1
Acetone	5.3
Acetone-d_6	5.7
Cyclohexanone	3.6

emission from the bipy and substituted bipy complexes of all three metals in EPA at 77°K were reported (203). Maximum intensities for the emissions of the bipy complexes occur between 10 and 15 kK. From an investigation of the effect of substituents in the bipyridine ligand on the spectral distribution of the emission, a tentative assignment of a triplet MLCT (L = bipy) excited state was made for the emitting level. Kaizu et al. (203) suggested that carbonyl complexes are in general nonluminescent because they have lowest lying ligand field excited states that decay via fast dissociative reactions. Complexes such as $M(CO)_4$(bipy), having MLCT (L ≠ CO) states of lowest energy, were predicted to be relatively photostable and luminescent.

Emission quantum yields and lifetimes were also reported for $Mo(CO)_4X$ and $W(CO)_4X$ (X = bipy and related ligands) in EPA at 77°K and as pure solids at room temperature (4). In contrast to the earlier results (203) on these systems no emission could be observed from the chromium analogs. The quantum yields (0.02 - 0.09) and lifetimes (~10 μs) are similar for both the tungsten and molybdenum complexes.

Luminescence was also observed from $W(CO)_4$(ethylenediammine) and cis-$W(CO)_4X_2$ where X is an aliphatic amine or pyridine (4). Assignments of ligand field spin-forbidden transitions were suggested for the emissions of these systems.

c. $ClRe(CO)_3X$. The compounds $ClRe(CO)_3X$, (X = phen and related ligands) provide more examples of luminescent CO complexes whose emitting levels have been assigned to the lowest MLCT (L ≠ 0) excited state (41). The photostability and luminescent nature of these complexes is in agreement with the projection of Kaizu et al. (203). Assignment of the radiative transitions was based on an assignment of the lowest energy absorption bands to MLCT (L = X) transitions. Substantial triplet character in the emitting states is suggested by the lifetimes, the lack of emission from analogous manganese complexes, and by sensitization experiments (Section III-D-1).

The most remarkable aspect of the $ClRe(CO)_3X$ systems is that their luminescence is fairly intense even in room temperature fluid solutions (Table 12). These are the first examples of metal carbonyl complexes known to luminesce under these conditions. Since such systems are extremely useful in the study of intermolecular as well as intramolecular relaxation of excited state complexes, their discovery is very significant. The similarly luminescent and photostable ionic complex, $Ru(bipy)_3^{2+}$, has been used extensively in such studies (204).

4. Metallocenes. In 1961 Scott and Becker (11) observed a long-lived (τ = 2 s) emission from a sample of ferrocene in EPA at 77°K. The structured emission band (15-21 kK) could be observed with excitation that produced upper excited states but not with irradiation into the lowest energy absorption band. Apparently the same emission was observed from ferrocene in low-temperature matrices (14). For example, in a methane matrix the maximum intensity of the emission band occurred at 20 kK. The emission lifetimes depended on the matrix material but were similar in magnitude to that observed earlier.

Aside from these observations, a number of negative results have been reported (13,205-208). For instance, Müller-Goldegg and Voitländer found that the weak emission observed from a ferrocene sample in ethanol was even weaker after the sample was further purified (207). Excitation of a single crystal of ferrocene with laser sources also produced no observable phosphorescence (208). With 355 nm excitation, however, four emission lines (402,418,547,574 nm), identified as fluorescence, were observed. The source of this emission is not yet certain (208). At this time the balance of experimental evidence certainly indicates that the long-lived emission previously observed from ferrocene samples was due to some impurity. These results are further substantiated by theoretical considerations (209).

Although the previously observed phosphorescence of ferrocene now appears to have been artifactual, a fairly strong luminescence has been observed from the analogous ruthenium(II) compound (53,210). The emission spectrum of ruthenocene, measured at low temperatures either from the pure solid or from glassy media, appears as a rather broad but highly structured band centered around a maximum at about 17 kK. The lifetime of the emission from the solid is strongly temperature dependent. From 90 to 32°K the lifetime increases fairly exponentially from 10 to 100 μs. At temperatures lower than this a limiting value of $\tau \sim$ 130 μs is reached and maintained from 16 to 4°K.

Different explanations of this temperature dependence have been proposed. Crosby et al. (53) showed that the behavior is consistent with substantial spin-orbit splitting of the emitting state, suggested to be the $^3E_1$ term in D_4 symmetry (assigned by Sohn et al. (9) to the lowest d,d level). The splittings between the four sublevels and their individual decay constants were obtained from a computer analysis of the experimentally determined lifetime vs. temperature plot (Fig. 10). Rapid Boltzmann relaxation between the levels compared with deactivation to the ground state produces the observed temperature effect.

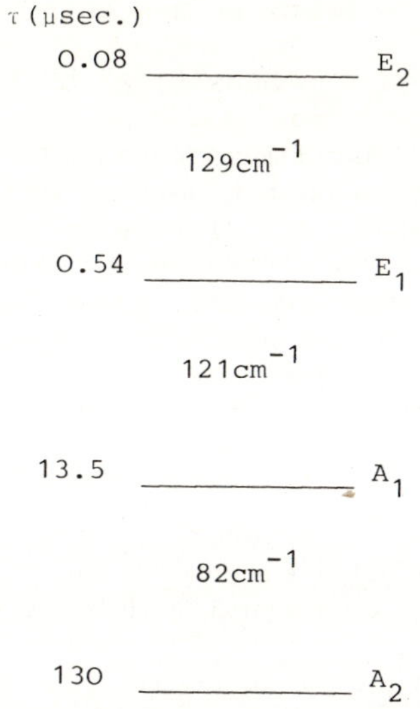

FIGURE 10. Calculated splittings and component decay constants of the luminescent excited state of ruthenocene (53).

Wrighton et al. (210) studied the temperature dependence of the emission quantum yield as well as of the lifetime. The relative quantum yield was found to vary in the same manner as the lifetime. This indicates that the ratio of the absolute quantum yield to the lifetime, Φ/τ, and thus the radiative decay rate of ruthenocene is essentially independent of temperature. The temperature effects on the lifetime and quantum yields, therefore, arise mainly from changes in the nonradiative decay rates.

The consistency of these results with the interpretation of Crosby et al. is at this point uncertain. It seems unlikely that the radiative decay constants of the spin-orbit components of the emitting level would be equal while the nonradiative rates show relatively large variations. However, since the changes in the nonradiative rates themselves are less than an order of magnitude (28-60°K) (210), the variations in both sets of decay constants in the model proposed by Crosby et al. would be predicted to be small. A thorough analysis of more accurate emission quantum yield and lifetime data in terms of the Crosby model should clarify this uncertainty.

At low temperatures photoluminescence has also been observed from the cyclopentadienides of rhodium(III) and iridium(III) (53). Nickelocene, however, is reported not to show any emission (11).

D. Quenching and Sensitization Processes

Sensitization and quenching processes are well-known tools for obtaining information on the natures and behaviors of excited state organic molecules (211,212). These techniques have also been applied in photochemical studies of coordination compounds (204). However, only very recently have low valent complexes been used very much in these studies. Information on the techniques (204,211,212) and complications (204,213) of photosensitization and quenching processes is available elsewhere.

1. Energy Transfer.

a. OTM complex donors. The generally high photoreactivity of OTM complexes has precluded extensive use of these compounds as sensitizers. Several exceptional complexes of ruthenium(II), in particular RuL_3^{2+} and <u>cis</u>-RuL_2CN_2 (L = phen and bipy), luminesce in room temperature fluid media, are photo-inert, and have been used in sensitization and quenching studies both as energy donors (204,214,215) and as electron donors (Section III-D-2).

A number of luminescent osmium(II) and iridium(III) complexes have also been successfully used as photosensitizers. The optical properties of these compounds are given in Table 11. In methanol solutions at room temperature, the osmium compounds are photostable but those of iridium tend to decompose upon photolysis. However, all six complexes efficiently sensitize the photooxidation of tetramethylethylene and trimethylethylene in oxygen saturated methanol solutions (193). The important primary step involved in these reactions is quenching of the excited complex by oxygen. The singlet excited oxygen produced in the quenching reaction then reacts with the organic substrates. Similar results have been obtained using ruthenium(II) complexes as sensitizers (214).

Other OTM compounds that are indicated by preliminary studies to be efficient triplet sensitizers include $ClRe(CO)_3L$ (Table 12). In dichloromethane at room temperature these complexes are strongly luminescent and photo-inert (41). Under the same conditions, the emissions are quenched by oxygen, anthracene, and <u>trans</u>-stilbene. The <u>trans</u> → <u>cis</u> isomerization of <u>trans</u>-stilbene is efficiently sensitized by several of these species. Quantitative sensitization experiments indicated that the luminescing excited states of $ClRe(CO)_3L$ are produced with 100 percent

TABLE 11. Optical properties of some OTM complex sensitizers in methanol at 21° (193).

Complex	E_o(kK)[a]	τ(μs)
$Os(bipy)_3^{2+}$	14.9	0.049
$Os(phen)_3^{2+}$	15.3	0.183
$Os(phen)_2(diphenylphen)^{2+}$	15.0	0.212
$Os(phen)_2[(SO_3Ph)_2phen]^{2+}$	14.8	0.093
$Ir(bipy)_3^{3+}$	22.7	2.4
$Ir(phen)_3^{3+}$	22.5	2.9

[a] Estimated excitation energy of thermally equilibrated emitting state.

efficiency by UV-VIS photoexcitation. The excitation energies of these species are estimated to be about 50 kcal/mol.

The luminescence of $W(CO)_5(py)$, which may be observed by photoexcitation of the complex in 1-pentene at 77°K, is completely quenched in a cis-1,3-pentadiene glass at the same temperature (49). Since the triplet energy of $W(CO)_5(py)$ is about 1 kcal/mol greater than the energy of the diene's lowest triplet, energy transfer is the suggested quenching mechanism. Interestingly, the photochemistry of $W(CO)_5(py)$ is not quenched in cis-1,3-pentadiene at room temperature (49). From these experiments, the lifetime of the reactive state in the room temperature solutions was estimated to be $\tau \leq 20$ ps. This is about five orders of magnitude smaller than the lifetime of the emitting state at 77°K in methylcyclohexane. The emission and reaction are proposed to occur from the same excited state. The large variations in the lifetime of this species were attributed to changes in the rates of ligand dissociation.

b. OTM complex acceptors. Of greater interest has been the use of OTM complexes as quenchers of excited molecules. Ferrocene, for example, was shown to be an efficient quencher of triplet excited state organic compounds in 1966 (216). Donors with triplet energies (E_T) between 66 and 42 kcal/mol were quenched at a nearly constant rate ($k_Q \approx 5 \times 10^9$ M^{-1} s^{-1}) in benzene

TABLE 12. Emission properties of $ClRe(CO)_3L$ in dichloromethane at 25° (41).

L	ν_{max}(kK)	τ(μs)	Φ
1,10-Phenanthroline	17.33	0.3	0.036
2,2'-Bipyridine		0.6	
5-Methyl-1,10-phenanthroline	17.01	≤ 0.65	0.030
4,7-Diphenyl-1,10-phenanthroline	17.24	0.4	
5-Chloro-1,10-phenanthroline	17.12	≤ 0.65	
5-bromo-1,10-phenanthroline	17.12	≤ 0.65	0.02

solutions at room temperature. As the quenching efficiency showed no dependence on the triplet energy of the donor, however, non-energy transfer mechanisms could not be excluded; thus the actual mode of quenching was undetermined.

KiKuchi et al. (217) extended the earlier work by studying the ferrocene quenching of several triplet donors with $E_T < 42$ kcal/mol. Although the quenching rate constants in this range are less than the diffusion limited value, they do not correlate with the donor excitation energies as expected for classical energy transfer (218). A very rough correlation between the quenching rate constants and calculated free energy changes for redox quenching suggested the involvement of such a process. Since flash studies provided no evidence for this mechanism, it was concluded that the products of the quenching reactions involving the less energetic sensitizers are very short-lived CT complexes.

The quenching of low-energy triplet sensitizers by ferrocene was more extensively investigated by Herkstroeder (55). The dependence of the quenching rate constant on the excitation energy of the donor was determined for a series of donors with triplet energies between 43 and 23 kcal/mol. The quenching rate constants correlated with the triplet energy of the donor in this range, however, not in the manner predicted for endothermic energy transfer which is given below (211):

$$\frac{\Delta(\log k_Q)}{\Delta E_T} = \frac{-1}{2.303\ RT}$$

In agreement with the results of KiKuchi et al. (217), the measured quenching rate constants were much greater than those predicted by this equation.

Activation energies were also measured by Herkstroeder for some of the quenching reactions. For anthrathrene (E_T = 33.8 kcal/mol) and tetracene (E_T = 29.3 kcal/mol) the activation energies are only approximately 1 kcal/mol and approximately 3 kcal/mol, respectively. These results were interpreted in terms of a nonvertical energy transfer quenching process. In this mechanism the geometry of the sensitized acceptor (excited ferrocene) is thought to be distorted from that of the corresponding ground state molecule (211). Large distortions in thermally equilibrated excited state metallocenes are in fact suggested by the large Stoke's shift observed in ruthenocene (Section III-C-4).

Triplet-triplet energy transfer was also proposed for the quenching of the triplet donors benzil and $Ru(bipy)_3^{2+}$ by d^6 metallocenes of iron, ruthenium, osmium, and cobalt (210). The quenching efficiencies did not correlate with the atomic weight of the central metal but instead with the energy of the longest wavelength absorption band of the metallocene.

The quenching of triplet fluorenone by a series of organometallic complexes including ferrocene has been investigated (219). The quenchers which were studied and the corresponding rate constants are given in Table 13. The quenching rate constants did not correlate with the ionization potentials of the quenchers. Flash photolysis of fluorenone with 10^{-3}M dibenzenchromium in cyclohexane and with 10^{-3}M ferrocene in ethanol or acetonitrile produced no further transient absorbance after the fluorenone triplet was quenched. Thus quenching by reversible electron transfer was suggested to be unimportant in these systems. The quenching was accompanied in some cases by sensitized reactions of the quencher molecules. It was thus concluded that the quenching reactions involve triplet-triplet energy transfer. Since the triplet energy of fluorenone is 53.3 kcal/mol (213), all of the quenchers studied should have excited triplet states lying lower in energy than this value.

The *cis* → *trans* isomerization of *cis*-styrylferrocene has been sensitized by benzophenone (220). The sensitized quantum yield was found to be the same as that obtained with direct excitation at 313 nm ($\Phi_{c \to t}$ = 0.005). The direct quantum yield for 366 nm irradiation, however, was only $\Phi_{c \to t}$ = 0.001. The isomerization was proposed to occur from an upper triplet of *cis*-styrylferrocene. It was suggested that while high-energy excitation either from benzophenone sensitization or from direct irradiation produced this upper triplet with unit efficiency, the yield of this state was less efficient with the lower-energy

TABLE 13. Rate constants for the quenching of triplet fluorenone by OTM compounds (Q) (219).

Q	$k_q \times 10^{-9}$ (M^{-1} s^{-1})
Dibenzenechromium	6.5
Ferrocene	5.5
Ferrocene dimethyldicarboxylate	5.3
Butadiene tricarbonyl iron	3.5
Bis(butadiene)monocarbonyl iron	4.5
Dimethylfumarate tetracarbonyl iron	3.2
Chalcone tricarbonyl iron	7.5

direct excitation. A high probability of internal conversion from the upper state to a lower unreactive triplet was the suggested explanation for the low quantum yields of isomerization.

The photooxidation of ferrocene in halocarbon containing solvents has been sensitized by napthalene (221). In 50 percent chloroform/ethanol the limiting quantum yield for the production of ferricenium ions is unity. As the Stern-Volmer constants for the reaction and for napthalene fluorescence quenching are almost equal ($K_{SV} \approx 200$ M^{-1}), a singlet-singlet energy transfer mechanism was proposed. In this case the quencher was suggested to be the ferrocene-chloroform complex. Charge transfer excited states of the complex were proposed to be the reactive species produced in the energy transfer process.

In contrast to what is usually observed in organic photochemistry, exothermic energy transfer involving transition metal complexes does not always occur at diffusion controlled rates (204). Several investigations employing well-characterized aromatic triplet donors and organometallic acceptors have been aimed at elucidating the factors that affect the efficiencies

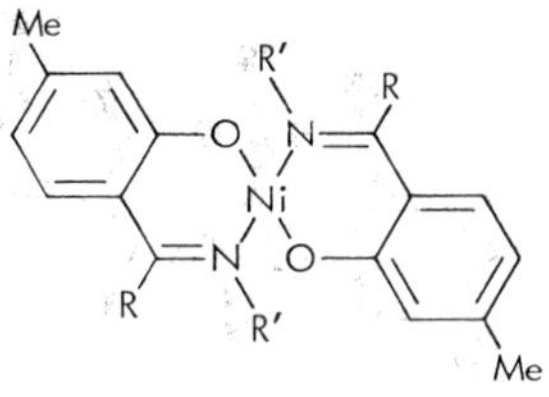

9

	R	R'
a)	Me	OH
b)	$C_{11}H_{23}$	OH
c)	Me	Bu
d)	$C_{11}H_{23}$	H
e)	Me	$(CH_2)_2OH$

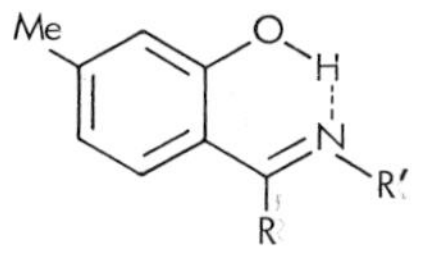

10

	R	R′
a)	Me	OH
b)	$C_{11}H_{23}$	OH
c)	Me	OBu
d)	Me	Bu

of such processes (222-224). In one such study, the quenching of organic triplet sensitizers by some planar nickel(II) complexes, [9], and by the related protonated ligands, [10], was investigated (223).

The quenching efficiencies of the free ligands fall off rapidly as the excitation energy of the donor (E_T) decreases from 70 to 60 kcal/mol. The quenching rate constants of the nickel(II) complexes show a slight decrease in this region but are then fairly constant ($k_q \approx 3 \times 10^9 M^{-1} s^{-1}$) down to E_T = 33 kcal/mol. The quenching of sensitizers with $E_T < 33$ kcal/mol was not investigated.

It was proposed that the high-energy sensitizers produce intraligand or CT excited states in the chelates, while the low-energy donors produce lower lying LF states. The relatively high-energy transfer efficiencies for producing ligand field states are suggested to arise from the high accessibility of d-orbitals in these planar complexes.

Supporting evidence for energy transfer quenching of the low-energy donors by the planar nickel(II) complexes was obtained by investigating the quenching ability of analogous palladium(II) compounds (222). In contrast to the nickel compounds, the quenching efficiency of the palladium(II) compounds decreased rapidly as the triplet energy of the donor became less than 50 kcal/mol. Since corresponding ligand field excited states are higher in energy in the palladium(II) compounds than in the nickel(II) analogs, the results are consistent with the proposed mechanism of energy transfer from the low energy donors to LF states of the chelates.

The importance of orbital overlap in energy transfer processes involving OTM compounds was further investigated with some paramagnetic tetrahedral and octahedral nickel(II) complexes [11], [12] (223). High energy sensitizers are quenched by these at near diffusion controlled rates while less energetic donors are quenched much less efficiently, even though all of the reactions studied are exothermic. As in the case of the

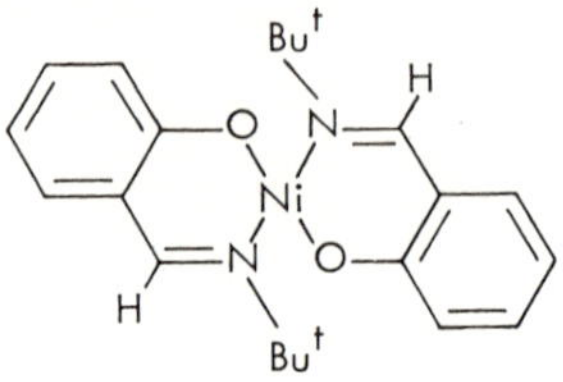

11

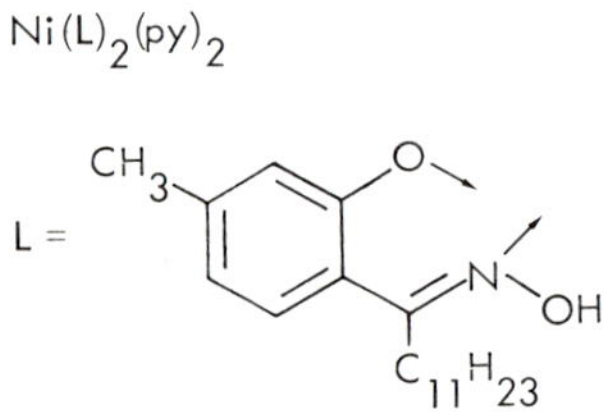

12

planar complexes, IL excited states are suggested to be produced with the high-energy donors and LF excited states with the lower-energy sensitizers. The fact that the tetrahedral and octahedral complexes quenched the low-energy donors much less efficiently than the planar complexes was taken as further support for the notion that favorable orbital overlap is required for efficient energy transfer.

The photosensitized conversion of $Cr(CO)_6$ to $Cr(CO)_5py$ has been investigated, but the mechanistic details of the reactions are not yet certain. In one experiment a two-step process was used to effect the net substitution (82). Benzophenone (E_T = 69 kcal/mol) (213) was photolyzed in the presence of $Cr(CO)_6$ in deaerated benzene to produce an intermediate complex, apparently $Cr(CO)_5$(benzophenone) rather than the proposed $Cr(CO)_5$ (Section III-B-1). The subsequent addition of pyridine to the irradiated solutions resulted in formation of $Cr(CO)_5py$. The limiting quantum yield based upon the amount of $Cr(CO)_5py$ produced was found to be $\Phi \approx 1$.

More recently, sensitization of the same net substitution process has been investigated using $W(CO)_6$ as well as $Cr(CO)_6$ (90). In these experiments various sensitizers were employed in cyclohexane solutions also containing pyridine. Benzophenone efficiently sensitized the formation of the $M(CO)_5py$ complexes, but with quantum yields somewhat lower than that previously found: M = Cr, $\Phi = 0.67$, M = W, $\Phi = 0.75$. However, when triphenylene (E_T = 66.5 kcal/mol) (225) was used as a sensitizer, the substituted product from $Cr(CO)_6$ was formed only very inefficiently. The factors, other than the energy requirement, that determine the efficiency of the overall reaction have not as yet been ascertained.

Sensitization and quenching of the intramolecular cis → trans isomerization of bis(glycinato)platinum(II) has been investigated (226). Pyrazine (E_T = 74.9 kcal/mol) and xanthone (E_T = 74.3 kcal/mol) are effective sensitizers, producing the limiting quantum yields $\Phi_{c \to t} = 0.03$ and $\Phi_{c \to t} = 0.1$, respectively. With the lower-energy sensitizers thioxanthone (E_T = 65.5 kcal/mol), quinoline (E_T = 62.9 kcal/mol), napthalene (E_T = 60.6 kcal/mol), and biacetyl (E_T = 54.9 kcal/mol), no sensitized isomerization was detected ($\Phi_{c \to t} \leq 10^{-3}$). As cis-bis(glycinato)platinum(II) did not quench the fluorescence of the relatively efficient sensitizer pyrazine, the sensitization occurs almost certainly via triplet-triplet energy transfer.

Quenching of the cis → trans isomerization obtained by direct photolysis of $Pt(gly)_2$ was efficiently effected by Ni^{2+}_{aq} (E_T = 24.3 kcal/mol). However, no quenching by Mn^{2+} (whose lowest excited state, which is a quartet, has an energy of 54 kcal/mol above the ground state) was observed. It was suggested that the excited state responsible for the isomerization is considerably distorted relative to the ground state, and that while the vertical excitation energy to this state is between 66 and 74 kcal/mol, the vertical deactivation energy is only 26 to 54 kcal/mol.

2. Chemical Quenching. Many photochemical reactions of organic compounds are known to occur via intermolecular interactions of the excited species with reactant molecules in solution. Common examples which may involve this mechanism include the photoreduction of ketones by hydrogen donor solvents and many photodimerizations (227). The scarcity of quenching reactions of metal complexes that have been attributed to chemical processes has already been discussed by Balzani et al. (204). With the increasing interest in the mechanistic aspects of OTM complex photochemistry, however, many more examples should be discovered.

An associative ligand photosubstitution reaction is also a chemical quenching process if the excited reactant complex is quenched in the association step:

$$ML_n + h\nu \rightarrow ML_n^*$$

$$ML_n^* + X \rightarrow ML_nX$$

$$ML_nX \rightarrow ML_{n-1}X + L$$

A part of the net photosubstitution reaction of $Mn(CO)_4NO$, in which CO is replaced by PPh_3, occurs via the scheme above and may thus be considered as a chemical quenching process (Section III-B-1).

The majority of the reported chemical quenching reactions of transition metal complexes involve electron transfer. The extent of such investigations, however, is small compared to the studies of electron transfer and charge transfer interactions of excited organic compounds (cf. 228-233).

Most of the work discussed here concerns reactions of the luminescent CT excited state of $Ru(bipy)_3^{2+}$, $Ru(bipy)_3^{2+*}$. The exceptional luminescence and photochemical stability of this species in ambient temperature fluid media have facilitated these investigations (204). The ability of $Ru(bipy)_3^{2+*}$ to act as a donor of triplet excitation energy is well documented (204, 234,235). Recent investigations have also confirmed the electron donating ability of this excited species.

It has been suggested that the quenching of $Ru(bipy)_3^{2+*}$ by some cobalt(III) complexes occurs by direct electron transfer (236-238):

$$Ru(bipy)_3^{2+*} + Co^{(III)}L_6 \rightarrow Co^{2+} + 6L + Ru(bipy)_3^{3+}$$

However, evidence for an energy transfer mechanism in these cases has also been presented (239,240). Recent investigations have demonstrated that $Ru(bipy)_3^{2+*}$ can be quenched in direct electron transfer reactions and is in fact a rather strong reductant.

Quenching of $Ru(bipy)_3^{2+*}$ by the organic cations <u>trans</u>-1,2-bis(N-methyl-4-pyridyl)ethylene^{2+} and 1,1'-dimethyl-4,4'-bipyridine^{2+} (paraquat^{2+}) and by the transition metal ion complexes $Fe(H_2O)_6^{3+}$ and $Ru(NH_3)_6^{3+}$ occurs at near diffusion controlled rates (241). In all four cases flash studies indicated that electron transfer is a principal component of the quenching mechanism. Selective excitation of $Ru(bipy)_3^{2+}$ in the presence of these quenchers (Q_{ox}) produced the oxidized ruthenium complex, $Ru(bipy)_3^{3+}$, and the reduced quencher, Q_{red}. The quenching reaction was followed by a rapid thermal process which regenerated $Ru(bipy_3{}^{2+}$ and the oxidized form of the quencher, Q_{ox}:

$$Ru(bipy)_3{}^{2+} + h\nu \rightarrow Ru(bipy)_3{}^{2+*}$$

$$Ru(bipy)_3{}^{2+*} + Q_{ox} \rightarrow Ru(bipy)_3{}^{3+} + Q_{red}$$

$$Ru(bipy)_3{}^{3+} + Q_{red} \rightarrow Ru(bipy)_3{}^{2+} + Q_{ox}$$

Other investigations have confirmed the proposed electron transfer nature of the quenching reaction involving $Ru(bipy)_3{}^{2+*}$ and $Fe(H_2O)_6{}^{3+}$ (238,242,243). From a kinetic analysis of quenching data involving measurements of the steady-state concentrations of $Ru(bipy)_3{}^{3+}$ and $Fe(H_2O)_6{}^{2+}$ produced under constant illumination, the fraction of total quenching events resulting in formation of these products was estimated to be 0.81 ± 0.16:

$$Ru(bipy)_3{}^{2+*} + Fe(H_2O)_6{}^{3+} \xrightarrow{81\%} Ru(bipy)_3{}^{3+} + Fe(H_2O)_6{}^{2+}$$

A model system, from which a means of converting light energy into storable chemical energy might ultimately be derived, has been based upon the highly efficient redox quenching reaction of $Ru(bipy)_3{}^{2+*}$ by paraquat^{2+}(P^{2+}) (244). In this case an electron donor (Red) (e.g., NPh_3 or $NHPh_2$) is involved in the photolyzed system and "captures" the oxidized ruthenium complex:

$$Ru(bipy)_3{}^{2+} \xrightarrow{h\nu} Ru(bipy)_3{}^{2+*}$$

$$Ru(bipy)_3{}^{2+*} + P^{2+} \longrightarrow Ru(bipy)_3{}^{3+} + P^{+}$$

$$Ru(bipy)_3{}^{3+} + Red \longrightarrow Ru(bipy)_3{}^{2+} + Ox$$

$$P^{+} + Ox \xrightarrow{k} P^{2+} + Red$$

Unfortunately, the back reaction between the reduced paraquat(P^+) and the oxidized amines (Ox) are very efficient ($k \approx 5 \times 10^9 M^{-1} s^{-1}$) and thus no energy is stored. Otherwise, the system is very similar to the basic processes which occur in photosynthesis (see ref. 3 for a recent review of related processes).

Quenching of $Ru(bipy)_3{}^{2+}$ by Tl^{3+} in aqueous solutions also occurs by direct electron transfer (242). Photolysis of $Ru(bipy)_3{}^{2+}$ in solutions containing Tl^{3+} produced $Ru(bipy)_3{}^{3+}$ with the limiting quantum yield $\Phi = 2.0 \pm 0.4$. Stern-Volmer constants for the quenching of the emission from $Ru(bipy)_3{}^{2+*}$ and for the production of $Ru(bipy)_3{}^{3+}$ were the same within experimental error. The intermediate Tl^{2+} initially produced in the quenching reaction was shown to oxidize $Ru(bipy)_3^{2+}$ thereby producing a quantum yield of $\Phi = 2$ for the net photooxidation.

A more detailed picture of electron transfer processes involving $Ru(bipy)_3^{2+*}$ was provided by a study of the quenching of this excited species by a series of nitrobenzene derivatives (245). The dependence of the quenching rate constant on the reduction potential of the quencher is shown in Fig. 11. The observed dependence is in excellent agreement with that expected for quenching which involves oxidation of the excited ruthenium complex.

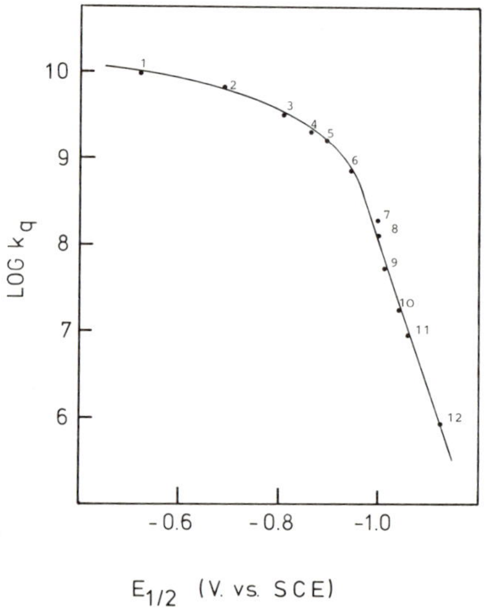

FIGURE 11. Quenching of $Ru(bipy)_3^{2+}$ by nitrobenzene derivatives in acetonitrile solutions. Dependence of the observed emission quenching rate constant, k_Q, on the reduction potential of the quencher (245). 1) *para*-Nitronitrosobenzene, 2) *para*-Dinitrobenzene, 3) *ortho*-Dinitrobenzene, 4) *para*-Nitrobenzaldehyde, 5) *meta*-Dinitrobenzene, 6) Methyl 4-Nitrobenzoate, 7) *cis*-4,4'-Dinitrostilbene, 8) 4,4'-Dinitrobiphenyl, 9) 3-Nitrobenzaldehyde, 10) Methyl 3-Nitrobenzoate, 11) 4-Chloronitrobenzene, 12) 4-Fluoronitrobenzene.

No transient free ion formation was observed by conventional flash photolysis of $Ru(bipy)_3^{2+}$ in acetonitrile solutions containing these quenchers. The mechanism proposed is analogous to the model used by Rehm and Weller for singlet exciplex formation (230,246) (Fig. 12).

The lack of significant dissociation of the ion-pair ($Ru(bipy)_3^{3+}\cdots Q^-$) was attributed to the combined effect of the electrostatic attraction between the partners in the ion-pair and a high rate constant for back electron transfer (k_{30}). From a kinetic analysis of the quenching data a value for the oxidation potential of $Ru(bipy)_3^{2+*}$, that is, for the reaction

$$Ru(bipy)_3^{2+*} \longrightarrow Ru(bipy)_3^{3+} + e^-$$

was estimated to be $E_{1/2} = +0.81 \pm 0.02$ V versus SCE.

Oxidation of photoexcited ferrocenes by N_2O has also been investigated (247). When ferrocene and some substituted ferrocenes were irradiated with 254 nm light in cyclohexane or ethanol solutions containing N_2O, evolution of nitrogen was observed. Irradiation at wavelengths longer than 313 nm had no effect. Plots of $1/\Phi N_2$ versus $1/[N_2O]$ were linear and in all cases the limiting quantum yield of N_2 production was significantly less than unity. Mechanisms involving photoejection of electrons from ferrocene were shown to be unlikely by flash studies. Thus, it was concluded that an excited state of ferrocene, other than that obtained initially by photolysis at 254 nm, is quenched by electron transfer to N_2O.

It should be noted that similar "wavelength effects" have been observed for the photooxidation of ferrocene in halocarbon-containing solvents (Section III-B-5). In these cases, however, the reaction is initiated by the excitation of ferrocene-solvent donor-acceptor complexes. As clear evidence against a similar mechanism was not presented for the ferrocene photooxidation involving N_2O, this appears to remain a possibility.

The limited results presently available indicate that intermolecular electron transfer can be an important deactivation process of excited metal complexes. However, the important factors which determine the probability of such reactions have not been extensively investigated. As is the case with ground state electron transfer (248), thermodynamic considerations alone cannot in general predict rates for these quenching processes (238,242, 245). A further complication with electron transfer quenching reactions is that in some cases the initial products can undergo a very fast thermal back reaction which regenerates the ground-state starting materials. These processes may then be undetectable by conventional flash photolysis.

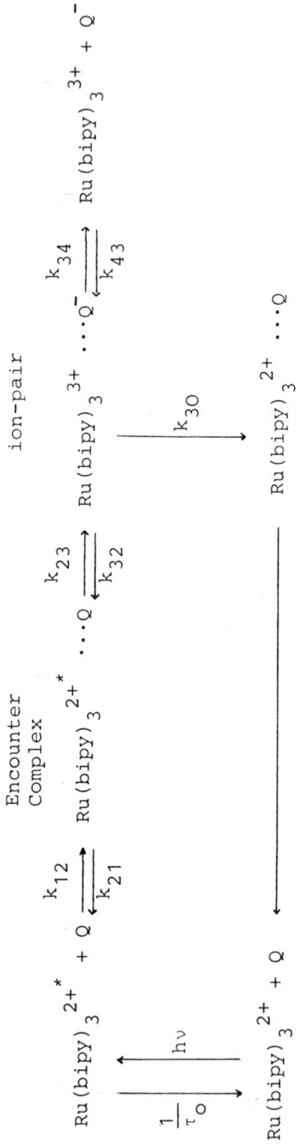

FIGURE 12. Kinetic scheme proposed for electron transfer quenching of $Ru(bipy)_3{}^{2+*}$ by some nitrobenzene derivatives (245).

3. Exciplex and Excimer Formation. The formation of excited complexes (excimers and exciplexes) in quenching reactions of excited organic compounds is a well-known phenomenon (cf. 70,230, 231,249). Evidence for the intermediacy of exciplexes is most readily obtained when exciplex emission is observed. However, other techniques, including flash absorption spectroscopy (250-252) and kinetic analysis of both steady-state and dynamic quenching rate data (253) have also been used to obtain evidence for the existence of these transients.

The instability of exciplexes has been attributed to charge transfer interactions in many cases (70), but other factors may also play important roles (252). Possible exciplex decay pathways other than emission include dissociation into the products of an electron transfer process (230,231), as well as various

$$D^* + A \rightleftharpoons (DA)^* \longrightarrow D^+ + A^-,$$

other physical and chemical decay modes (70).

Quenching of photoexcited cis-$IrCl_2(phen)_2^+$ by naphthalene and by 2-acetonapthone occurs by exciplex formation (254). The emission of the iridium complex is quenched by napthalene in deaerated DMF with $K_{SV} \approx 2M^{-1}$. Concurrent with the substrate emission quenching, a new emission band is produced which has characteristics typical of exciplex bands (broad and structureless and red shifted from the substrate emission). Since the emitting state in the iridium complex is characterized by increased electron density on the phenanthrolines (Section III-C-2), the exciplexes may have considerable charge transfer (bound phenanthroline → napthalene) character.

Both excimer and exciplex formation have been proposed for some quenching reactions involving triplet excited state metalloporphyrins (250-252,255). The concentration quenching effect observed for the triplet state lifetimes of zinc and platinum etioporphyrin and palladium mesoporphyrin in both degassed mineral oil and THF solutions was suggested to be due to the formation of triplet excimers (255). The quenching data are consistent with the following scheme in which S = the metalloporphyrin ground state, S^* = singlet excited metalloporphyrin, T^* = triplet excited metalloporphyrin, and $(ST)^*$ = triplet excimer:

$$S + h\nu \longrightarrow S^* \longrightarrow T^*$$

$$T^* + S \rightleftharpoons (ST)^*$$

$$T^* \longrightarrow S$$

Direct observation of the triplet excimers was prevented by their relatively low concentrations, which were estimated for the conditions employed to be only a small fraction of the concentration of unassociated excited triplet molecules.

Triplet exciplexes formed from the triplet excited states of zinc and magnesium etioporphyrin I (ZnEtio I and MgEtio I) and zinc octaethylporphyrin (ZnOEP) with acceptors such as nitroaromatics and organochloro compounds were directly observed by conventional flash spectroscopy (250-252). The nature and decay properties of the exciplexes are sensitive to quencher concentration and medium composition. For example, as the concentration of para-nitrotoluene (PNT) was increased in benzene solutions of ZnEtioI, the lifetime of the transient observed after flash excitation of the metal complex proportionately increased up to a maximum and then decreased (250). The first effect was attributed to the formation of long-lived exciplexes involving two PNT molecules as shown in the scheme below, where D = metalloporphyrin and A = PNT:

$$D^{3*} + A \longrightarrow (DA)^{3*}$$

$$(DA)^{3*} + A \rightleftharpoons (DAA)^{3*}$$

$$(DA)^{3*} \longrightarrow D + A$$

The latter was shown to be probably due to a medium effect on the exciplex lifetime. Addition of small amounts of polar solvents to the solutions produced similar results. With further increases in the concentrations of polar additives, the transient half-life increased to a constant value. The transitions in decay kinetics from first-order to second-order and in transient absorption spectra indicated that a new decay process of the exciplex was dominant:

$$(DA)^{3*} \longrightarrow D^{+} + A^{-}$$

Thus the exciplex was suggested to change from being a delocalized excited complex in the nonpolar medium to a system described as an ion-pair in polar solvents.

The limited extent of charge transfer interaction between the exciplex components in the nonpolar benzene solutions is indicated by plots of the logarithm of k_Q, the observed rate constant for the quenching of the metalloporphyrin triplet, versus the quencher electrochemical reduction potential. Although the slopes of such plots were linear for the nitroaromatics and for the chloro compounds, the slopes differed for the two groups of quenchers. In no case was the slope of the plot as steep as that predicted for purely charge transfer exciplexes (230,246) (Section III-D-2). The proposed nature of these exciplexes, suggested to be analogous to weak ground-state donor-acceptor systems, is such that the donor excited state is only slightly perturbed.

4. Heavy Atom Effects. Organometallic compounds have also been used to investigate "external-heavy-atom" quenching. The deactivation of anthracene singlets and triplets by $Hg(CH_3)_2$, $Pb(CH_3)_4$, and $Sn(CH_3)_4$ was studied in paraffin, ethanol, and benzene solutions (256). The singlet quenching by $Hg(CH_3)_2$ is essentially diffusion controlled in ethanol and benzene. Rate constants for the singlet quenching by $Pb(CH_3)_4$ and $Sn(CH_3)_4$ were one order of magnitude and two orders of magnitude less than the diffusion controlled limit, respectively. Quenching of anthracene fluorescence by these compounds in paraffin solutions was compared to enhancement of the yield of anthracene triplets. From these experiments it was concluded that all of the anthracene singlet quenching occurs by catalyzed intersystem crossing. Similar results were obtained using $HgPh_2$ as quencher in ethanol and toluene solutions (257).

The quenching of singlet excited molecular oxygen ($^1\Delta_g$) by ferrocene was also proposed to arise from the external heavy atom effect (258). The bimolecular rate constant for this quenching reaction is $k_Q = (9.0 \pm 2.0) \times 10^6\ M^{-1}s^{-1}$.

5. Intramolecular Interactions. An interesting case of intramolecular energy transfer is provided by the <u>cis</u> → <u>trans</u> photoinduced isomerization of some azastilbenes and related compounds coordinated to zinc and magnesium porphyrin complexes (259). Excitation that initially produces porphyrin localized excited states in the <u>cis</u>-1-(1-napthyl)-2-(4-pyridyl)-ethylene-zinc Etioporphyrin I complex (ZnEtioI(NPE)) results in efficient <u>cis</u> → <u>trans</u> isomerization of NPE, even though the energy of the lowest triplet state of <u>cis</u>-NPE is spectroscopically predicted to be approximately 15 kcal/mol greater than the lowest porphyrin localized triplet. The fact that the isomerization occurs at all is not the most remarkable aspect of the results. In degassed benzene solutions containing excess <u>cis</u>-NPE the quantum yield of total <u>cis</u> → <u>trans</u> isomerization is $\Phi = 6.6$. The following mechanism, involving a quantum chain process, has been proposed (where M = metalloporphyrin, C = <u>cis</u>-NPE, and t = <u>trans</u>-NPE).

$$M\text{-}C \xrightarrow{h\nu} {}^1M^*\text{-}C \longrightarrow {}^3M^*\text{-}C$$

$$^3\overset{*}{M}\text{-}C \longrightarrow M\text{-}{}^3C^*$$

$$M\text{-}^3C^* \longrightarrow {}^3M^*\text{-}t + {}^3M^*\text{-}C$$

$${}^3M^*\text{-}t + C \longrightarrow {}^3M^*\text{-}C + t$$

$${}^3M^*\text{-}t(C) \longrightarrow M\text{-}t(C)$$

The seemingly unfavorable energy migration from the porphyrin centered excited state to the coordinated ligand was interpreted as a "nonvertical" energy transfer process. Preferential quenching of the triplet cis-olefin to the trans-isomer and rapid exchange of the ligands during the lifetime of the triplet porphyrin produce the large quantum yields observed in some cases.

Intramolecular quenching and sensitization processes have also been investigated with nonconjugated covalently linked organometallic compounds such as the bichromophoric "double porphyrins" [13] (260). From a study of the emission characteristics of these double porphyrins the following conclusions were

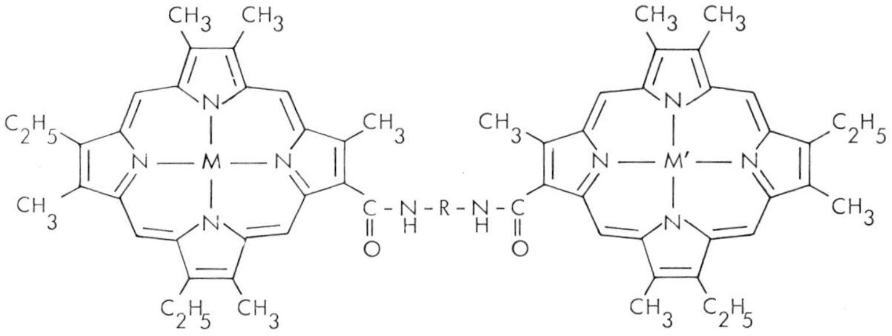

13

a) $M = Zn$, $M' = Cu$, $R = C_2H_4$

b) $M = Zn$, $M' = Cu$, $R = C_6H_4$

c) $M = Zn$, $M' = Co$, $R = C_2H_4$

d) $M = Zn$, $M' = Co$, $R = C_6H_4$

made concerning their excited state interactions:

1. Singlet-singlet energy transfer does not occur in any case.
2. The triplet level of the zinc moiety is quenched in the ethylene linked compounds but not in those linked by phenylene.
3. In [13a] the triplet excited zinc portion is only partially quenched while in [13c] it is essentially completed quenched.

The different quenching efficiences shown by the cobalt and copper units were suggested to arise from different mechanisms. In [13a], a triplet-triplet exchange interaction requiring orbital overlap is the proposed quenching process. The very efficient quenching observed in [13c] occurs via formation of a charge transfer intermediate, in which the zinc moiety is reduced by the cobalt system. The lifetime of this intermediate is estimated to be $10^{-9} < \tau < 10^{-5}$s.

IV. CLOSING REMARKS

Synthetic applications of the photosensitivity of OTM compounds have fluorished. Until recently, however, investigation and interpretation of the primary photobehavior exhibited by these species have been rather limited. The slow pace of progress in this area has been due, at least partially, to the generally complex electronic structure and spectra of OTM compounds. For example, applications of flash photolysis techniques and correlations of spectroscopic properties with photoreactivity have been limited by the scarcity of complexes which have well resolved, well assigned optical absorption spectra. Because of the relatively small percentage of OTM compounds that are known to luminesce, spectroscopic studies of excited state species, as well as investigations of quenching and sensitization processes have been hindered also.

Economics as well has probably played a role in creating the time-lag between the synthetic work and the investigation of primary processes. While the discovery of novel and efficient photosyntheses may have almost immediate practical or industrial utility, investigations of the latter sort are comparatively long-term investments.

Future progress may, however, be accelerated by recent and forthcoming developments. Recent flash photolysis studies of metal-carbonyls and metal-metal bonded systems should provide a foundation for more extensive development of this area. Applications of picosecond and nanosecond flash systems to a wide variety of problems is expected. Such techniques should be particularly useful for investigation of the photodynamics of nonluminescent complexes (261). The development of IR detection systems for use in conjunction with flash apparatus would facilitate structural characterization of primary photoproducts and other short-lived reaction intermediates (262). The knowledge gained from investigations of the available luminescent complexes may be systematically applied to the development of more luminescent and photoinert species which then could be used in quenching and sensitization studies. Further application of matrix isolation techniques should also provide much information regarding the primary photoprocesses of OTM compounds.

ACKNOWLEDGMENTS

C.R.B. gratefully acknowledges the inspiration and encouragement which he received from the late Dr. Ernst A. Koerner von Gustorf, whose further engagement with this work was precluded by his own sudden and tragic death. All errors, omissions, and so on, are therefore the full responsibility of C.R.B.

C.R.B. is grateful to the Max-Planck-Gesellschaft for financial support and particularly to Vincenzo Balzani for reading the manuscript and offering critical suggestions. C.R.B. thanks Margaret Y. Hancox, friend and typist, for her patience and efficient work; his wife, Alice, for helping in many ways; Pierre Kirsch for collecting some of the literature; Frau Ilse Schneider for preparing illustrations; and Robert H. Grubbs, Hans-Werner Frühauf, Andrew Wojcicki, Richard Blackborrow, and Roland Severson for proofreading and making helpful suggestions.

Abbreviations

aca	acetylacetonate
bipy	2,2'-bipyridine
Cp	η^5-cyclopentadienyl
cx	cyclohexyl
diphenylbipy	4,4'-diphenyl-2,2'-bipyridine
diphenylphen	4,7-diphenyl-1,10-phenanthroline
DMF	dimethylformamide
DMSO	dimethylsulfoxide
dmtc	dimethyldithiocarbamate
dpae	1,2-bis(diphenylarsine)ethane
dppe	1,2-bis(diphenylphosphino)ethane
dtp	diethyldithiophosphate
en	ethylenediamine
Et	ethyl
gly	glycinate
hfaca	hexafluoroacetylacetonate
Me	methyl
ox	oxalate
Ph	phenyl
phen	1,10-phenanthroline
PMM	poly(methyl methacrylate)
py	pyridine
pyz	pyrazine
$(SO_3Ph)_2$phen	disulfonated diphenylphen
tfaca	trifluoroacetylacetonate
THF	tetrahydrofuran
tripy	2,2',2"-tripyridine

REFERENCES

(1) E. Koerner von Gustorf and F.-W. Grevels, Fortschr. Chem. Forsch. (Topics in Current Chemistry), 13, 366 (1969).

(2) G. B. Porter, V. Balzani, and L. Moggi, in "Advances in Photochemistry", (J.N. Pitts, Jr., G.S. Hammond, and K. Gollnick, Eds.), Vol. 9, Interscience, New York, 1974, p. 147.

(3) L.H.G. Leenders, Ingrid Fischler, R. N. Perutz, and E. A. Koerner von Gustorf, in "Advances in Inorganic Chemistry and Radiochemistry", (H.J. Emeléus and A.G. Sharpe, Eds.), Vol. 19, Academic Press, New York.

(4) M. Wrighton, Chem. Rev., 74, 401 (1974).

(5) F.-W. Grevels and E. Koerner von Gustorf, Liebigs. Ann. Chem., 1975, 547.

(6) A. Vogler, Photochemistry of Carbonyl Complexes, in "Concepts of Inorganic Photochemistry", (A.W. Adamson and P.D. Fleischauer, Eds.), Wiley, New York, 1975, p. 269.

(7) V. Balzani and V. Carassiti, "Photochemistry of Coordination Compounds", Academic Press, New York, 1970.

(8) A. W. Adamson and P. D. Fleischauer, Eds., "Concepts of Inorganic Photochemistry", Wiley, New York, 1975.

(9) Y. S. Sohn, D. N. Hendrickson, and H. B. Gray, J. Am. Chem. Soc., 93, 3603 (1971).

(10) D. R. Scott and R. S. Becker, J. Organometal. Chem., 4, 409 (1965).

(11) D. R. Scott and R. S. Becker, J. Chem. Phys., 35, 516 (1961).

(12) Y. S. Sohn, D. N. Hendrickson, J. H. Smith, and H. B. Gray, Chem. Phys. Lett., 6, 499 (1970).

(13) A. T. Armstrong, F. Smith, E. Elder, and S. P. McGlynn, J. Chem. Phys. 46, 4321 (1967).

(14) J. J. Smith and B. Meyer, J. Chem. Phys., 48, 5436 (1968).

(15) M. Dartiguenave, Y. Dartiguenave, and H. B. Gray, Bull. Soc. Chim. Fr., 1969, 4223.

(16) A.B.P. Lever, "Inorganic Electronic Spectroscopy", Elsevier, Amsterdam, 1968.

(17) E. P. Kündig, M. Moskovits, and G. A. Ozin, Angew. Chem., 87, 314 (1975); Angew. Chem. Int. Ed., 14, 292 (1975).

(18) G. A. Ozin and A. Vander Voet, Acc. Chem. Res., 6, 313 (1973).

(19) K. Fukui, Fortschr. Chem. Forsch. (Topics in Current Chemistry), 15, 1 (1970).

(20) F. Basolo and R. G. Pearson, "Mechanisms of Inorganic Reactions", 2nd ed., Wiley, New York, 1967.

(21) V. J. Koester, Chem. Phys. Lett., 32, 575 (1975).

(22) G. M. Tom, Carol Creutz, and H. Taube, J. Am. Chem. Soc., 96, 7827 (1974).

(23) R. W. Callahan, G. M. Brown, and T. J. Meyer, J. Am. Chem. Soc., 96, 7829 (1974).

(24) G. C. Allen and N. S. Hush, in "Progress in Inorganic Chemistry", (F.A. Cotton, Ed.), Vol. 8, Interscience, New York, 1967, p. 357.

(25) N. S. Hush, in "Progress in Inorganic Chemistry", (F.A. Cotton, Ed.), Vol. 8, Interscience, New York, 1967, p. 391.

(26) A. Vogler and H. Kunkely, Ber. Bunsenges. Phys. Chem., 79, 83 (1975).

(27) F. A. Cotton, W. T. Edwards, F. C. Rauch, M. A. Graham, R. N. Perutz, and J. J. Turner, J. Coord. Chem., 2, 247 (1973).

(28) C. K. Jørgensen, Acta Chem. Scand., 11, 53 (1957).

(29) C. D. Flint, Coord. Chem. Revs., 14, 47 (1974).

(30) D. Sutton, "Electronic Spectra of Transition Metal Complexes", McGraw-Hill, London, 1968.

(31) M. S. Wrighton, D. L. Morse, and L. Pdungsap, J. Am. Chem. Soc., 97, 2073 (1975).

(32) P. D. Fleischauer and P. Fleischauer, Chem. Rev., 70, 199 (1970).

(33) P. Krumholz, O. A. Serra, and M. A. De Paoli, J. Inorg. Nucl. Cham., 37, 1820 (1975).

(34) J. F. Endicott, G. J. Ferraudi, and J. R. Barber, J. Phys. Chem., 79, 630 (1975).

(35) H. tom Dieck and I. W. Renk, Chem. Ber., 105, 1419 (1972).

(36) H. tom Dieck, K.-D. Franz, and F. Hohmann, Chem. Ber., 108, 163 (1975).

(37) H. tom Dieck and I. W. Renk, Angew. Chem., 82, 805 (1970); Angew. Chem. Int. Ed. 9, 793 (1970).

(38) I. W. Renk and H. tom Dieck, Chem. Ber., 105, 1403 (1972).

(39) D. Walther, Z. Anorg. Allg. Chem., 396, 46 (1973).

(40) P. Ford, De F. P. Rudd, R. Gaunder, and H. Taube, J. Am. Chem. Soc., 90, 1187 (1968).

(41) M. Wrighton and D. L. Morse, J. Am. Chem. Soc., 96, 998 (1974).

(42) M. Wrighton, H. B. Gray, and G. S. Hammond, Mol. Photochem., 5, 165 (1973).

(43) J. I. Zink, Mol. Photochem., 5, 151 (1973).

(44) J. I. Zink, Inorg. Chem., 12, 1018 (1973).

(45) J. I. Zink, Inorg. Chem., 12, 1957 (1973).

(46) J. I. Zink, J. Am. Chem. Soc., 96, 4464 (1974).

(47) J. I. Zink, J. Am. Chem. Soc., 94, 8039 (1972).

(48) J. I. Zink, Inorg. Chem., 14, 446 (1975).

(49) M. Wrighton, G. S. Hammond, and H. B. Gray, Mol. Photochem., 5, 179 (1973).

(50) P. C. Ford, Inorg. Chem., 14, 1440 (1975).

(51) A. W. Adamson, Pure Appl. Chem., 20, 25 (1969).

(52) R. A. Pribush, C. K. Poon, C. M. Bruce, and A. W. Adamson, J. Am. Chem. Soc., 96, 3027 (1974).

(53) G. A. Crosby, G. D. Hager, K. W. Hipps, and M. L. Stone, Chem. Phys. Lett., 28, 497 (1974).

(54) K. W. Hipps and G. A. Crosby, Inorg. Chem., 13, 1543 (1974).

(55) W. G. Herkstroeter, J. Am. Chem. Soc., 97, 4161 (1975).

(56) F. S. Richardson, D. D. Shillady, and A. Waldrop, Inorg. Chim. Acta, 5, 279 (1971).

(57) J. K. Burdett, J.C.S. Faraday II, 70, 1599 (1974).

(58) R. G. Pearson, Chem. Phys. Lett., 10, 31 (1971).

(59) D. P. Keeton and F. Basolo, Inorg. Chim. Acta, 6, 33 (1972).

(60) M. K. DeArmond, Acc. Chem. Res., 7, 309 (1974).

(61) M. Gouterman, J. Chem. Phys., 36, 2846 (1962).

(62) G. J. Goldsmith, F. V. Shallcross, and D. S. McClure, J. Mol. Spectrosc., 16, 296 (1965).

(63) W. M. Gelbart, K. F. Freed, and S. A. Rice, J. Chem. Phys., 52, 2460 (1970).

(64) J. E. Hillis and M. K. DeArmond, J. Lumin., 4, 273 (1971).

(65) W. J. Mitchell and M. K. DeArmond, J. Lumin., 4, 137 (1971).

(66) D. J. Robbins and A. J. Thomson, Mol. Phys., 25, 1103 (1973).

(67) R. G. Pearson, Fortschr. Chem. Forsch. (Topics in Current Chemistry), 41, 75 (1973).

(68) P. Natarajan and A. W. Adamson, J. Am. Chem. Soc., 93, 5599 (1971).

(69) N. Rösch, R. P. Messmer, and K. H. Johnson, J. Am. Chem. Soc., 96, 3855 (1974).

(70) J. B. Birks, "Photophysics of Aromatic Molecules", Interscience, London, 1970.

(71) M. H. Hui, P. De Mayo, R. Suau, and W. R. Ware, Chem. Phys. Lett., 31, 257 (1975).

(72) J. B. Birks, Ed., "Organic Molecular Photophysics", Vol. 2, Wiley, London, 1975.

(73) M. Wrighton, Inorg. Chem., 13, 905 (1974).

(74) P. Kirsch, R. C. Kerber, R. Rumin, and E. A. Koerner von Gustorf, private communication.

(75) P. Kirsch, J. Buchkremer, O. Jaenicke, R. Knoesel, R. Rumin, J. Shields, and E. Koerner von Gustorf, Meeting of the Photochemistry Group, German Chemical Society, Göttingen, 1973, Abstract 5.

(76) P. P. Zarnegar and D. G. Whitten, J. Am. Chem. Soc., 93, 3776 (1971).

(77) P. P. Zarnegar, C. R. Bock, and D. G. Whitten, J. Am. Chem. Soc., 95, 4367 (1973).

(78) Y. Ohashi, K. Yoshihara, and S. Nagakura, J. Mol. Spectrosc., 38, 43 (1971).

(79) G. A. Crosby, K. W. Hipps, and W. H. Elfring, Jr., J. Am. Chem. Soc., 96, 629 (1974).

(80) W. Strohmeier and K. Gerlach, Chem. Ber., 94, 398 (1961).

(81) W. Strohmeier, Angew. Chem., 76, 873 (1964); Angew. Chem. Int. Ed., 3, 370 (1964).

(82) A. Vogler, Z. Naturforsch., 25b, 1069 (1970).

(83) J. A. McIntyre, J. Phys. Chem., 74, 2403 (1970).

(84) J. Nasielski, P. Kirsch, and L. Wilputte-Steinert, J. Organometal. Chem., 29, 269 (1971).

(85) I. W. Stolz, G. R. Dobson, and R. K. Sheline, J. Am. Chem. Soc., 84, 3589 (1962).

(86) I. W. Stolz, G. R. Dobson, and R. K. Sheline, J. Am. Chem. Soc., 85, 1013 (1963).

(87) J. M. Kelly and A. Morris, Rev. Latinoamericana Quim., 2, 163 (1972).

(88) J. M. Kelly, H. Hermann, and E. Koerner von Gustorf, Chem. Commun., 1973, 105.

(89) J. M. Kelly, D. V. Bent, H. Hermann, D. Schulte-Frohlinde, and E. Koerner von Gustorf, J. Organometal. Chem., 69, 259 (1974).

(90) J. Nasielski and A. Colas, Seventh Int. Conf. on Organometal. Chem., Venice, 1975, Abstr. 64.

(91) M. S. Wrighton, D. S. Ginley, M. A. Schroeder, and D. L. Morse, Pure Appl. Chem., 41, 671 (1975).

(92) N. Serpone and D. G. Bickley, in "Progress in Inorganic Chemistry", (J. O. Edwards, Ed.), Vol. 17, Interscience, New York, 1972, p. 391.

(93) F. Scandola, O. Traverso, V. Balzani, G. L. Zucchini, and V. Carassiti, Inorg. Chem. Acta, 1, 76 (1967).

(93a) C. J. Ballhausen, N. Bjerrum, R. Dingle, K. Ericks, and C. R. Hare, Inorg. Chem., 4, 514 (1965).

(94) F. Bolletta, M. Gleria, and V. Balzani, Mol. Photchem., 4, 205 (1972).

(95) L. Moggi, G. Varani, N. Sabbatini, and V. Balzani, Mol. Photochem., 3, 141 (1971).

(96) S. H. Mastin and P. Haake, Chem. Commun., 1970, 202.

(97) P. Haake and T. A. Hylton, J. Am. Chem. Soc., 84, 3774 (1962).

(98) K. Öfele and M. Herberhold, Z. Naturforsch., 28b, 306 (1973).

(99) K. Öfele and M. Herberhold, Angew. Chem., 82, 775 (1970).

(100) M. Pankowski and M. Bigorgne, J. Organometal. Chem., 19, 393 (1969).

(101) R. K. Pomeroy, R. S. Gay, G. O. Evans, and W. A. G. Graham, J. Am. Chem. Soc., 94, 272 (1972).

(102) P. R. Brookes and B. L. Shaw, Chem. Commun., 1968, 919.

(103) P. R. Brookes, C. Masters, and B. L. Shaw, J. Chem. Soc. A, 3756 (1971).

(104) G. S. Kovalenko, I. B. Baranovskii, and A. V. Babaeva, Russian J. Inorg. Chem., 16, 148 (1971).

(105) J. Jeffery and R. J. Mawby, J. Organometal. Chem., 40, C42 (1972).

(106) M. S. Wrighton and D. S. Ginley, J. Am. Chem. Soc., 97, 2065 (1975).

(107) M. S. Wrighton and D. S. Ginley, J. Am. Chem. Soc., 97, 4246 (1975).

(108) D. S. Ginley and M. S. Wrighton, J. Am. Chem. Soc., 97, 4908 (1975).

(109) R. A. Levenson, H. B. Gray and G. P. Ceasar, J. Am. Chem. Soc., 92, 3653 (1970).

(110) J. L. Hughey IV, C. R. Bock, and T. J. Meyer, J. Am. Chem. Soc., 97, 4440 (1975).

(111) M. Wrighton and D. Bredesen, J. Organometal. Chem., 50, C35 (1973).

(112) J. L. Hughey, IV, and T. J. Meyer, Unpublished.

(113) R. A. Levenson and H. B. Gray, J. Am. Chem. Soc., 97, 6042 (1975).

(114) S. A. Hallock and A. Wojcicki, J. Organometal. Chem., 54, C27 (1973).

(115) J. L. Hughey, IV, and T. J. Meyer, Seventh Int. Conf. on Organometal. Chem., Venice, 1975, Abstr. 65.

(116) A. Hudson, M. F. Lappert, and B. F. Nicholson, J. Organometal. Chem., 92, C11 (1975).

(117) C. L. Kwan and J. K. Kochi, J. Organometal. Chem., 101, C9 (1975).

(118) A. G. Osborne and M. H. B. Stiddard, J. Chem. Soc., 634 (1964).

(119) E. O. Fischer and W. A. Herrmann, Chem. Ber., 105, 286 (1972).

(120) A. R. Burkett, T. J. Meyer, and D. G. Whitten, J. Organometal. Chem., 67, 67(1974).

(121) R. J. Haines, R. S. Nyholm, and M. H. B. Stiddard, J. Chem. Soc. (A), 43 (1968).

(122) J. L. Hughey, IV, and T. J. Meyer, Inorg. Chem., 14, 947 (1975).

(123) D. S. Ginley and M. S. Wrighton, J. Am. Chem. Soc., 97, 3533 (1975).

(124) R. J. Klingler, W. Butler, and M. D. Curtis, J. Am. Chem. Soc., 97, 3535 (1975).

(125) J. P. Fawcett, R. A. Jackson and A. Poe, Chem. Commun., 1975, 733.

(126) G. L. Geoffrey, H. B. Gray, and G. S. Hammond, J. Am. Chem. Soc., 96, 5565 (1974).

(127) M. A. Graham, R. N. Perutz, M. Poliakoff, and J. J. Turner, J. Organometal. Chem., 34, C34 (1972).

(128) M. A. Graham, A. J. Rest, and J. J. Turner, J. Organometal. Chem., 24, C54 (1970).

(129) M. A. Graham, M. Poliakoff, and J. J. Turner, J. Chem. Soc. (A), 2939 (1971).

(130) J. K. Burdett, R. N. Perutz, M. Poliakoff, and J. J. Turner, Chem. Commun., 1975, 157.

(131) J. K. Burdett, M. A. Graham, R. N. Perutz, M. Poliakoff, A. J. Rest, J. J. Turner, and R. F. Turner, J. Am. Chem. Soc., 97, 4805 (1975).

(132) R. N. Perutz and J. J. Turner, J. Am. Chem. Soc., 97, 4791 (1975).

(133) R. N. Perutz and J. J. Turner, J. Am. Chem. Soc., 97, 4800 (1975).

(134) M. J. Newlands and J. F. Ogilvie, Can. J. Chem., 49, 343 (1971).

(135) M. Poliakoff and J. J. Turner, Sixth International Conference on Organometallic Chemistry, Amherst, 1973, Abstr. 45.

(136) M. Poliakoff and J. J. Turner, J.C.S., Dalton, 1973, 1351.

(137) M. Poliakoff and J. J. Turner, J.C.S., Faraday II, 70, 93 (1974).

(138) M. Poliakoff, J.C.S., Dalton, 1974, 210.

(139) M. Poliakoff and J. J. Turner, J. Chem. Soc. A, 2403 (1971).

(140) A. J. Rest, J. Organometal. Chem., 40, C76 (1972).

(141) A. J. Rest and J. J. Turner, Chem. Commun. 1969, 1026.

(142) O. Crichton, M. Poliakoff, A. J. Rest, and J. J. Turner, J.C.S., Dalton, 1321 (1973).

(143) J. K. Burdett, Chem. Commun. 1973, 763.

(144) J. D. Black, M. J. Boyland, P. S. Braterman, and W. J. Wallace, J. Organometal. Chem., 63, C21 (1973).

(145) J. D. Black and P. S. Braterman, J. Organometal. Chem., 63, C19 (1973).

(146) W. J. R. Tyerman, M. Kato, P. Kebarle, S. Masamune, O. P. Strausz, and H. E. Gunning, Chem. Commun., 1967, 497.

(147) O. L. Chapman, J.Pacansky, and P. W. Wojtkowski, Chem. Commun., 1973, 681.

(148) M. Poliakoff, Seventh Int. Conf. on Organometal. Chem., Venice, 1975, Abstr. 62.

(149) O. Crichton and A. J. Rest, Sixth Int. Conf. on Organometal. Chem., Amherst, 1973, Abstr. 238.

(150) O. Crichton and A. J. Rest, Chem. Commun., 1973, 407.

(151) J. C. D. Brand and W. Snedden, Trans. Faraday Soc., 53, 894 (1957).

(152) O. Traverso, S. Sostero, and G. A. Mazzocchin, Inorg. Chim. Acta, 11, 237 (1974).

(153) O. Traverso and F. Scandola, Inorg. Chim. Acta, 4, 493 (1970).

(154) P. Borrell and E. Henderson, Inorg. Chim. Acta, 12, 215 (1975).

(155) T. Akiyama, Y. Hoshi, S. Goto, and A. Sugimori, Bull. Chem. Soc. Japan, 46, 1851 (1973).

(156) T. Akiyama, A. Sugimori, and H. Hermann, Bull. Chem. Soc. Japan, 46, 1855 (1973).

(157) Y. Hoshi, T. Akiyama, and A. Sugimori, Tetrahedron Lett., 1970, 1485.

(158) G. B. Porter and H. L. Schläfer, Ber. Bunsenges., Phys. Chem., 68, 316 (1964).

(159) F. E. Lytle, Appl. Spectrosc., 24, 319 (1970).

(160) G. A. Crosby, Acc. Chem. Res., 8, 231 (1975).

(161) J. P. Paris and W. W. Brandt, J. Am. Chem. Soc., 81, 5001 (1959).

(162) G. A. Crosby, W. G. Perkins, and D. M. Klassen, J. Chem. Phys., 43, 1498 (1965).

(163) F. Zuloaga and M. Kasha, Photochem. Photobiol., 7, 549 (1968).

(164) D. M. Klassen and G. A. Crosby, Chem. Phys. Lett., 1, 127 (1967).

(165) D. M. Klassen and G. A. Crosby, J. Chem. Phys., 48, 1853 (1968).

(166) G. A. Crosby, J. Chim. Phys., 64, 160 (1967).

(167) J. Van Houten and R. J. Watts, J. Am. Chem. Soc., 97, 3843 (1975).

(168) J. N. Demas and G. A. Crosby, J. Mol. Spectrosc., 26, 72 (1968).

(169) F. E. Lytle and D. M. Hercules, J. Am. Chem. Soc., 91, 253 (1969).

(170) J. N. Demas and G. A. Crosby, J. Am. Chem. Soc., 93, 2841 (1971).

(171) R. J. Watts and G. A. Crosby, J. Am. Chem. Soc., 94, 2606 (1972).

(172) R. J. Watts and G. A. Crosby, J. Am. Chem. Soc., 93, 3184 (1971).

(173) R. J. Watts, R. W. Harrigan, and G. A. Crosby, Chem. Phys. Lett., 8, 49 (1971).

(174) R. W. Harrigan, G. D. Hager, and G. A. Crosby, Chem. Phys. Lett., 21, 487 (1973).

(175) R. W. Harrigan and G. A. Crosby, J. Chem. Phys., 59, 3468 (1973).

(176) D. C. Baker and G. A. Crosby, Chem. Phys., 4, 428 (1974).

(177) K. W. Hipps and G. A. Crosby, Abstracts, 169th National Meeting of the American Chemical Society, Philadelphia, Pa., April, 1975, Phys. 89.

(178) G. A. Crosby, D. E. Lacky and B. J. Pankuch, Abstracts, 169th National Meeting of the American Chemical Society, Philadelphia, Pa., April, 1975, INOR 69.

(179) D. H. W. Carstens and G. A. Crosby, J. Mol. Spectrosc., 34, 113 (1970).

(180) M. K. DeArmond and J. E. Hillis, J. Chem. Phys., 54, 2247 (1971).

(181) G. A. Crosby and D. W. H. Carstens, in "Molecular Luminescence", (E. C. Lim, Ed.), Benjamin, New York, 1969, p. 309.

(182) J. N. Demas and G. A. Crosby, J. Am. Chem. Soc., 92, 7262 (1970).

(183) W. Halper and M. K. DeArmond, Chem. Phys. Lett., 24, 114 (1974).

(184) W. Halper and M. K. DeArmond, J. Lumin., 5, 225 (1972).

(185) M. K. DeArmond and J. E. Hillis, J. Chem. Phys., 49, 466 (1968).

(186) P. E. Hoggard and H.-H. Schmidtke, Ber. Bunsenges. Phys. Chem., 77, 1052 (1973).

(187) T. R. Thomas and G. A. Crosby, J. Mol Spectrosc., 38, 118 (1971).

(188) T. R. Thomas, R. J. Watts, and G. A. Crosby, J. Chem. Phys., 59, 2123 (1973).

(189) R. J. Watts and J. Van Houten, J. Am. Chem. Soc., 96, 4334 (1974).

(190) J. E. Hillis and M. K. DeArmond, Chem. Phys. Lett., 10, 325 (1971).

(191) G. S. Arnold, W. L. Klotz, W. Halper, and M. K. DeArmond, Chem. Phys. Lett., 19, 546 (1973).

(192) C. M. Flynn, Jr. and J. N. Demas, J. Am. Chem. Soc., 96, 1959 (1974).

(193) J. N.Demas, E. W. Harris, C. M. Flynn, Jr., and D. Diemente, J. Am. Chem. Soc., 97, 3838 (1975).

(194) R. J. Watts, G. A. Crosby and J. S. Sansregret, Inorg. Chem. 11, 1474 (1972).

(195) G. A. Crosby, R. J. Watts, and D. H. W. Carstens, Science, 170, 1195 (1970).

(196) R. J. Watts, J. Am. Chem. Soc., 96, 6186 (1974).

(197) K. R. Wunschel, Jr., and W. E. Ohnesorge, J. Am. Chem. Soc., 89, 2777 (1967).

(198) R. J. Watts and G. A. Crosby, Chem. Phys. Lett., 13, 619 (1972).

(199) R. Ballardini, G. Varani, L. Moggi, V. Balzani, K. R. Olson, F. Scandola, and M. Z. Hoffman, J. Am. Chem. Soc., 97, 728 (1975).

(200) M. Wrighton, G. S. Hammond, and H. B. Gray, J. Am. Chem. Soc., 93, 4336 (1971).

(201) M. Wrighton, G. S. Hammond, and H. B. Gray, Inorg. Chem., 11, 3122 (1972).

(202) G. W. Robinson and R. P. Frosch, J. Chem. Phys., 37, 1962 (1962).

(203) Y. Kaizu, I. Fujita, and H. Kobayashi, Z. Phys. Chem. (Frankfurt am Main), 79, 298 (1972).

(204) V. Balzani, L. Moggi, M. F. Manfrin, F. Bolletta, and G. S. Laurence, Coord. Chem. Revs., 15, 321 (1975).

(205) J. P. Guillory, C. F. Cook, and D. R. Scott, J. Am. Chem. Soc., 89, 6776 (1967).

(206) A. M. Tarr and D. M. Wiles, Can. J. Chem., 46, 2725 (1968).

(207) A. Müller-Goldegg and J. Voitländer, Z. Naturforsch., 23a, 1236 (1968).

(208) R. Schandry and J. Voitländer, Z. Naturforsch., 26a, 1772 (1971).

(209) R. Krieger and J. Voitländer, Z. Naturforsch., 27a, 1444 (1972).

(210) M. S. Wrighton, L. Pdungsap, and D. L. Morse, J. Phys. Chem., 79, 66 (1975).

(211) A. A. Lamola, in "Energy Transfer and Organic Photochemistry", Vol. 14, Technique of Organic Chemistry, (P. A. Leermakers and A. Weissberger, Eds.), Interscience, New York, 1969, p. 17.

(212) P.J. Wagner and G. S. Hammond, in "Advances in Photochemistry", (W. A. Noyes, Jr., G. S. Hammond and J. N. Pitts, Jr., Eds.), Vol. 5, Interscience, New York, 1968, p. 21.

(213) Paul Engel and Bruce Monroe, in "Advances in Photochemistry", (J. N. Pitts, Jr., G. S. Hammond, and W. A. Noyes, Jr., Eds.), Vol. 8, Interscience, New York, 1971, p. 245.

(214) J. N. Demas, D. Diemente, and E. W. Harris, J. Am. Chem. Soc., 95, 6864 (1973).

(215) J. N. Demas and J. W. Addington, J. Am. Chem. Soc., 96, 3663 (1974).

(216) A. J. Fry, R. S. H. Liu, and G. S. Hammond, J. Am. Chem. Soc., 88, 4781 (1966).

(217) M. KiKuchi, K. KiKuchi, and H. Kokubun, Bull. Chem. Soc. Japan, 47, 1331 (1974).

(218) K. Sandros, Acta Chem. Scand., 18, 2355 (1964).

(219) A. Gilbert, J. M. Kelly, and E. Koerner von Gustorf, Mol. Photochem., 6, 225 (1974).

(220) J. H. Richards and N. Pisker-Trifunac, J. Paint Technol., 41, 363 (1969).

(221) O. Traverso, F. Scandola, and V. Carassiti, Inorg. Chim. Acta, 6, 471 (1972).

(222) S. R. Allsopp and F. Wilkinson, Chem. Phys. Lett., 19, 535 (1973).

(223) A. Adamczyk and F. Wilkinson, J.C.S., Faraday II, 68, 2031 (1972).

(224) F. Wilkinson, Pure Appl Chem., 41, 661 (1975).

(225) Steven L. Murov, "Handbook of Photochemistry", Dekker, New York, 1973.

(226) F. Bolletta, M. Gleria, and V. Balzani, J. Phys. Chem., 76, 3934 (1972).

(227) J. G. Calvert and J. N. Pitts, Jr., "Photochemistry", Wiley, New York, 1966.

(228a) H. Leonhardt and A. Weller, Z. Phys. Chem. Neue Folge, 29, 277 (1961).

(228b) H. Leonhardt and A. Weller, Ber. Bunsenges. Phys. Chem., 67, 791 (1963).

(229) H. Knibbe, D. Rehm, and A. Weller, Ber. Bunsenges. Phys. Chem., 72, 257 (1968).

(230) D. Rehm and A. Weller, Ber. Bunsenges. Phys. Chem., 73, 834 (1969).

(231) M. Ottolenghi, Acc. Chem. Res., 6, 153 (1973).

(232) N. Mataga and O. Tanimoto, Theoret. Chim. Acta, (Berl.), 15, 111 (1969).

(233a) K. H. Grellmann, A. R. Watkins, and A. Weller, J. Lumin. 1,2, 678 (1970).

(233b) K. H. Grellmann, A. R. Watkins, and A. Weller, J. Phys. Chem., 76, 469 (1972).

(234) J. N. Demas and A. W. Adamson, J. Am. Chem. Soc., 93, 1800 (1971).

(235) M. Wrighton and J. Markham, J. Phys. Chem., 77, 3042 (1973).

(236) H. Gafney and A. Adamson, J. Am. Chem. Soc., 94, 8238 (1972).

(237) J. N. Demas and A. W. Adamson, J. Am. Chem. Soc., 95, 5159 (1973).

(238) G. Navon and N. Sutin, Inorg. Chem., 13, 2159 (1974).

(239) P. Natarajan and J. F. Endicott, J. Phys. Chem., 77, 971 (1973).

(240) P. Natarajan and J. F. Endicott, J. Phys. Chem., 77, 1823 (1973).

(241) C. R. Bock, T. J. Meyer, and D. G. Whitten, J. Am. Chem. Soc., 96, 4710 (1974).

(242) G. S. Laurence and V. Balzani, Inorg. Chem., 13, 2976, (1974).

(243) C. T. Lin and N. Sutin, J. Am. Chem. Soc., 97, 3543 (1975).

(244) R. C. Young, T. J. Meyer, and D. G. Whitten, J. Am. Chem. Soc., 97, 4781 (1975).

(245) C. R. Bock, T. J. Meyer, and D. G. Whitten, J. Am. Chem. Soc., 97, 2909 (1975).

(246) D. Rehm and A. Weller, Isr. J. Chem., 8, 259 (1970).

(247) J. A. Powell and S. R. Logan, J. Photochem., 3, 189 (1974).

(248) W. L. Reynolds and R. W. Lumry, "Mechanisms of Electron Transfer", Ronald Press, New York 1966.

(249) B. Stevens, in "Advances in Photochemistry", (J. N. Pitts, Jr., G. S. Hammond, and W. A. Noyes, Jr., Eds.), Vol. 8, Interscience, New York, 1971, p. 161.

(250) J. K. Roy and D. G. Whitten, J. Am. Chem. Soc., 93, 7093 (1971).

(251) J. K. Roy and D. G. Whitten, J. Am. Chem. Soc., 94, 7162 (1972).

(252) J. K. Roy, F. A. Carroll, and D. G. Whitten, J. Am. Chem. Soc., 96, 6349 (1974).

(253) C.Lewis and W. R. Ware, Mol. Photochem., 5, 261 (1973).

(254) R. Ballardini, G. Varani, L. Moggi and V. Balzani, J. Am. Chem. Soc., 96, 7123 (1974).

(255) J. B. Callis, J. M. Knowles, and M. Gouterman, J. Phys. Chem., 77, 154 (1973).

(256) E. Vander Donckt and J. P. Van Bellinghen, Chem. Phys. Lett., 7, 630 (1970).

(257) E. Vander Donckt, M. Lootens, D. Swinnen, and C. Haquin, Bull. Soc. Chim. Belg., 84, 77 (1975).

(258) A. Farmilo and F. Wilkinson, Photochem. Photobiol. 18, 447 (1973).

(259) D. G. Whitten, P. D. Wildes, and Carold A. DeRosier, J. Am. Chem. Soc., 94, 7811 (1972).

(260) F. P. Schwarz, M. Gouterman, Z. Muijiani, and D. H. Dolphin, Bioinorg. Chem., 2, 1 (1972).

(261) D. Magde, B. A. Bushaw, and M. W. Windsor, Chem. Phys. Lett., 28, 263 (1974).

(262) L. Y. Tan, A. M. Winer, and G. C. Pimentel, J. Chem. Phys., 57, 4028 (1972).

Intramolecular Proton Transfer in Electronically Excited Molecules

WALTER KLÖPFFER
Battelle-Institut
Frankfurt Am Main
West Germany

CONTENTS

I. INTRODUCTION

In 1955 Weller (1) reported very large Stokes shifts in the fluorescence spectra of salicyclic acid and its methyl ester. Both compounds show intramolecular hydrogen bonds from the phenolic hydroxyl group to the carbonyl group. The unusual spectral shift has been explained as a consequence of intramolecular proton transfer in the first excited singlet state (1,2), which is possible due to the pronounced difference between excited state and ground state acidity (3,4). Cohen, Hirschberg, and Schmidt (5) showed in 1957 that the photochromic behavior (reversible color change) of salicyclidene anilines is also due to proton transfer in the excited state. The molecules showing the effect have intramolecular hydrogen bonds. Both discoveries have been followed by many papers showing that intramolecular proton transfer takes place in many molecules in their excited states. The great photochemical stability of derivatives of 2-hydroxy benzophenone - again molecules with intramolecular hydrogen bonds - was also established in the 1950s (6) and has been attributed to intramolecular proton transfer followed by very rapid radiationless deactivation (7,8).

From a phenomenological point of view the compounds showing intramolecular excited state proton transfer can be divided roughly into three groups: fluorescent, photochromic, and nonluminescent molecules. Prototypes are methyl salicylate (I), salicylindeneaniline (II), and 2-hydroxy benzophenone (III).

The first group is characterized by relatively slow radiationless deactivation so that radiative deactivation (fluorescence) can compete. In the second group the intermediate state can be stabilized by a geometrical rearrangement into a photochromic isomer. In the third group of substances the radiationless deactivation is so rapid that no* luminescence is observed unless the hydrogen bond is broken by polar solvents (8). It

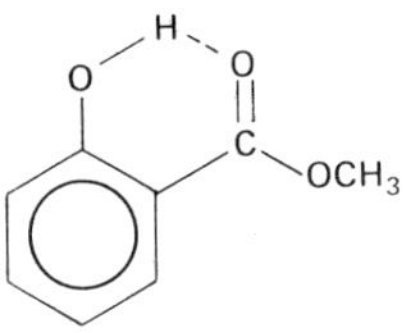

I

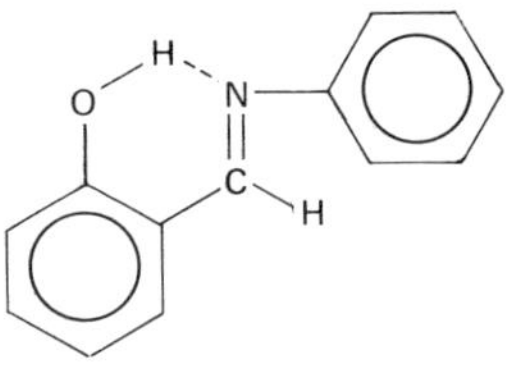

II

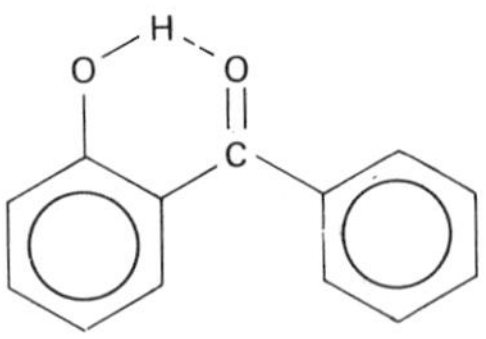

III

---

*or only very weak.

should be noted that in the last group the mechanism is not yet understood in detail and that other mechanisms in addition to proton transfer have been proposed (9). In analogy to the behavior of the first two groups, however, proton transfer in group III is a likely mechanism. Common to all three groups is the existence of an intramolecular hydrogen bond forming a chelate ring, as evidenced by infrared, NMR, and the general physical and chemical behavior of the substances.

Following brief discussion of the acidity of excited states (Section II), the spectral properties of the three groups are discussed in Sections III to V. Not all substances possibly showing this effect will be covered. The discussion is rather restricted to compounds on which experiments have been performed which contribute to the elucidation of the mechanism of intramolecular proton transfer. In the last section the three groups will be discussed together.

II. ACID-BASE PROPERTIES OF EXCITED STATES

Studies on the acid-base properties of molecules in their excited states are closely related to intermolecular proton transfer. This topic has been reviewed by Weller (4) and Parker (10), and only some basic phenomena necessary for the understanding of the following sections will be discussed here.

In view of the differences of electron distribution, molecules in an excited state and in the ground state may be regarded as different molecules as far as their chemical reactivity is concerned. In the case of molecules with acidic or basic groups these differences are reflected in their dissociation constants. Experimentally the formation of deprotonated or protonated species is often observed at pH values not causing any ionization in the ground state (3,4,11-13).

The different excited species can be detected and studied at varying pH by their fluorescence, phosphorescence, or transient absorption spectra. If dissociation occurs fast enough, equilibrium dissociation constants of the excited states may be obtained. The corresponding ground state constants are obtained spectroscopically from the changes in the absorption spectra as a function of the pH of the solvent.

The dissociation constants are defined as follows:

$$\frac{[A^-]\,[H^+]}{[AH]} = K_G \qquad (1)$$

$$\frac{[A^{-*}]\ [H^+]}{[AH^*]} = K_{S^*} \text{ or } K_{T^*} \qquad (2)$$

$$pK = -\log K \qquad (3)$$

K_G, $K_S{}^*$, $K_T{}^*$ denote, respectively: equilibrium dissociation constants in the ground state, the first excited singlet state, and the first triplet state. As usual, the strength of bases in their ground and excited states is expressed by the pK and pK^* values of their conjugate acids.

2-Naphthol is a good example to demonstrate the different dissociation behavior in the ground and excited states. The low frequency absorption maximum of the undissociated molecules is at 30,300 cm^{-1}, that of the naphtholate ion at 28,800 cm^{-1} (pK_G = 9.46) (14). The corresponding fluorescence maxima, measured in 0.15m $HClO_4$ and 0.02m NaOH are at 28,200 cm^{-1} and at 23,600 cm^{-1}, respectively (4). The fluorescence decay times of the two forms are 7.6 ± 0.3 ns and 8.0 ± 0.3 ns (14).

In Fig. 1, the relative fluorescence intensities of ROH^* and RO^{-*} are shown as a function of $[H^+]$. As can be seen, the fluorescence of RO^{-*} starts to appear at about pH = 2, where

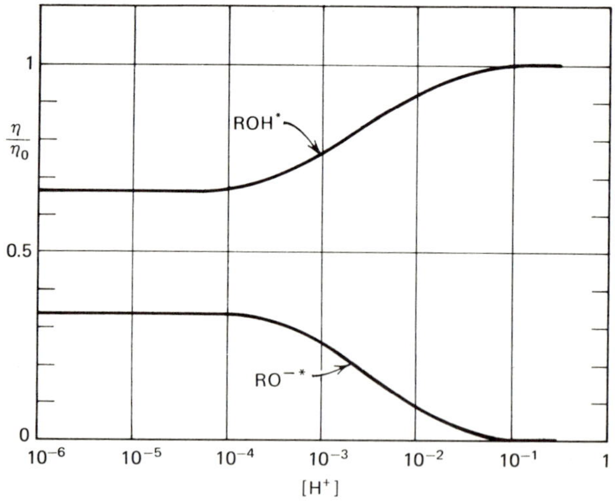

FIGURE 1. Relative quantum efficiencies of 2-naphthol in aqueous $HClO_4$ solutions at 25°C (after Weller[12]).

nearly all molecules are in the undissociated form ROH in the ground state. The transformation of ROH^* into RO^{-*} is not complete, since the dissociation equilibrium is not completely established during the lifetime of the excited states.

The exact evaluation of this experiment yields $pK_S^* = 2.81 \pm 0.03$ (14); hence the acidity of 2-naphthol in its first excited singlet state is by more than six orders of magnitude larger than in its ground state.

A simple estimate of pK^* has been proposed by Weller (4), based on the so-called Förster cycle (3) (Fig. 2). The difference between the ground state and excited state dissociation enthalpy is given by the difference of the energy levels, as measured by the absorption or emission spectra of the protonized (ΔE) and deprotonized ($\Delta E'$) species.

$$\Delta H - \Delta H^* = \Delta E - \Delta E' \qquad (4)$$

Assuming equal reaction entropies in the two states, we have

$$\ell n \frac{K^*}{K} = \frac{\Delta E - \Delta E'}{RT} = \frac{hc}{kT} \Delta\nu' \qquad (5)$$

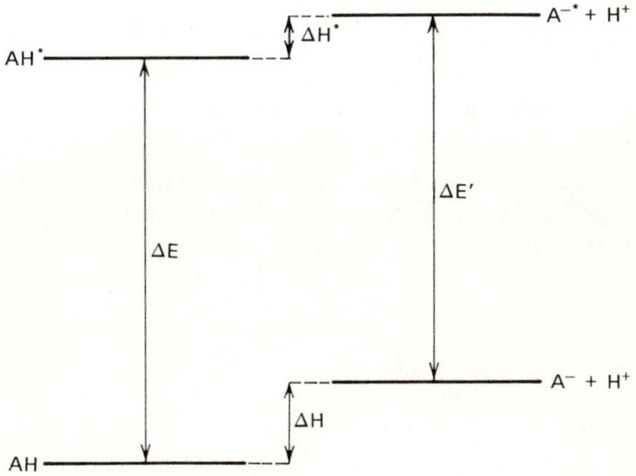

FIGURE 2. Förster cycle[15] of ground and excited state dissociation neglecting vibrational levels; ΔE and ΔE' correspond to 0-0 bands or to the average of absorption and emission bands.

(since $\Delta G = \Delta H - T\Delta S = -RT \ln K$), or

$$pK^* = pK - \frac{0.625}{T}\Delta\nu' \tag{6}$$

if the wave number difference is given in cm^{-1}.

Since no assumption concerning the electronic nature of the excited states and the rate of the dissociation and association processes is made in this thermodynamic approach, eq. 6 can be applied to triplet states and singlet states as well, and is useful regardless of whether the states of equilibrium are reached during the excited state lifetimes or not. The agreement of the pK^* values according to eq. 6 with values obtained by kinetic measurements is quite good and justifies the simplifying assumptions.

Since phosphorescence is generally not observed in liquid solutions, pK_T^* values are determined from flash transient absorption measurements (13) or by means of eq. 6 from phosphorescence spectra in acidic and basic solid solutions (13,16).

The pK, pK_S^*, and pK_T^* values of several compounds are summarized in Table 1.

The following conclusions can be drawn from these and other data not included in Table 1 (4,10,13,16):

1. Phenolic compounds are much stronger acids in their excited singlet state than in the ground state ($pK_S^* \ll pK_G$).
2. Aromatic carboxylic acids are much weaker acids in S_1 ($pK_S^* \gg pK_G$).
3. In both groups the triplet state acidity is much more similar to ground state acidity than to excited singlet state acidity.

These observations have been explained qualitatively (15) by the pronounced charge transfer character of the first excited singlet states, for example,

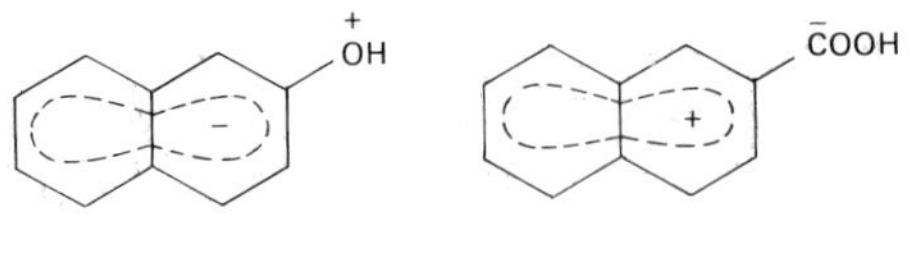

(after Godfrey, Porter, and Suppan (15)), whereas the corresponding triplet states show much less distinct charge transfer (13).

TABLE 1. Dissociation of selected organic substances in the ground and in the excited state.

		pK_S^*		pK_T^*		
Substance	pK_G	a)	b)	a)	b)	Ref.
Phenol	10.0		5.7			(17)
1-Naphthol	9.23	(2.5)	2.0±0.3			(14)
2-Naphthol	9.46	2.81±0.03	2.5±0.3	8.1	7.7	(14,13)
1-Naphthoic acid	3.7	10 - 12		3.8	4.6	(13)
2-Naphthoic acid	4.2	10 - 12		4.0	4.2	(13)
2-Naphthyl amine	4.1	-2		3.3	3.1	(13)
Benzophenone	-5.7		-3.6		-0.4	(16)
4,4-Dichloro-benzophenone	-6.5		-2.9		2.1	(16)
4,4-Dimethoxy-benzophenone	-4.4		4.2		5.5	(16)
4-Hydroxybenzo-phenone (C=O)	-5.0		0.0		2.3	(16)
4-Hydroxy-benzo-phenone (OH)	6.5		3.4		1.5	(16)

[a]From kinetic measurements.

[b]From spectral data using eq. 6.

4. Benzophenones are stronger bases in their excited states, especially in T_1 ($pK_T > pK_S > pK_G$).

This behavior has been explained (16) by the enhanced T_1-S_1 splitting of the $\pi - \pi^*$ states of the conjugate acid (Ar-COH$^+$-Ar), compared to that of the n-π^* states of the unprotonated benzophenone.

III. FLUORESCENT MOLECULES

A. Salicyclic Acid Derivatives

The phenolic OH group forms an intramolecular hydrogen bond to the CO group in salicyclic acid, its esters, and amides. The infrared and NMR data are included in Table 2. These data and the estimates of the enthalpy of formation derived therefore indicate relatively strong hydrogen bonding (ΔH = -6 to -8 kcal/mol).

The ultraviolet spectroscopic properties of salicyclic acid (SA), methyl salicyclate (MSA), and salicyclic anilide (SAN) have been studied in some detail. Since SA itself is more complex due to the possibility of dimerization and the existence of two acid hydrogen atoms we shall first discuss MSA and SAN.

MSA shows two strong absorption bands in the near ultraviolet (Fig. 3 and Table 2) whose spectral position and intensity are rather insensitive to solvent polarity. The blue fluorescence of MSA in methyl cyclohexane shows a very strong Stokes shift of 10,000 cm^{-1} compared to the long-wavelength absorption band (2).

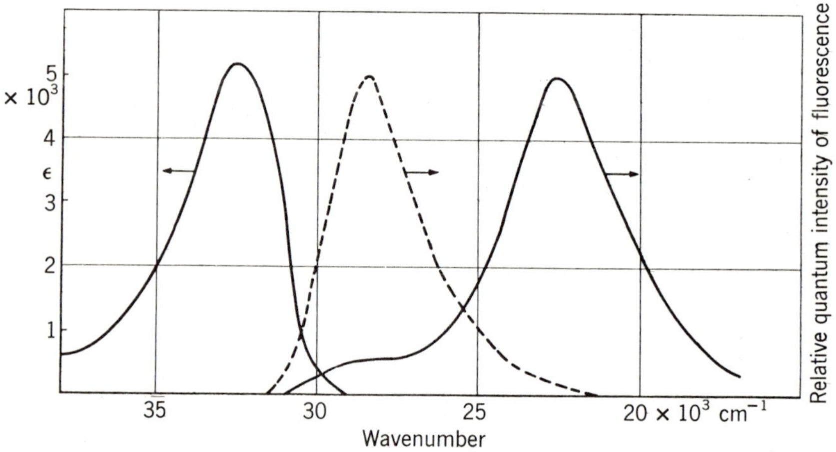

FIGURE 3. Absorption of fluorescence spectra of MSA in methyl cyclohexane (solid lines) and fluorescence spectrum of methyl 2-methoxy benzoate in the same solvent (dashed); fluorescence spectra after Weller[2].

A second band in the ultraviolet region is indicated as a shoulder. The anomalously large Stokes shift of the blue band is not detected in methyl 2-methoxy benzoate (2) whose fluorescence has a maximum in the region of the weak ultraviolet band of MSA (Fig. 3). This behavior has been explained by Weller (1,2) on the basis of intramolecular proton transfer in the first excited singlet state according to the following:

TABLE 2. Spectroscopic data of salicyclic acid derivatives[a] (cm^{-1}).

	IR		NMR	UV - VIS		
	O-H stretching	C=O stretching	Chem. shift (ppm)	Absorption (ε)	Fluorescence	$-\Delta H$ (kcal/mol)
Salicyclic acid		1696(mon.)(20) 1662(dim.)(20)	No sharp peak	33,000(4000) in CH_3OH	22,100(10^{-3}m)(1) 24,800(10^{-5}m)(25) in CH_3OH	6.8[b]
Methyl salicylate	3191 (19) 3200 (18)	1683 (20) 1680 (19)	10.6(25)	32,500(4260)(38) 42,000(9600)(38)	22,600(max.) ∿28,000(shoulder)(2)	6.9(19)
Methyl-2-methoxy-benzoate	-	1737(20)[c]	-	34,100 in CH_3OH(1)	28,400(2)	-
Salicylic acid-anilide	3070 (21)	1659 (22)	12.0(25)	31,850(10,000)(21) ∿38,000(∿10,000)	21,300(21)	8.0(21)
Phenyl salicylate	3240 (18)		10.53(25) ($CDCl_3$)	33,000 (34)[d]	22,200(34)[d]	5.8(34)

[a] If not stated otherwise, the solvents are CCl_4 for IR and NMR and aliphatic hydrocarbons for UV-VIS; the chemical shift is relative to TMS and refers to the H-bonded hydrogen. Reference numbers are given in parentheses.

[b] Estimated from ν'_{CO} data (20) and the emipirical eq. 2 in ref. 19.

[c] Methyl-2-methoxy-6-methylbenzoate.

[d] No solvent stated.

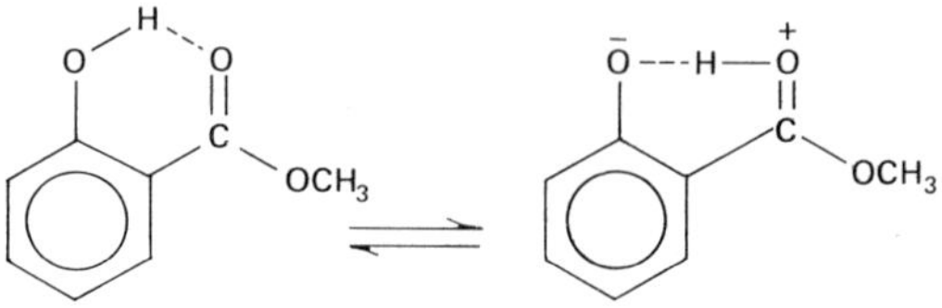

A* B*

Since the exact electron distribution in S_1 is not known both for the phenolic and for the phenolate-carboxonium form, the formulas should be regarded as more or less arbitrary mesomeric structures. Intermolecular proton transfer is not likely in nonpolar solvents; furthermore, the emission of MSA in 0.1M NaOH/CH_3OH has its maximum at 24,000 cm^{-1} (23), compared to the observed maximum at 22,600 cm^{-1}.

The ultraviolet fluorescence band could be due to:

1. a second ground state species.
2. emission from A^*, before B^* is formed.
3. emission from A^* in equilibrium with B^*.

Two ground state species are improbable in a pure nonpolar solvent. Possibilities 2 and 3 could be distinguished by fluorescence decay measurements, which are not yet available. Quenching experiments using CS_2 showed that the two components of the fluorescence spectrum are quenched to the same degree pointing to equal lifetimes of A^* and B^* and thus to equilibrium conditions (2). Lowering the temperature decreases the intensity of the ultraviolet component. The enthalpy of $A^* \rightarrow B^*$ has been estimated from the temperature dependence of the relative intensities of the ultraviolet and the blue components to be $\Delta H^* = -1.0 \pm 0.5$ kcal/mol (4) (-0.7 kcal/mol) (24). The ground state enthalpy ($A \rightarrow B$) is then $\Delta H = +13.5 \pm 0.5$ kcal/mol, based on the difference in the emission frequencies $\Delta\nu' = 5100$ cm^{-1} (=14.5 kcal/mol).

The activation energy of proton transfer has been estimated to be $E \leq 2$ kcal/mol (4) according to

$$k_{A^*B^*} = \nu_{OH} \times \exp(-E|RT)$$

where ν_{OH} is the stretching frequency, $k > 1/\tau_o$ ($>10^8 s^{-1}$) and $T = 77^oK$ where proton transfer is still observed. From 4^oK

experiments an even smaller limit of $E \leq 0.1$ kcal/mol is derived, suggesting proton tunneling as transfer mechanism (24).

Since no quantum efficiencies have been reported, the role of intersystem crossing or internal conversion in A^* and B^* is not known. MSA shows no phosphorescence in solid methyl cyclohexane (25) so that T_1 is either not populated or, more likely, very short-lived.

Methyl 5-ethoxy salicylate also shows two fluorescences (24), but in this case the short wavelength component (25,300 cm^{-1}) is the stronger one at room temperature. The enthalpy has been determined to be $\Delta H^* = +0.9$ kcal/mol (24). Thus, increasing the temperature has the opposite effect on the A^*/B^* ratio as in MSA (24).

In alcoholic solvents the behavior of MSA is more complex. Although it has been stated (2,4) that in alcohols the ultraviolet component is not observed, a more recent paper by Weller's group (24) reports an ultraviolet fluorescence of MSA in ethanol. Investigations at our laboratories showed (23) that in methanol, ethanol, water, and acetic acid there are two strong emissions with maxima at 22,500 and 28,200 cm^{-1}, whose relative intensity depends on the excitation wavelength. This points to different ground state species, a chelated (A) and a broken hydrogen bond (C) form (the fluorescence maximum of methyl 2-methoxy benzoate in CH_3OH is at 29,300 cm^{-1}) (1). This effect is known from 2-hydroxy benzophenone derivatives (see Section V). NMR measurements (23) indicate the presence of the C form in quantities of a few percent in methanol. It should be noted that two strong fluorescences have been observed in the vapor spectrum of MSA at 110°C (17).

A third species has been detected in methanol solution (23), showing an emission maximum at 24,000 and a corresponding excitation maximum at 28,600 cm^{-1}. This is due to traces of the phenolate ion (D) generated by dissociation of MSA.

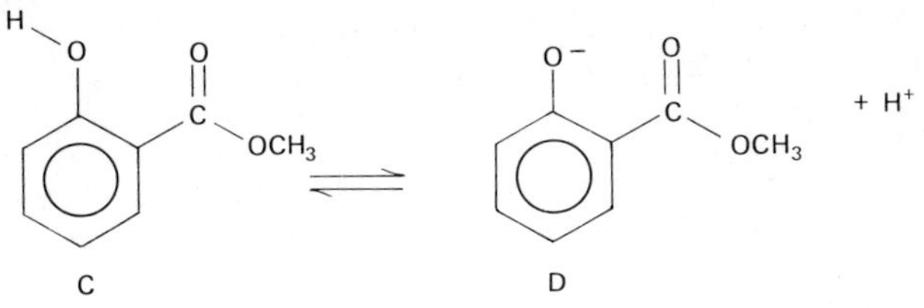

Salicylic anilide (SAN) has been investigated by Naboikin and co-workers (21). The spectral behavior of SAN is much more solvent dependent than that of MSA.

The absorption spectrum in heptane (Table 2) is characteristic of the structure with intact intramolecular H bond (A).

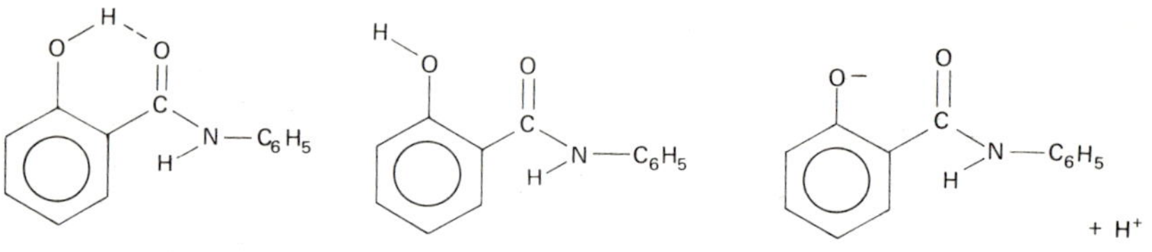

A C D

In ethanol, the 31,850 cm^{-1} maximum (A) is reduced to a shoulder near 33,000 cm^{-1}. Whereas in heptane a normal "Weller shift" of fluorescence indicating intramolecular proton transfer is observed, in alcohols we note a fluorescence band at higher frequency with a maximum at 23,800 cm^{-1}. The latter is due to the anion (D) and corresponds to the absorption maximum of D at 29,400 cm^{-1}. The intensity of this emission increases with increasing pH and remains constant above pH = 9 (in ethanol). The anilide of 2-methoxy benzoic acid shows a ultraviolet fluorescence that was too weak to be recorded (21). Low quantum efficiency may, therefore, be the reason that the fluorescence of C has not been detected. The only evidence of the presence of C is the close similarity of the absorption spectra of SAN and the methoxy derivative in ethanol (21).

SA, too, shows anomalous fluorescence (1,2), but a straightforward interpretation is not possible in this case. The long wavelength absorption maximum is close to that of the MSA (Table 2), and a fluorescence with a maximum at 22,100 cm^{-1} (CH_3OH) has been reported by Weller (1,2). Recent experiments of our group have shown that this fluorescence depends critically on the concentration of SA (25). The emission maximum in methanol is shifted from 22,700 (10^{-2}M) to 22,900 (10^{-3}), 24,600 (10^{-4}) and 24,800 ($\leq 10^{-5}$ M). Although a very small dissociation constant of the carboxylic group of SA in methanol (pK = 7.9 (26), compared to pK = 3 in water) has been reported, the concentration dependence is due to dissociation into salicylate ions. This is evidenced by measurements of the decay times and quantum efficiencies (25).

The absorption and emission maxima of SA in different states of protonation are compared in Table 3. The data show that Stokes shifts of about 6000 cm^{-1} are usual for compounds in which intramolecular proton transfer does not occur. Concentrated solutions of the acids where dimerization can occur show very large shifts of about 11,000 cm^{-1}, regardless of whether intramolecular hydrogen bonds are formed or not. (The association of SA has not been studied in alcoholic solvents; in benzene there is considerable association even at 80°C (apparent $\bar{M}/M$ = 1.4 in 0.01 molar solution)(33).

Intramolecular proton transfer in salicylic derivatives can be understood qualitatively considering the increased acidity of phenolic OH groups and the increased basicity of carboxyl and carbonyl groups in the first excited singlet state (1,2). From the shift of the absorption spectrum of MSA in alcoholic NaOH compared to that of a neutral solution ($\Delta\nu'$ = 2,900 cm^{-1}), ΔpK = 6.09 (6). With pK_G (OH) = 10.19 we have $pK_S{}^*$ = 4.1. Since ΔH^* is very small (2), $pK_S{}^*$ (CO) should have approximately the same value.

B. Hydroxy Naphthoic Acids and Derivatives

The following hydroxy naphthoic acids and their esters have been investigated: 3-hydroxy-2-naphthoic acid (29,35-37), 4-hydroxy-1-naphthoic acid (29) and the corresponding methoxy derivatives. The spectral data are summarized in Table 4. The infrared (and NMR) data suggest a relatively strong intramolecular hydrogen bond in 3-hydroxy-2-naphthoic acid and the corresponding methyl ester. The ultraviolet absorption spectra of the derivatives showing this bond are red shifted compared with those lacking the hydrogen bond (37-39). A theoretical explanation of this shift has been attempted (40) using the PPP method and allowing for partial covalency of the hydrogen bond. The same method reproduced satisfactorily the shifts observed in the salicylic derivatives (40). The influence of solvent polarity on the spectra of hydrogen-bonded esters is insignificant (38). The spectra of the acids are subject to a stronger solvent influence (38).

The fluorescence of methyl 3-hydroxy-2-naphthoate has been investigated by Naboikin and co-workers (37). They found two emissions in hydrocarbon solvents with maxima near 23,800 and 15,400 cm^{-1}. The former should be due to a normal emission, analogous to $A^* \rightarrow A$ (see Section III-A). The Stokes shifts amounts to 3200 cm^{-1}, relative to the long wavelength absorption maximum. A shift of 4000 cm^{-1} is observed in methyl 3-methoxy-2-naphthoate. The red fluorescence (15,400 cm^{-1}) is rather weak,

TABLE 3. Absorption and fluorescence maxima of salicylic acid and 2-methoxy salicylic acid (cm^{-1}).

Species	Solvent	ν'(Abs.) (cm^{-1})	ν'(Fl.) (cm^{-1})	$\Delta\nu'$
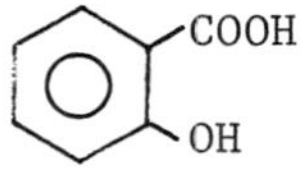	CH_3OH	33,000 (1)	22,100 (10^{-3}m) (1) 24,800 (10^{-5}m) (25)	10,900
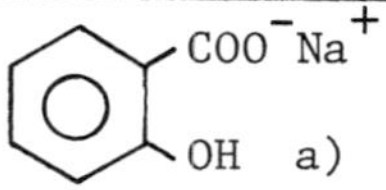	C_2H_5OH	33,600 (28)	24,800 ($2x10^{-3}$m) (28)	8,800
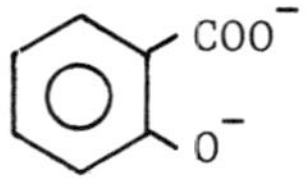	6mKOH/H_2O	31,100 (1)	25,500 (1)	5,600
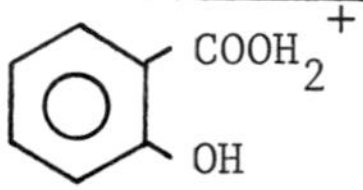	conc. H_2SO_4	30,400 (1)	24,700 (1)	5,700
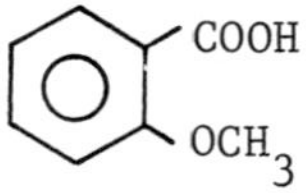	CH_3OH	33,800 (2)	27,800 (10^{-3}m) (2) 23,700 (conc.) (29)	6,000 10,000
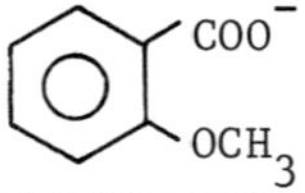	KOH/ CH_3OH	35,800 (32)	24,700 (29) 1m $NaOH/H_2O$	∿11,100

[a]Decay time in ethanol saturated with air τ_F = 8.8 ns (30), quantum yield in water η_F = 0.28 (31).

so that it can be observed only at high concentrations. Since it has been claimed that impurities have been excluded, there remain three possible explanations for this band: formation of dimers or excimers, or intramolecular proton transfer. The formation of dimers has been refuted since no corresponding changes in the absorption spectra (VIS and IR) have been found. Excimers are also unlikely, because the red fluorescence has been observed in solid solution in polystyrene where diffusion is restricted. As the most probable explanation there remains intramolecular proton transfer, as in MSA. Unfortunately, no attempt has been made to show whether the blue and red emitting levels are in equilibrium or not. The 15,400 cm^{-1} band has not been observed by Hirota (36) and Ware (35) in either the methyl ester or in the free acid (see below). It may therefore be due to an artifact. The anion of the ester emits at 18,200 cm^{-1} in alcoholic NaOH, corresponding to an absorption maximum near 25,000 cm^{-1} (37).

The free acid shows an anomalous Stokes shift only in proton-accepting solvents (acetone, acetonitrile) or mixtures such as toluene/pyridine (36,35), whereas in benzene only normal fluorescence with its maximum near 23,700 cm^{-1} is observed. The system 3-hydroxy-2-naphthoic acid in toluene plus small amounts of pyridine has been studied in some detail by Hirota (36) and Ware (35). Two fluorescence emissions have been observed near 24,000 and 19,000 cm^{-1}. The two bands have been interpreted in analogy to those of MSA as being due to excited state equilibrium $A^* \rightleftarrows B^*$ (Section III-A) and $\Delta H^* = -3.1$ kcal/mol has been calculated from the temperature dependence of the relative intensity (36). This interpretation has been criticized by Ware (35) on the basis of a detailed kinetic study using steady-state and flash techniques. According to Ware, pyridine acts as static quencher by formation of a ground state complex. The temperature dependence observed by Hirota should be given by the enthalpy of formation of this complex, which has been measured to be $\Delta H = -3.2$ kcal/mol.

The proposed mechanism of the anomalous fluorescence of this acid consists in the formation of the hydrogen-bonded complex, followed by absorption of a photon and intramolecular proton

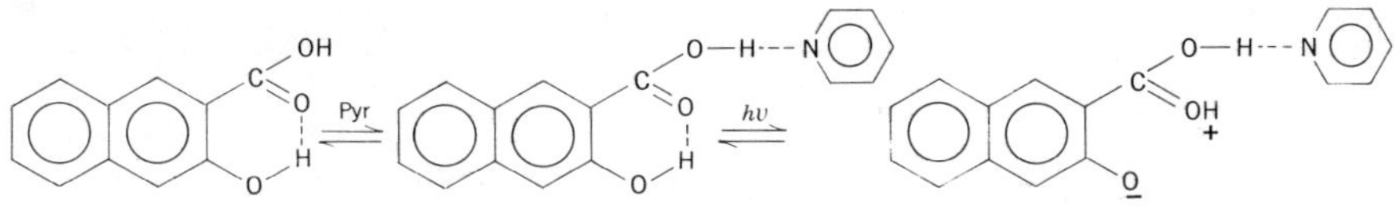

TABLE 4. Spectral data of hydroxy naphthoic acids and their derivatives[a] (cm^{-1}).

	IR		NMR	UV - VIS			
	OH stretching	C = O stretching	OH chem. shift (ppm)	Absorption(ε)	Fluorescence	τ_F (ns)[b]	$-\Delta H$ (H-bond) (kcal/mol)
3-Hydroxy-2-naphthoic acid	3230(29) (-OH)	1678(41)	-	27,200(2600)(38)D	23,800(29) 23,700(36) 24,400(35) (15,400)(29)	15.8	6.3[c]
3-Methoxy-2-naphthoic acid	3366(62) 3395(29) 3359(39) (-COOH)[d]	7151(39)	-	28,600(1650)(39) 30,500(1350)(61)D	23,200(29)	12.1	3.2(34) 2.8(62) 0.5 in S_1(62)
Methyl-3-hydroxy-2-naphthoate	3268(19,62) 3280(77)	1674(19)	10.11(6)[e]	27,350(2700)(38) 27,000(2630)(39)D shoulder 26,000	23,800(37) (15,400(37))	16.4	5.7(19,62) 5.0(37) 12.9 in S_1(62)
Methyl-3-methoxy-2-naphthoate	-	-	-	29,600(1740)(38) 29,400(1690)(39)D	25,400(37)		
4-Hydroxy-1-naphthoic acid	-	-	-	31,800(19,000)(61)D	27,000(29)		
4-Methoxy-1-naphthoic acid	-	1677	-	32,200(10,000)(61)D	No or very weak (29)		

[a]Solvent: Carbon tetrachloride for IR and NMR, hydrocarbon solvent for UV + VIS; D = dioxane. Reference numbers are given in parentheses.
[b]Toluene, 25°C (35).
[c]Estimated according to ref. (19).
[d]COOH forms a weak H-bond to $-OCH_3$.
[e]8.68 ppm rel. to C_6H_{12} (60).

transfer. At high concentration of the acid a complex spectral behavior is observed due to association and quenching by H^+ ions (29).

4-Hydroxy-1-naphthoic acid does not show anomalous fluorescence shifts, nor does 2-hydroxy-1-naphthoic acid.

C. Hydroxyphenyl Benzothiazole and Related Compounds

2-(2'hydroxyphenyl)-benzothiazole, the corresponding benzoxazole, and several derivatives show anomalous spectroscopic properties similar to those reported in the previous sections (Flavian and Cohen (42), Williams and Heller (43).) The intramolecular hydrogen bond is formed between the phenolic 2'OH group and the aza nitrogen group of the heterocycle.

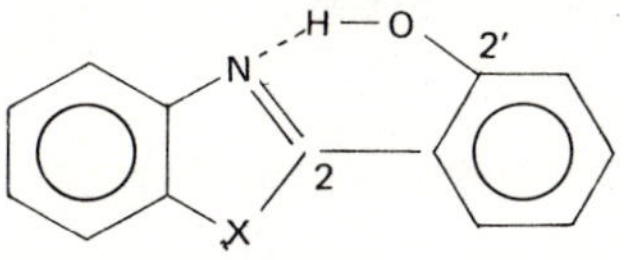

There is evidence by NMR that the proton is located at the indicated position in the ground state (τ = 12.06 ppm, X = S; τ = 11.15 ppm, X = O) in $CDCl_3$ (51).

The absorption and fluorescence of the thiazole has been studied in non-hydrogen bonding and hydrogen bonding solvents and in the solid state. The discussion will therefore mainly be concerned with this substance, although many derivatives have been studied; spectral data of some of them are compiled in Table 5.

In hydrocarbon solution, the thiazole shows absorption maxima at 30,300 and 31,250 cm^{-1} and fluorescence with a maximum near 20,000 cm^{-1} (19,350 in the solid) (42). In ethanol, the intensity of these absorption bands decreases and a new, concentration-dependent absorption near 25,600 cm^{-1} appears. Together with this change, an additional fluorescence band appears at 27,000 cm^{-1}. The excitation spectrum of the long-wavelength fluorescence coincides with the long-wavelength absorption and that of the short-wavelength fluorescence with the absorption which is also observed in the nonpolar solvent (42). This behavior points to two different ground state species in ethanol, as has been found in several salicyl anilines (see Section IV).

The study of solid state fluorescence and photochemical stability (43) showed the following results:

1. Only those derivatives having 2'-OH or 2'-OD are fluorescent.
2. The quantum efficiencies depend strongly on the substituents, η_F = 0.01 to 0.6.
3. Deuteration of the phenolic OH reduces η_F.
4. Photochemical stability is greatly enhanced in the derivatives with intramolecular hydrogen bonds.
5. Fluorescence decay times are smaller than 10 ns.
6. No phosphorescence has been found in solids and in plastic solutions.
7. Fluorescence spectra seem to consist of two overlapping emissions in most derivatives.

The following reaction course has been proposed as the predominant one (43):

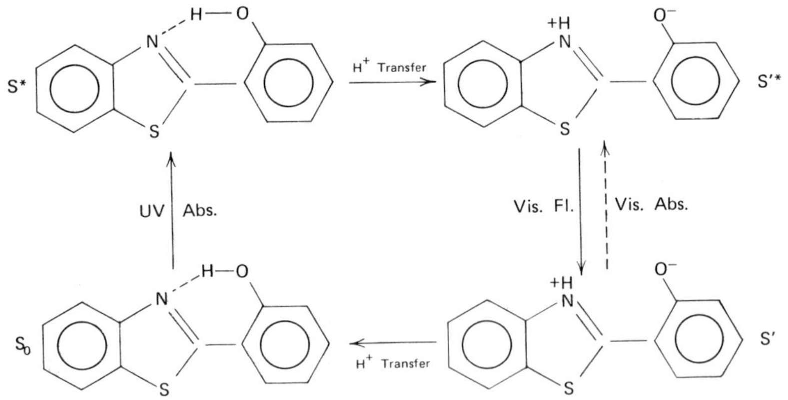

Other quinoid and ionic structures, both cis and trans forms by rotation about 2(C)-1'(C) may be involved to a minor extent. For comparison, a model compound of S' has been prepared

TABLE 5. Spectral properties of azoles[a].

Substance	Solvent	Absorption (cm^{-1})	Fluorescence (cm^{-1})	η_F	Photo-chem.[b] stability
2-Phenyl benzothiazole	Solid			0	
2'Methoxy phenyl benzothiazole	Solid			0	
2'-Hydroxy phenyl benzothiazole	Solid		19,350	0.31	10
	Trimethyl-hexane	28,600(sh)+ 29,600 (51) (16,500)	17-22,000(42)		
	Ethanol	25,600(2870)(42) 30,400(13000)(42)[c]	20,400(42)+ 27,000 (77°K)		
	Chloroform		∿19,400 + 18,500		
2'Deuterohydroxy phenyl benzothiazole	Solid		19,400	0.21	
2'Hydroxy-5'-methoxy phenyl benzothiazole	Solid		17,400	0.13	36
2'-Deutero-hydroxy 5'-methoxy phenyl benzothiazole	Solid		17,350	0.08	
2'-Hydroxy-5' methyl phenyl benzothiazole	Solid		18,800	0.57	97
2'-O^-, 3-N^+-CH_3[d] phenyl benzothiazole	Solid dry		16,550	0.05	0
2'Hydroxy-phenyl benzoxazole	Solid		19,750	0.42	38
	Ethanol	27,000(42)	20,400+ (42) 27,000		
	Trimethyl-hexane	29,800 + (51) 31,300(17,400)			

TABLE 5 (cont'd)

2'-Deutero hydroxy benz-oxazole	Solid	19,900	0.34
2'-Hydroxyphenyl-benzimidazole	Solid	21,700	0.035

[a]After Williams and A. Heller (43) if no other reference; NMR see text.

[b]Percent residual brightness after 26 h UV illumination (75 mW/cm^2).

[c]ε at 5.33×10^{-6} mole/l; depends on concentration (42).

[d]Model compound, see text.

which exists in a yellow, wet-stable and an orange, dry-stable modification (43):

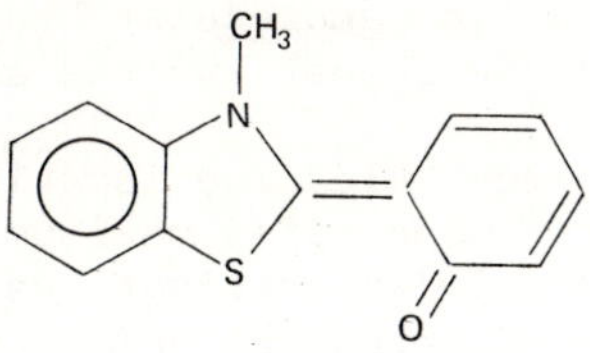

(1) Yellow

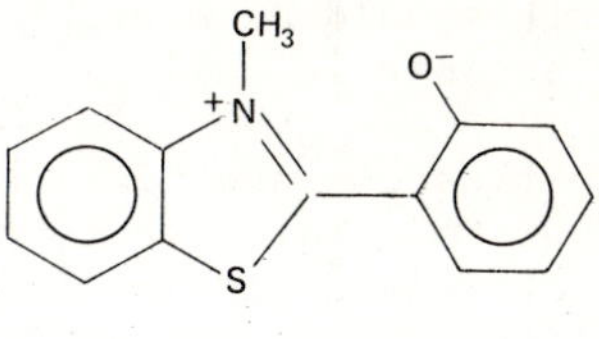

(2) Orange

The assignment of the above structures is somewhat tentative; in the yellow modification an IR band at 1635 cm^{-1} is found, which points to a quinoid structure; this band is missing in very dry samples. Form (1) fluoresces yellow-green, (2) has a maximum at 16,550.

Deuteration of 2'-OH reduces η_F (Table 5). If no impurity effects are involved, this would indicate that either the rate of $S^* \rightarrow S'^*$ is reduced or that radiationless deactivation is enhanced, the latter effect differing from the usual effect of deuteration. No measureable activation energy is needed to pass from S^* to S'^*, since η_F at 7^oK is decreased only by about 35 percent both in the H and the D product. Proton tunneling is, therefore, proposed as the mechanism for interconversion.

D. Azines

Weller and Wolf (44) investigated a great number of azines, derived from aromatic o-hydroxy aldehydes and o-hydroxy-ketones. We shall discuss only the aldazine derived from salicylic aldehyde.

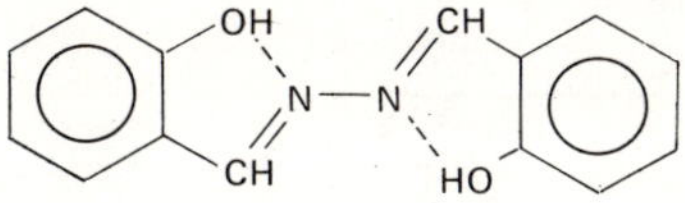

Salicyl aldazine

The intramolecular hydrogen bond is indicated by a red shift of the ultraviolet absorption maximum, compared to the 2,2'dimethoxy derivative (Table 6). In the absorption spectrum of the latter in toluene, an $n-\pi^*$ band is found as a shoulder around 25,000 cm^{-1} which vanishes in alcohol due to the usual blue shift. The same shift should be brought about by intramolecular hydrogen bonding so that in salicyl aldazine the lowest excited state is $\pi-\pi^*$.

The fluorescence maximum of salicyl aldazine is shifted by about 9000 cm^{-1} relative to the first absorption maximum. The quantum efficiency is small ($\eta_F < 2 \times 10^{-3}$); derivatives without intramolecular hydrogen bonds are nonfluorescent unless they are protonated.

In order to estimate the excited state basicity of the azine group, the absorption and fluorescence spectra have been measured in $HClO_4/CH_3OH$ (Table 6). A calculation using eq. 6 showed that $pK_G - pK_S^* = -8.4$; this corresponds to a greatly enhanced basicity in the excited state of benzaldazine and 2,2' dimethoxy-benzaldazine. In salicyl aldazine the bathochromic shift upon acidification is smaller (2800 cm^{-1}) due to the already H-bond-lowered

TABLE 6. Salicylaldazine and reference substances (44).

Substance	Solvent	UV absorption (cm^{-1}) (ε)	Fluorescence (cm^{-1})
Benzaldazine	Cyclohexane	33,500	No
	0.1m $HClO_4/CH_3OH$	29,300	23,600
Salicylaldazine	Toluene	27,600 (20,000)	18,700[a]
	0.5m $HClO_4/CH_3OH$	25,400	20,300
2,2'-Methoxy benzaldazine	Toluene	29,150 (29,500)	No
	0.05m $HClO_4/CH_3OH$	25,300	19,600
1,4-Diphenyl-butadiene	Toluene	30,000	26,300
1,4-Bis(2'-hydroxyphenyl) butadiene	Toluene	28,500 (39,000)	24,500
	OH^-/CH_3OH	25,300	

[a]($\eta_F < 2 \times 10^{-3}$ with η_F (diphenyl butadien) = 0.44 (45).

absorption maximum. A corresponding experiment with 1,4-bis(2' hydroxy phenyl) butadiene showed that the acidity of the OH group is increased by 6.8 pK units. From these values it can be concluded that the equilibrium constant for intramolecular proton transfer is changed by 10 to 15 orders of magnitude in favor of the N-protonated form (44).

IV. PHOTOCHROMIC ANILS

A. Introduction

Photochromism has been observed in salicylidene aniline, in many of its derivatives and Schiff's bases derived from other orthohydroxy aldehydes (46). This has been investigated extensively by Cohen, Hirschberg, and Schmidt (5,47-50,42).

It is not intended here to deal with all photochromic anils, but rather to discuss particular experiments that lead to a better understanding of the proton transfer process. In this context, the papers by Cohen and Flavian (42,51), Ottolenghi, McClure, and Potashnik (52-54), and Becker and Richey (55,63) deserve particular attention.

The reversible color change (mostly from yellow to red) has been observed in the crystalline state, in solid solutions and, if flash techniques (56) are used, also in liquid solutions. There is no clear-cut distinction between the photochromic anils and the molecules discussed in Section III since fluorescence can be observed in most compounds. The main difference seems to be the stabilization of an intermediate (produced by proton transfer) by geometrical rearrangement in the excited state, which is possible in the photochromic anils but not in the "fluorescent" compounds.

B. Ground State Properties

In typical anils the hydrogen bond is formed from the phenolic OH group to the nitrogen. This is the case, for example, in the parent compound N-salicylidene aniline (SAA) and in 4-chloro and 4-bromo-N-salicylidene aniline (42). In some derivatives, however, proton transfer occurs in the ground state to give quinoid tautomers, in equilibrium with the phenolic forms (42).

The formation of quinoid tautomers is favored by hydrogen bonding and polar solvents. As an example, tha absorption of N-(2-hydroxy-1-naphthylidene) aniline (HNA) in cyclohexane and in ethanol is shown in Fig. 4. The long-wavelength band in ethanol is attributed to the quinoid tautomer (B), which is present only in a small amount in non-polar solvents (50,58,51). A similar tautomerism has been observed in 1-(phenylazo)-2-hydroxy

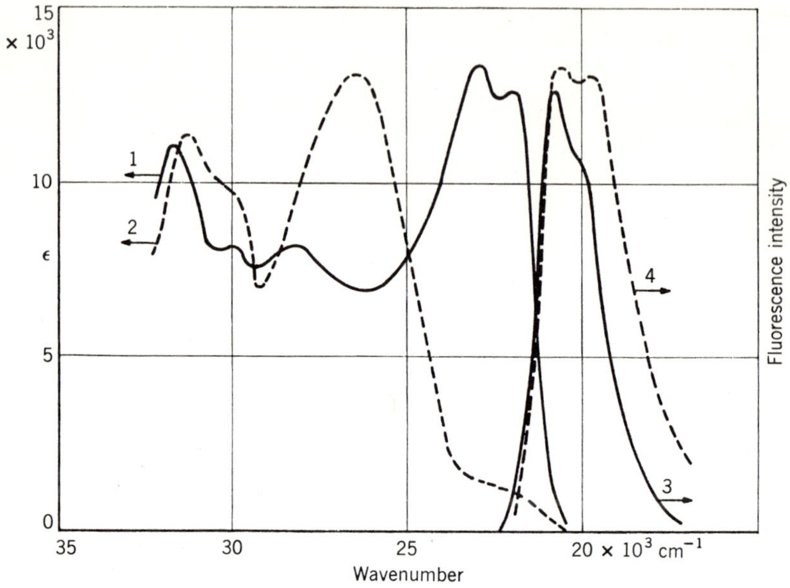

FIGURE 4. Absorption and fluorescence spectra of HNA, curve 1: absorption in absolute ethanol[58]; curve 2: absorption in cyclohexane[58]; curve 3: fluorescence in ethanol, 77°K, excitation at 25,600 cm^{-1} [42]; curve 4: fluorescence in 3-methylpentane, 77°K, excitation at 26,000 cm^{-1} [63]; the fluorescence spectra are not corrected.

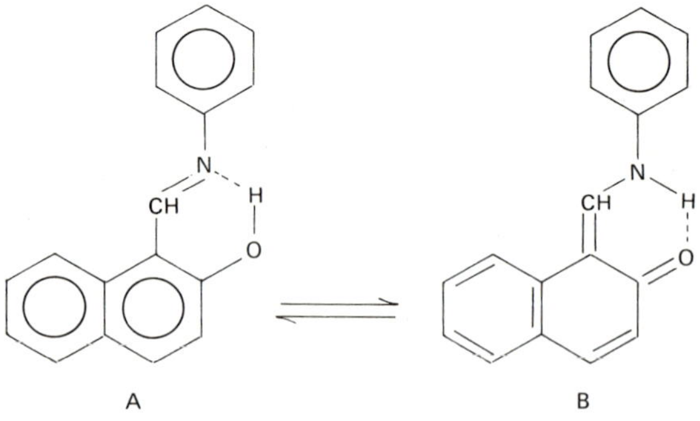

napthalene (64), whereas 2-hydroxy-azobenzene exists in the "enol" form only.

Dudek and Dudek (58) determined the equilibrium constants of the tautomerism in different solvents by NMR, using N^{15}-H spin-coupling. Since the concentration of the two isomers is known, it was possible to calculate the extinction coefficients (Table 7). It has been shown that in CCl_4 14 percent of HNA is present in the cis-quinoid form, compared to 37 percent in $CDCl_3$ (29.5°C). No quinoid form of SAA has been detected in $CDCl_3$, but 15 percent of SAA exists as quinone at 0°C in ethanol.

The proton chemical shifts of OH (A) and NH (B) show that the two tautomers are chelates (Table 7). Although the proton exchange is rapid on the time scale of the NMR experiment, the tautomers exist as separate molecules, as is shown by the ultraviolet absorption spectra. The potential of the proton in the hydrogen bond, therefore, has two minima (58).

The enthalpy of the reaction A $\rightleftarrows$ B (HNA) has been deduced from the temperature dependence of the equilibrium constant (58); ΔH = -1.5 ($CDCl_3$); -0.5 (CCl_4); -0.08 (CH_3CN) kcal/mol. These data would indicate that the "enol" form has the higher energy. This is, however, not consistent with the findings by Ottolenghi (54) which indicate that the enol form is favored at low temperatures in HNA and SAA. His absorption spectra showed that ΔH = +0.92 kcal/mol (HNA in methyl cyclohexane/isopentane). As a possible explanation of this discrepancy it has been suggested that due to the rather high concentrations necessary in NMR experiments, dimers may play a role in these measurements, whereas the ultraviolet work is done at very low concentrations and should give the true equilibrium constants (58). At low temperatures dimers have been found in absorption measurements, too (species Q_B in ref. 58).

The partly quinoid nature of SAA in ethanol is indicated by the absorption spectrum, compared to that in nonpolar solvents (Fig. 5). The concentration of the quinoid form in methyl cyclohexane is $\ll$ 1 percent (Fig. 5, curve 2). The spectra of the quinoid isomers are important for the elucidation of the mechanism of proton transfer, since similar structures are created in the excited state of the photochromic anils. It should be noted that intramolecular hydrogen bonding is not necessary for the formation of quinoid ground state tautomers, as has been found by Ledbetter (65) who observed that anils derived from p-hydroxybenzaldehyde show tautomerism in solution. It has been postulated that the stabilizing effect of the solvent is due to hydrogen bonding to the amine group of B.

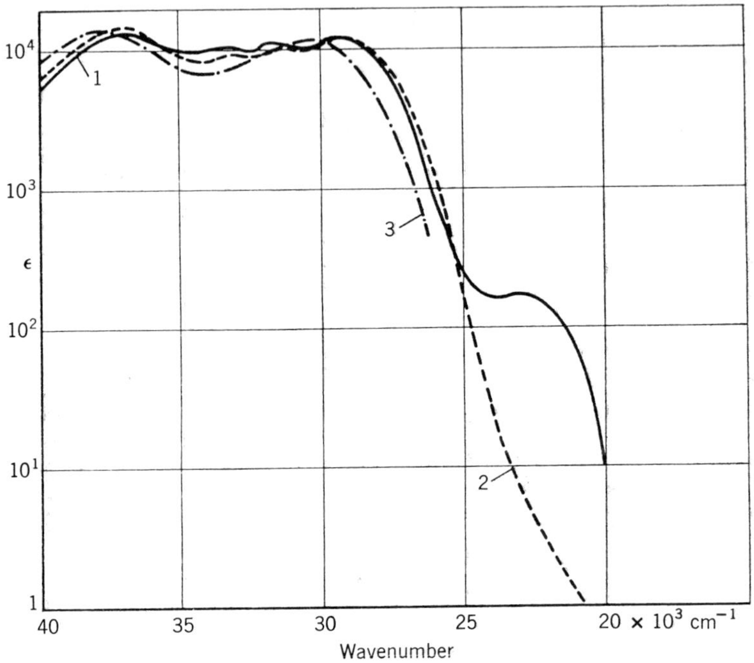

FIGURE 5. Absorption spectra of zone-refined SAA in ethanol (1) and methyl cyclohexane (2)[25]; the formation of tautomers is disregarded in the calculation of the molar absorption coefficient (ε); curve 3: N(2-methoxy benzylidine)aniline in methanol[52].

C. Absorption and Fluorescence Spectra

The absorption spectra of SAA and HNA have already been presented in Figs. 4 and 5 and in Table 7. The electronic transitions of SAA have been correlated with those of structurally related substances such as trans-benzylidene aniline (BA), trans-azobenzene, and trans-stilbene by Ottolenghi and McClure (52).

The lowest-energy $\pi-\pi^*$ transition has its maximum in the 32,000 to 33,000 cm^{-1} region in stilbene, azobenzene, and BA. It has been shown that this (long-axis polarized) absorption in stilbene is due to a transition between the seventh and eighth molecular orbitals ($^1A_o \rightarrow {}^1B_1$) (66). The decreased symmetry in BA

TABLE 7. Spectroscopic properties of selected anils.

Substance	IR - OH Stretching	NMR Chem. Shift (ppm)	UV absorption (ε)[a]	Fluorescence
N-benzylidene aniline	-	-	32,200(16,900)isooctane(57) 32,200(8,530)ethanol (57)	
N-Salicylidene aniline (SAA)	∿2800(25,69)[b]	O - H; 13.22(58),13.27(59)	29,400(11,200)isooctane(57)	18,500 in trimethyl hexane 19,000+18,000(sh)in methyl cyclohexane, 77°K(52)
		in $CDCl_3$ (25)	29,500(12,800)ethanol(57)	19,336 in 3 - MP (63)
		12.8 in CCl_4 N^{15}-H: 15.02 in	23,000(14,600)ketoform in methanol(58)	19,200+18,400(sh) in ethanol, -78°C (54)
		C_2H_5OH, 0°C (58)		19,600+18,500(sh)in EPA 77°K (52) 19,412 in EPA (63)
N-(2-Methoxy benzyl-dine)aniline	-	-	30,600(12,100)in methanol (52)	
N-Salicylidene-o-toluidine (SAT)			29,000 in 3-methyl pentane 77°K ("enol") (55)	19,000 + 18,200 (55)
			22,200 in 3-methyl pentane + propionic acid ("cis-ketoform") (55)	20,100 (55)
			21,300 in 3-methyl pentane ("trans-ketoform") (55)	17,850 + 17,000 (55)
N-(2-hydroxy-1-naphthylidene aniline (HNA)		N^{15}-H: 15.03 in CCl_4 0°C(58); 15.58 in $CDCl_3$, 0°C (58)	26,400 (15,800)("enol") 23,000 (22,400) + 22,000 (19,500)("keto") in CCl_4 (58)	20,700+20,200 (sh) in ethanol at 77°K (42) 21,000 + 20,000(sh) in ethanol at -78°C (54)

[a]Maximum of the long-wavelength band only.
[b]C-H peaks superposed.

and SAA allows additional transitions which are forbidden in stilbene and azobenzene (Table 7, Fig. 5). There is a bathochromic shift in the long-wavelength band from BA to 2-methoxy benzylidene aniline. The formation of the chelate ring in SAA includes a further red shift. The strong 29,400 cm^{-1} band in SAA is due to a transition analogous to the allowed stilbene transition indicated above (51,52). A shoulder near 28,000 cm^{-1} in the BA spectrum in ethanol has tentatively been assigned to an $n-\pi^*$ transition which is found at 23,800 cm^{-1} in azobenzene (52). Since hydrogen bonding increases the energy of such transitions, in SAA the lowest excited singlet state is most probably $\pi-\pi^*$ (63). The lowest triplet state in SAA has recently been reported to be located near 16,000 cm^{-1} (92); this is in agreement with the estimate of the lowest $\pi-\pi^*$ triplet (52). In stilbene, T_1 is found at 17,400 cm^{-1} (52), in agreement with the calculated position.

The fluorescence of SAA is roughly a mirror image of the absorption of the _cis_-quinoid isomer (Table 7). Minor blue shifts have been observed if the hydrocarbon solvent is replaced by an alcoholic solvent, but the evidence is not sufficient to permit postulating an $n-\pi^*$ fluorescent state (63). Evidence of the $\pi-\pi^*$ character of the emitting state is the high absorption coefficient of the quinoid form of the order of $\varepsilon = 10^4$. Furthermore, fluorescence from $n-\pi^*$ levels of ketonic compounds is unusual. The fluorescence decay time has been measured to be $\tau_F = 20 \pm 10$ ns (54). The decay time $\tau = 80$ μs published earlier (53) only corresponds to a few percent of the total intensity and is due to delayed fluorescence rather than $n-\pi^*$ phosphorescence (54).

The large Stokes shift of the fluorescence shows that proton transfer occurs in the excited state of SAA. As has been pointed out by Richey and Becker (63), the emissions of the excited SAA in its phenolic form and of the _cis_-quinoid form are similar but not identical. From this and photochemical evidence (see Section IV-D) it is concluded that the species produced by excited state proton transfer from the phenol and by absorption of a photon from the quinoid tautomer are both excited _cis_-quinones of slightly different structure.

The fluorescence of salicylidene-o-toluidene (SAT) has been investigated by Becker and Richey (55). There is a 10,000 cm^{-1} Stokes shift between the first absorption maximum and the fluorescence maximum in hydrocarbon solvents (Table 7). Addition of small amounts of propionic acid to the 3-methylpentane solution of SAT produced some _cis_-quinoid tautomer whose fluorescence maximum is at 20,100 cm^{-1}, blue shifted by about 1000 cm^{-1}, compared to that of SAT "phenol." The _cis_-quinoid tautomer could further be produced by controlled heating of a solution of the photochromic isomer (presumably the trans-quinoid form) and recooling to 77°K. The emission of this species is nearly identical to that of the "phenol."

The fluorescence spectra of HNA in different solvents are illustrated in Fig. 4. In this case, too, the emission mirrors the absorption of the quinoid species. The emission is, however, dependent on the excitation wavelength (63). Curve 4 in Fig. 4 shows the fluorescence spectrum of HNA upon excitation in the phenol absorption region. Excitation of the cis-quinoid form produces a fluorescence spectrum in the same spectral region but of different shape (63).

In summary, anils show fluorescence from a state (most probably π-π^*) which is similar to the cis-quinoid tautomer observed in appropriate solvents even in the ground state. This state can only be produced by intramolecular proton transfer in the excited state. Since the energy difference between the phenolic and quinoid ground state isomers is rather small in the anils (see Section IV-B), the increase in (-OH) acidity and/or (=N-)basicity needs not be as large as in the case of salicylic esters.

D. Mechanism of the Photochromic Process

Ultraviolet irradiation of the anils of o-hydroxy aldehydes in solid solutions produces a change in color - mostly from yellow to red (5,47-51). In the solid state, a "soft" crystal structure, allowing some molecular movement is necessary in order to allow photochromism (47). The initial form is regenerated under the action of heat or visible light (eradication). In fluid solutions this process is so fast that photochromism can only be studied by flash techniques (56,67). The nature of the photocolored species and the mechanism by which it is produced and eradicated has been studied extensively but so far no complete and unambiguous mechanism has been proposed. The following section deals with the spectral properties of the photoproduced species, followed by a discussion of the different mechanisms proposed in the literature.

1. Absorption and fluorescence of the photoproducts. Irradiation of SAA and SAT with ultraviolet light in solid solutions produces a new absorption band with a maximum at 21,300 cm^{-1} (5,49,53,55,69). The 0-0 peak of this transition has been found at 17,550 cm^{-1} in a crystalline dibenzyl matrix at 4°K (53). The extreme sharpness of this spectrum, not expected for internally hydrogen-bonded molecules, is an indication of the trans-quinoid nature of the photoproduct (53).

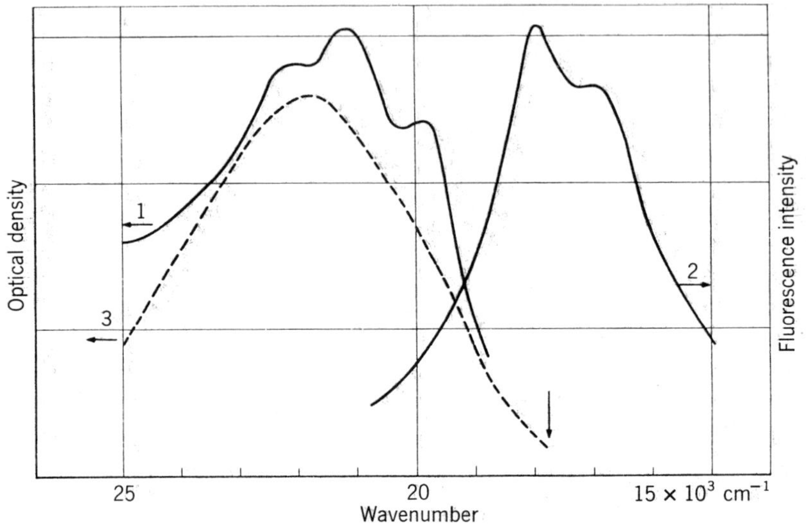

FIGURE 6. Spectra of the photoproducts of anils; absorption of SAA photoproduct in methylcyclohexane isopentane 2:1, 77°K after[68] (1); fluorescence of SAT photoproduct in 3-methylpentane, 77°K after[55] (2); millisec transient of SAA in n-heptane after[56]; the arrow indicates 0-0 peak found for SAA photoproduct in dibenzyl at 4°K.

Figure 6 shows the absorption spectrum of irradiated SAA and the emission spectrum of irradiated SAT. As can be seen, the spectra are red-shifted compared to the corresponding spectra of the cis-quinoid tautomers by about 2000 cm^{-1}. The transient observed after flashing a liquid solution shows a spectrum similar, although structureless, to the photoproduct in solid solution (Fig. 6).

Irradiation of HNA in solid solutions with ultraviolet light yields a cis-quinoid isomer with an absorption spectrum as in Fig. 4, curve 1, and no further species is produced. The photoproduct can be eradicated by light and heat.

2. Proposed mechanisms. From the spectroscopic evidence presented in the foregoing sections the most simple mechanism of the photochromic reaction would be (47,42):

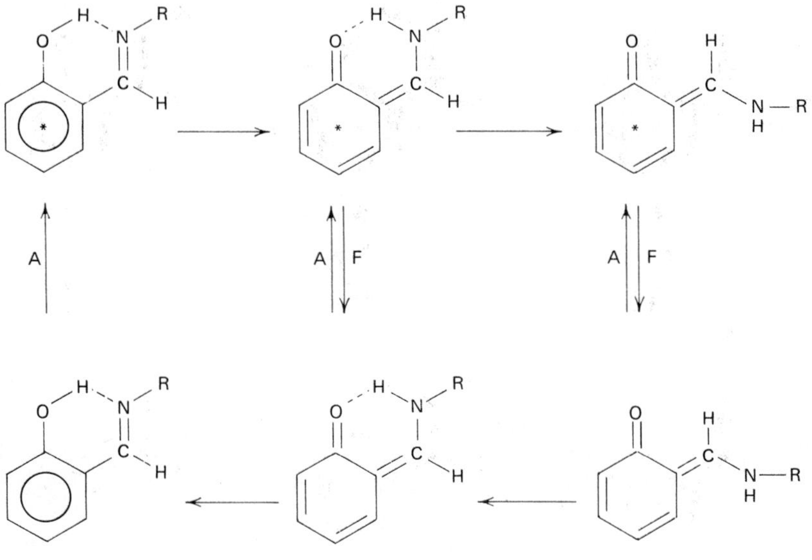

Phenol *cis*–quinone *trans*–quinone

It has been shown by Becker and Richey that this mechanism is too simple. Studying SAT (SAA yielded very similar results) these authors found that a cis-quinone is formed upon heating the photochemically produced transquinone just above the softening point of the solvent (3-methyl pentane), whereas heating to higher temperatures gives the phenol. Another method of producing the cis-quinone consists in adding a small amount of acid to the nonpolar solvent and cooling to low temperature. It has been shown that the cis-quinone in solid solution can neither be photochemically converted into the photoproduct (presumably the trans-quinone) nor can the phenol be regenerated photochemically from the cis-quinone. This proves that the excited cis-quinone cannot be an intermediate in the photochromic process.

The close similarity of the fluorescence spectra of phenol and cis-quinone, on the other hand, shows that the intermediate and the cis-quinone should have very similar structures.

Richey and Becker (63) later proposed that the red photoproduct, too, is a cis-quinoid species, since only a small activation is needed for the thermal photoproduct → cis-quinone process. Arguments against this view have been raised by Potashnik and Ottolenghi (54). They showed that 7-pnehyl (SAA)

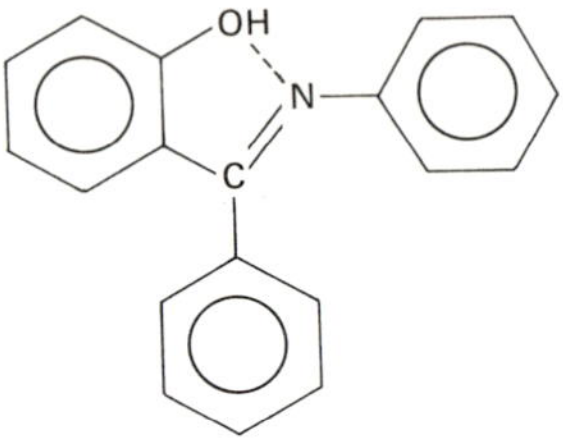

in which cis-trans isomerization is sterically hindered is not photochromic. Preventing the rotation about the C_7-N bond does not hinder photochromism, since 2-hydroxy phenyl benzothiazole is photochromic. In agreement with Becker, the authors maintain that competition between cis-quinone* fluorescence and isomerization is not the key to understanding the mechanism of SAA photochromism, since

1. the temperature effect on trans-quinone formation and fluorescence occur in different temperature ranges (53) and
2. the processes phenol $\xrightarrow{h\nu}$ trans-quinone (a) and cis-quinone $\xrightarrow{h\nu}$ trans-quinone (b) depend differently on the temperature and rigidity of the solvent; (b) occurs only in liquid solutions.

The authors conclude that the emitting state created after excitation of the phenol form is predominantly quinoid but not identical with the excited cis-quinone. It is speculated that the proton transfer process may not be a simple transfer between two minima of a hydrogen bond but rather involve another minimum in which the proton can be trapped. The same suggestion has been made previously by Shigorin (84) on the basis of theoretical arguments.

In liquid solution, it has been shown by Wettermark and Dogliotti (56) that a transient showing a similar absorption spectrum (Fig. 6) as the trans-quinone is formed, which decays in the millisecond range.

The decay is of first order and acid-catalyzed in polar solvents (56), but of second order in non-polar solvents (54).

In order to explain the kinetics of the acid catalysis, it is assumed (67) that a rapid equilibrium is established *trans*-quinone $\rightleftarrows$ *trans*-quinone$^-$ + H^+ and that neutralization of the anion again leading to "phenol" is the rate-determining step.

In non-polar solvents, the reverse reaction involves the formation of dimers (54), which may be identical with the dimers observed upon cooling.

The photochromism of HNA seems to be simpler than that of SAA, since no *cis*-*trans* isomerization is involved (63), the *cis*-quinone being the photoproduct. In this case, too, the emitting state which is produced by proton transfer from the excited phenol is similar but not identical with the excited *cis*-quinone. This is demonstrated by somewhat different fluorescence spectra obtained in the presence of ultraviolet and visible exciting light.

V. NONLUMINESCENT SUBSTANCES

A. Photochemical Stability and Proton Transfer

It has been observed that many substances showing an intramolecular hydrogen bond are photochemically very stable. The stability of the 2-hydroxy phenyl benzothiazole has been referred to in Section III-C, and some data are given in Table 5. The photochromism detected by flashing nonpolar solutions of 2-hydroxy phenylbenzothiazole (54) does not exclude photostability since it is a reversible process. Of prime importance as photostable substances are derivatives of 2-hydroxy-benzophenone (HB) and 2-(2'-hydroxy-phenyl)-benzotriazole.

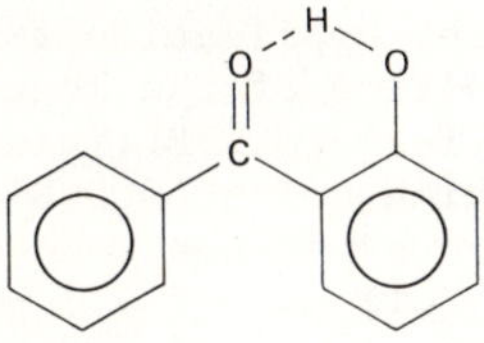

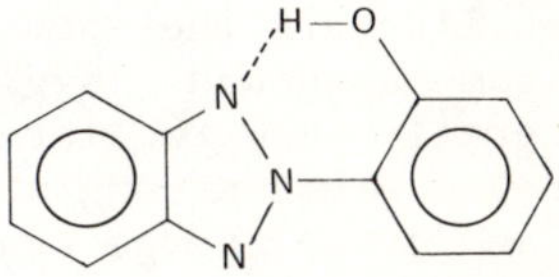

These substances are used as ultraviolet stabilizers (6,9,70-72) on account of their intense absorption in the near ultraviolet combined with high photostability and lack of photochemical sensitization. They are nonluminescent, at least in pure, nonpolar solvents, that is, if the intramolecular hydrogen bond is intact

(8).* It has been proposed that, in analogy to Weller's mechanism, proton transfer in the first excited singlet state is responsible for the fast nonradiative decay and high photostability (7,8). However, it is doubtful whether proton transfer alone can account for these effects (8,71,73,74). As in the foregoing sections, only a few substances will be discussed for which interesting spectroscopic or photochemical evidence is available.

B. Hydrogen Bonding

The infrared and NMR data are included in Table 8. They indicate moderately strong hydrogen bonds similar to those of salicylic acid derivatives. It has been suggested (82) that the stronger the hydrogen bond, as measured by the chemical shift, the better the ultraviolet stabilizing effect. H. J. Heller (9), on the other hand, observed that the hydrogen-accepting group should be weakly basic and that therefore the hydrogen bond should not be too strong. No free OH groups are found in solvents suitable for infrared measurements. Steric hindrance has been suggested as the reason for weaker chelation in 2,2'-dihydroxy benzophenone (75). Substitution in 4 position of HB by a methoxy group increases the stength of the hydrogen bond.

The acidity of HB and several derivatives has been measured by Hrdlovič et al. (83); pK_G (HB) = 9.37 in 5 percent ethanol/water. The excited state pK^* data (83) and their implications for the mechanism of radiationless deactivation will be discussed below.

C. Absorption Spectra

The ultraviolet spectra of these substances and many derivatives have been reported (6,9,81,83,79). Their shape and intensity is of utmost importance for the application as ultraviolet stabilizers. The spectral position of the long-wavelength maxima of some compounds is given in Table 8. Efficient stabilizers generally show two intense maxima below 40,000 cm^{-1} (9) (Fig. 7). The first strong absorption of HB has been reported to be due to an "ortho charge transfer" (CT) band (79); the weak $n-\pi^*$ band of benzophenone should be blue-shifted and hidden below the strong CT band (85). The assignment of the band near 30,000 cm^{-1} as ortho CT band is supported by quantum mechanical calculations (HMO-SCF) by Kysel' and Zahradnik (85). They show that in the first excited singlet state of HB a structure

*In a recent paper (94), very weak fluorescences were reported to occur at low temperatures.

TABLE 8. Spectral data of some benzophenones and benzotriazoles (cm^{-1}).

Substance	IR[a]		NMR[a]	UV - VIS	
	OH stretching	CO stretching	OH chem. shift	Absorption (ε)	Phosphorescence
2-Hydroxy-benzophenone	∿3100(18)	1628(76,25) 1633(79)[b]	11.9(25) 11.83(78)	29,760(5600)isooctane(81) 29,920,ethanol (81) 26,180(15,000),96% H_2SO_4 (86) 26,620(3,400) 0.2m NaOH (86)	23,900(0-0)ether/ethanol at 77°K (8) 21,850(max), ethanol 77°K(81) 20,100(max),96% H_2SO_4(81) 18,650(max),0.2m NaOH(81)
2-Hydroxy-4-methoxy-benzophenone	2960(75)	1620(25)	12.55 (75,25)[c]	30,860(10,000)isooctane (81) 31,000, ethanol (81) 27,150(31,600)96% H_2SO_4 (86) 27,320(6,600)0.2m NaOH(86)	23,600; 22,200; 20,800 ethanol/methanol at 77°K(80) 23,750(0-0) ether/ethanol at 77°K (81) 21,600(max)ethanol(81) 20,000(max), 92% H_2SO_4(81) 19,000(max), 0.2m NaOH(81)
2,2'-Dihydro-xybenzo-phenone	3200 (75)	1620-1640 (75)	10.4[75]		23,700 ether/ethanol at 77°K (8)
2-(2'-hydroxy-5'-methyl phenyl) benzotriazole		-	10.8[25])	29,400 (16,500) heptane(9)	

[a]In CCl_4 if not stated otherwise.

[b]CO (benzophenone): 1664 (77,79).

[c]2,4-Dihydroxy benzophenone: 12.6 ppm in isopropanol (73).

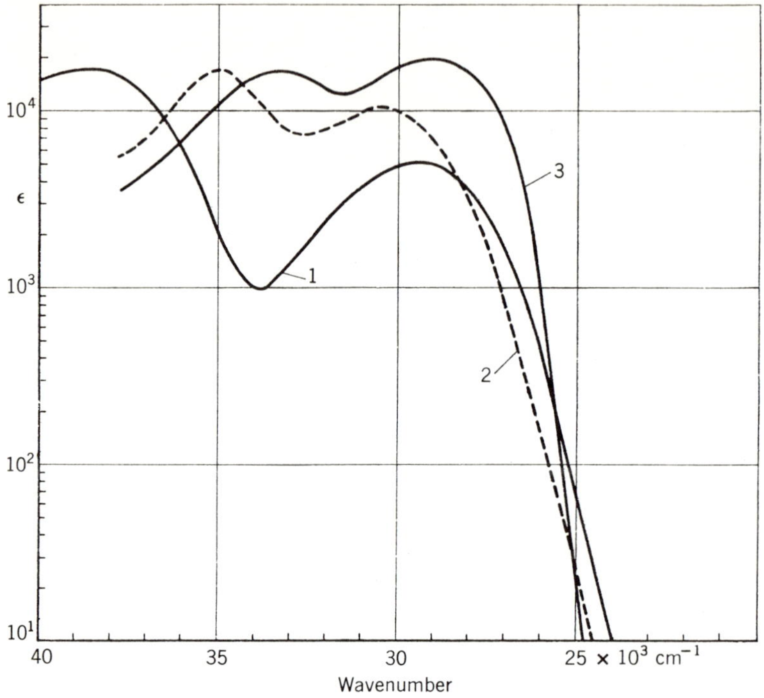

FIGURE 7. Absorption spectra of nonluminescent chelates; 2-hydroxy-benzophenone (1); 2-hydroxy-4-methoxy-benzophenone (2) and 2(2'hydroxy-5'methyl-phenyl)benzotriazole; (3) in CCl_4.

dominates in which the carbonyl oxygen group is more negative and the hydroxyl oxygen group more positive than in the ground state.

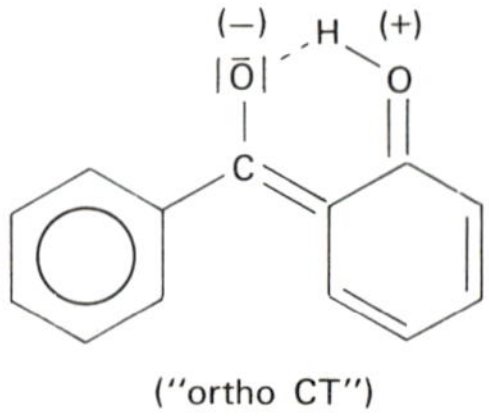

("ortho CT")

The predominance of this structure immediately suggests higher basicity of CO and increased acidity of OH. It has been shown from the spectra and by calculations that the first absorption band in all substituted HB derivatives studied has the same character as in the parent compound. Since the spectra of 4-methoxy and 4-hydroxy HB are very similar (85), the theoretical interpretation of the dihydroxy compound may be applied to the methoxy derivative as well.

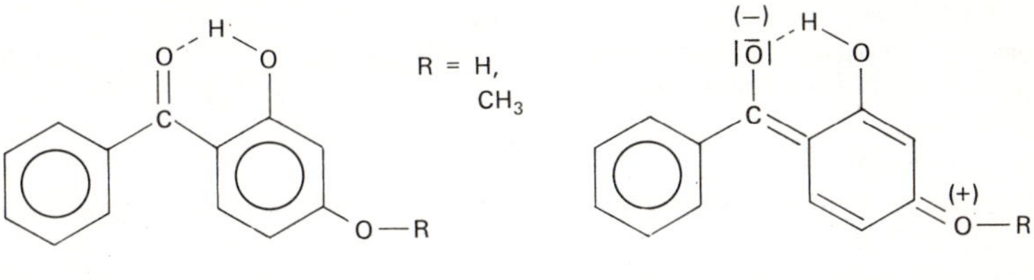

From the observation that the second band of 4-methoxy-HB (Fig. 7) corresponds closely to the first strong band of 4-hydroxybenzophenone and from calculations it has been concluded that these bands are due to a "para CT" transition (85) with an important contribution by a structure with a positively charged para oxygen.

Protonation and deprotonation of HB and derivatives causes bathochromic shifts (81,86) (Table 7), in accordance with HMO calculations. These shifts, too, indicate increased OH acidity and CO basicity in the first excited singlet state.

No theoretical calculations have been published which could elucidate the spectra of hydroxyphenylbenzotriazoles. The spectra of many derivatives in different solvents have been presented by H. J. Heller (9). These spectra demonstrate beyond any doubt that the long-wavelength band (Fig. 7) is connected with the six-membered chelate ring. This absorption vanishes in solvents destroying the intramolecular hydrogen bond and if bulky substituents possibly present in the molecule prevent a coplaner arrangement of the two aromatic rings (9). The strong shorter-wavelength band near 33,000 cm^{-1} has tentatively been attributed to a structure in which the two aromatic rings are vertical to one another (9). Such a structure has necessarily no intramolecular hydrogen bond. This, however, is contradicted by the fact that no free OH group is found by infrared and NMR measurements in carbon tetrachloride (25) (the same is true for HB and 4-methoxy HB). Furthermore, the intensity of the 33,000 cm^{-1}

band is not increased by a hydrogen-bond-breaking solvent (Fig. 13 in ref. 9). It is, therefore, more likely that this band is characteristic of 2-phenyl-benzotriazoles and is not significantly influenced by different substituents and solvents; the absorption maximum of 2-phenyl-benzotriazole is at 32,600 cm^{-1} in chloroform.

D. Emission Spectra

It was stated at the beginning that the stable molecules discussed here are nonluminescent. Now we must restrict this statement to solutions in which the intramolecularly hydrogen-bonded structure is intact, that is, in nonpolar solvents. In solid solvents containing alcohols or water, part of the hydrogen bonds are broken and phosphorescence is observed. In derivatives of HB, Sharp and Lamola (8) observed benzophenone-type phosphorescence in ether/ethanol, A. Heller and Wasserman in ethanol/methanol glasses (Table 8). Weak phosphorescences observed in hydrocarbon glasses (8,85) are very likely due to traces of water. It has been observed in our laboratory that solutions of HB in freshly distilled and red-di methylcyclohexane do not show any phosphorescence and that short standing of the (liquid) solution in air causes phosphorescence after solidification of the solution. Lamola and Sharp (8) observed that rapid cooling of solutions results in higher phosphorescence intensity compared to solutions which have been allowed to relax just above the glass freezing point. This experiment shows that at higher temperatures there are more broken H-bonds and that consequently this form has the higher energy.

The phosphorescence spectra of HB and derivatives thereof have been measured in the protonated and deprotonated forms (Table 8) (86). The spectral shifts are used below to calculate the acidities in the first triplet state.

E. Photochemical Studies

Data on the photochemical stability and stabilizing action of HB's, benzotriazoles, and related compounds have been reported on several occasions (6,9,70,71,7,82,87). In order to be a good stabilizer, it is not sufficient that a compound does not decompose if dissolved; for example, in a plastic, it is equally important that it does not act as a sensitizer. Investigations on the sensitizing action of HB's compared to that of benzophenone and nonchelated derivatives are, therefore, of interest. Such studies are included by Lamola (8), A. Heller (80), and Hammond (74). *Cis-trans* isomerization of piperylene (1,3-pentadiene) in benzene is easily sensitized by benzophenone via triplet transfer but not by 4-methoxy HB (74,8). In ethanol/piperylene glasses, the phosphorescence of HB's is quenched by energy transfer (8). In liquid ethanol, the yield of quenchable triplets is <0.03 (benzophenone 1.00) for HB, 4-methoxy HB and 2,2' dihydroxybenzophenone (8). The yield of isoprene dimers relative to benzophenone (50% isoprene) is found to be 0.03 in cyclohexane and 0.15 in ethanol if HB is used as sensitizer. The lack of piperylene sensitization even in ethanol indicates a very short lifetime of T_1 of HB even in its broken H-bond form (8).

From the sensitization of isoprene in liquid ethanol it can be concluded that the quantum efficiency of triplet formation of HB is at least 0.15 (8). No sensitization of Tb^{3+} or Eu^{3+} emission by 4-methoxy-HB has been found in polar solutions. This, too, points to short lifetimes of HB triplets.

It should be noted that the photostability of some dyes has been attributed to intramolecular hydrogen bonds (88,89). In the case of indigo, however, there seems to be no direct evidence for intramolecular H-bonds (90), which have been postulated by Wyman (89).

F. Possible Mechanism of Excited State Deactivation

Radiationless deactivation has to be very fast in order to account for the absence of any luminescence in the substances discussed in this section. It has been proposed that internal conversion ($S_1 \rightarrow S_o$) is so fast that the triplet system is not populated (73,8,94). It is, however, not possible to distinguish from existing spectroscopic and photochemical evidence whether the triplet states are not populated by intersystem crossing or whether they are only very short lived.

Excited state proton transfer is very likely according to the excited state acidities estimated according to eq. 6. The ground and excited state pK data of HB and 4-methoxy HB are compiled in Table 9. They follow the general order $pK_S^* < pK_T^* < pK_G$ for OH and the inverse order for the CO group.

Although pK values obtained with the aid of eq. 6 must be considered as estimates, the following trend may be derived: The stable position of the proton is at the phenolic oxygen atom in the ground singlet and in the first excited triplet state; at the carbonyl group, on the other hand, it is in the first excited singlet state ($pK_S^*(OH) < pK_S^*(CO)$). This behavior is very pronounced in the case of 4-methoxy HB. The following reaction sequence may therefore be tentatively proposed (see Fig. 8):

1. Absorption of a photon creates an excited singlet (CT) state of the "phenol".

TABLE 9. Ground and Excited State pK Data[a]

	2-Hydroxybenzophenone (HB)		2-Hydroxy-4-methoxy benzophenone	
	OH	CO	OH	CO
pK_G	9.37 (83)	-5^b	8.49 (83)	-4.8^b
pK_S^*	2.32 (83) 2.43	2.85	1.93 (83) 1.76	3.27
pK_T^*	2.65	-1.23	4.04	-1.44

[a]Relative to water; all data without reference have been calculated on the basis of eq. 6 and spectral data after Kysel' (81).

[b]Estimated from pK_G values of similar compounds: 4-hydroxy benzophenone: -5.03; 4-methoxy benzophenone: -4.93; 2,4-dihydroxy benzophenone: -4.82 (91).

2. The proton of the OH group is transferred to the carbonyl group.
3. The electronic structure changes to an ortho quinone structure, possibly of $n-\pi^*$ character.
4. Intersystem crossing of the quinone to T_1.
5. Since in the triplet state the stable structure is the phenol, the proton tunnels back.
6. Intersystem crossing leads to S_0 of the phenol.

This is, of course, only a hypothesis since several important questions are not yet answered by appropriate experiments:

First, it has not proved that S_1 (quinone) is a $n-\pi^*$ state. If this is the case, very fast intersystem crossing is likely.

Secondly, the exact positions of S_1 and T_1 (quinone) and T_1 (phenol) are not known.

Finally, it is not yet understood why intersystem crossing $T_1 \rightarrow S_0$ (phenol) is so fast.

As an alternative mechanism there remains the assumption of a very rapid internal conversion either in the phenol or, more probably, in the quinone structure, preventing triplet state population (94). It should be noted that for no substance showing

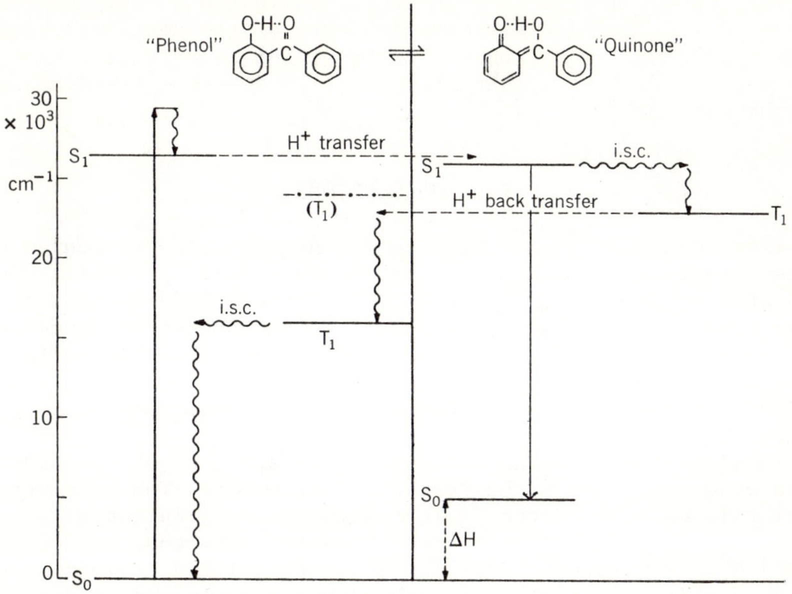

FIGURE 8. Tentative deactivation scheme of HB and similar compounds: S_1 (phenol) estimated from absorption maximum of HB (solid arrow); ΔH from MSA[2]; S_1 (quinone) from SAA[58,54]; T_1 (quinone) estimated for $n\pi^*$ state ($S_1 - T_1 \approx 3000$ cm^{-1}); T_1 (phenol) from SAA $S_0 \rightarrow T_1$ absorption[92]; this state could actually lie higher in HB but still below (T_1) of the form with broken H bond. The fluorescence of S_1 (quinone) is taken from[94], (2-hydroxy-4-dodecyloxy-benzophenone in 3 MP at 77K).

intramolecular proton transfer phosphorescence or any other indication of efficient triplet state formation has been reported. The approximate location of a phenol-type $S_o \rightarrow T_1$ absorption has been obtained in alkyl iodides (external heavy atom effect) for SAA (92) $S_o - T_1$ (00) $\approx$ 16,000 cm^{-1}, in good agreement with the calculated value. This value has been used for HB in Fig. 8.

The deactivation of benzotriazoles has tentatively been explained by H.J. Heller (71) as being due to the rapid change between H-bonded and non-bonded structures. Molecular vibrations of the flexible molecules should take up rapidly the electronic energy. The absence of free OH groups indicated above, however, favors a rigid, hydrogen-bonded structure of the benzotriazoles, at least in nonpolar solvents. Perhaps, the deactivation mechanism is similar to that proposed here for the orthohydroxybenzophenones (Fig. 8).

VI. CONCLUSIONS

Intramolecular proton transfer is an important mechanism of excited state deactivation. It is, however, only partially understood at present; especially the role of the triplet states is still unknown. This is due to the fact that phosphorescence has never been observed except in hydrogen-bond-breaking solvents. Intramolecular proton transfer is of practical importance in the field of stabilizers, optical brighteners, photochromic substances, and dyes.

In all three cases discussed the primary step of the deactivation sequence seems to be the proton transfer. The consecutive behavior is quite different. In the first group including fluorescent compounds the excited state after proton transfer seem to be π-π^*. Intersystem crossing (or internal conversion) is slow enough to allow fluorescence. No molecular fluorescence quantum efficiencies have been reported in literature. Unpublished measurements (25) suggest that they are of the order of 10^{-2} for salicylic acid derivatives and that radiationless processes dominate in this group. The same is true for the aldazines (Table 6). Solid state data for hydroxyphenyl benzothiazoles show high quantum efficiencies.

The second group covering photochromic aromatic ketones is related to the first group insofar as the electronic state created by proton transfer is π-π^* and able to fluoresce. The differences are:

1. The hydrogen bond is stronger owing to the higher basicity of the imine nitrogen.
2. Quinoid structures have an only slightly higher energy and can, therefore, be observed in some cases in the ground state (tautomers).
3. *Cis-trans* isomerization in the excited state provides a possibility for stabilization of a quinoid species.

The exact deactivation mechanism is still unknown. The energy and population of most triplet states is unknown and two (or even three) excited *cis*-quinones of slightly different structures are discussed.

The H-bond strength of the third group--nonluminescent molecules--is similar to that of the salicylic derivatives, and no ground state tautomers are observed. Tremendous pK changes are necessary in order to facilitate proton transfer in the excited state. The deactivation path is still a matter of speculation. It is worth noting that a substance of the fluorescent class (phenyl salicylate, Table 2) has been the first ultraviolet stabilizer (9) used on a technical scale, despite its poor absorption intensity (screening effect). This substance is photochemically converted into 2,2'-dihydroxy benzophenone (87) in a photo-Fries reaction. This can, however, not explain the total efficiency of this stabilizer (93).

As has been shown, the primary proton transfer is followed by a variety of secondary processes, depending on the structure of the substance and on the electronic nature of the excited states created by the proton transfer. It is, therefore, suggested that the proton transfer is primarily a very rapid "switching mechanism" between different electronic states (e.g., $\pi-\pi^* \rightarrow n-\pi^*$) and that the actual deactivation behavior is largely determined by the nature of the electronic states thus created. This is an alternative possibility to previous suggestions maintaining that the proton transfer itself is the decisive energy dissipating step especially in molecules showing enhanced photochemical stability.

REFERENCES

(1) A. Weller, Naturwissenshaften, 42, 175 (1955).

(2) A. Weller, Z. Elektrochem., 60, 1144 (1956).

(3) Th. Förster, Z. Elektrochem., 54, 43 (1950).

(4) A. Weller, Progr. React. Kinet. 1, 188 (1961).

(5) M. D. Cohen, Y. Hirschberg, and G.M.J. Schmidt, "Hydrogen Bonding", Symp. in Ljubljana, 1957 (D. Hadži, Ed.) Pergamon, New York (1959), p. 293.

(6) G. M. Gantz and W. G. Summer, Textile Res. J., 27, 244 (1957).

(7) J. N. Pitts, Jr., H. W. Johnson and T. Kuwana, J. Phys. Chem., 66, 2456 (1962).

(8) A. A. Lamola and L. J. Sharp, J. Phys. Chem., 70, 2634 (1966).

(9) H. J. Heller, Europ. Pol. J. Suppl., 1969, 105.

(10) C. A. Parker, "Photoluminescence of Solutions," Elsevier, Amsterdam 1968, p. 328.

(11) A. Weller, Z. Elektrochem., 56, 662 (1952).

(12) A. Weller, Z. Phys. Chem. NF., 3, 238 (1955).

(13) G. Jackson, G. Porter, Proc. Roy. Soc. (London), A260, 13 (1961).

(14) A. Weller, Z. Phys. Chem. NF., 17, 224 (1958).

(15) T. S. Godfrey, G. Porter, and P. Suppan, Disc. Far. Soc., 39, 194 (1965).

(16) J. F. Ireland and P.A.H. Wyatt, J. Chem. Soc., Faraday Trans. I, 69, 161 (1973).

(17) W. Bartok, P.J. Lucchesi, and N. J. Snider, J. Am. Chem. Soc., 84, 1842 (1962).

(18) R. A. Nyquist, Spectrochim. Acta, 19, 1655 (1963).

(19) B. A. Zadorozhnyi and I. K. Ishchenko, Opt. Spectrosc. (Engl. transl.) 19, 306 (1965).

(20) D. Peltier, A. Pichevin, P. Dizabo, M.-L. Josien, Compt. Rend. Acad. Sci., 248, 1148 (1959).

(21) Iu. V. Naboikin, E. N. Pavlova, and B. A. Zadorozhnyi, Opt. Spectrosc. (Engl. Transl.), 4, 231 (1959).

(22) D. Welti, Spectrochim. Acta, 22, 281 (1966).

(23) W. Klöpffer and G. Naundorf, J. Luminescence, 8, 457 (1974).

(24) H. Beens, K. H. Grellmann, M. Gurr, and A. H. Weller, Disc. Faraday Soc., 39, 183 (1965).

(25) W. Klöpffer, H. Bauser, G. Naundorf, and V. Betz, Unpublished results.

(26) L. J. Minnick and M. Kilpatrik, J. Phys. Chem., 43, 258 (1939).

(27) J. K. Marsh, J. Chem. Soc., 125, 418 (1924).

(28) I. B. Berlman, "Handbook of Fluorescence Spectra of Aromatic Compounds," 2nd ed., Academic Press, New York, 1971.

(29) Iu. V. Naboikin and B. A. Zadorozhnyi, Opt. Spectrosc. (Engl. transl.), 8, 347 (1960).

(30) S. F. Kilin and I. M. Rozman, Opt. Spectrosc. (Engl. transl.), 15, 266 (1963).

(31) G. Weber and F.W.J. Teale, Trans. Faraday Soc., 53, 646 (1957).

(32) S. G. Schulman and H. Gershon, 72, 3297 (1968).

(33) K. L. Wolf and G. Metzger, Liebigs Ann., 536, 157 (1949).

(34) Yu. V. Naboikin, V. A. Zadorozhnyi, and E. N. Pavlova, Bull. Acad. Sci. USSR (Engl. transl.), Phys. Ser., 23, 9 (1959).

(35) W. R. Ware, P. R. Shukla, P. J. Sullivan, and R.V. Bremphis, J. Chem. Phys., 55, 4048 (1971).

(36) K. Hirota, Z. Phys. Chem. N.F., 35, 222 (1962).

(37) Yu. V. Naboikin, B. A. Zadorozhnyi and E. N. Pavlova, Opt. Spectrosc. (Engl. transl.), 6, 312 (1959).

(38) E. D. Bergmann, Y. Hirshberg, and S. Pinchas, J. Chem. Soc., 1950, 2350.

(39) M. Oki, M. Hirota and S. Hirofuji, Spectrochim. Acta, 22, 1537 (1966).

(40) E. Razafindrakoto and S. Besnainon, Theor. Chim. Acta, 7, 321 (1967).

(41) L. J. Bellamy and L. Becher, J. Chem. Soc., 1954, 4487.

(42) M. D. Cohen and (Mrs.) S. Flavian, J. Chem. Soc. B, 1967, 317.

(43) D. L. Williams and A. Heller, J. Phys. Chem., 74, 4473 (1970).

(44) A. Weller and H. Wolf, Liebigs Ann., 657, 64 (1962).

(45) J. B. Birks and D. J. Dyson, Proc. Roy. Soc. (London), 275, 135 (1963).

(46) J. D. Margerum and L. J. Miller in "Photochromism", G. H. Brown, Ed., Wiley, New York, 1971, Chapter 6, p. 557.

(47) M. D. Cohen and G.M.J. Schmidt, J. Phys. Chem., 66, 2442 (1962).

(48) M. D. Cohen, G.M.J. Schmidt and S. Flavian, J. Chem. Soc., 1964, 2041.

(49) M. D. Cohen, Y. Hirschberg, and G.M.J. Schmidt, J. Chem. Soc., 1964, 2051.

(50) M. D. Cohen, Y. Hirschberg, and G.M.J. Schmidt, J. Chem. Soc., 1964, 2060.

(51) M. D. Cohen and S. Flavian, J. Chem. Soc. B, 1967, 321.

(52) M. Ottolenghi and D. S. McClure, J. Chem. Phys., 46, 4613 (1967).

(53) M. Ottolenghi and D. S. McClure, J. Chem. Phys., 46, 4620 (1967).

(54) R. Potashnik and M. Ottolenghi, J. Chem. Phys., 51, 3671 (1969).

(55) R. S. Becker and W. F. Richey, J. Am. Chem. Soc., 89, 1298 (1967).

(56) G. Wettermark and L. Dogliotti, J. Chem. Phys., 40, 1486 (1964).

(57) J. D. Margerum and J. A. Sousa, Appl. Spectrosc., 19, 91 (1965).

(58) G. O. Dudek and E. P. Dudek, J. Am. Chem. Soc., 88, 2407 (1966).

(59) R. W. Hay and P. P. Williams, J. Chem. Soc., 1964, 2270.

(60) D. N. Shigorin, N. N. Shapet'ko, A. P. Skoldinov, and T. S. Ryagchikova, Dokl. Akad. Nauk. SSSR, 148, 1141 (1963).

(61) A. E. Lutskii, L. A. Antropova, Russ. J. Phys. Chem., 40, 482 (1966).

(62) B. A. Zadorozhnyi, Zh. Fiz. Khim., 39, 1944 (1965). (Russ. J. Phys. Chem., 39, 1032 (1965)).

(63) W. R. Richey and R. S. Becker, J. Chem. Phys., 49, 2092 (1968).

(64) D. Gegiou and E. Fischer, Chem. Phys. Lett., 10, 99 (1971).

(65) J. W. Ledbetter, Jr., J. Phys. Chem., 70, 2245 (1966).

(66) D. L. Beveridge and H. H. Jaffé, J. Am. Chem. Soc. 87, 5340 (1965).

(67) D. G. Anderson and G. Wettermark, J. Am. Chem. Soc., 87, 1433 (1965).

(68) D. Lexa, M.-M. Duval, R. Viovy, and G. Chene, J. Chim. Phys., 65, 292 (1968).

(69) H. H. Freedman, J. Am. Chem. Soc., 83, 2900 (1961).

(70) H. Gysling and H. J. Heller, Kunststoffe, 51, 13 (1961).

(71) H. J. Heller and H. R. Blattmann, in "Chemical Transformations of Polymers" (R. Rado, Ed.), Butterworths, London, 1972, p. 145.

(72) O. Cicchetti, Adv. Polym. Sci., 7, 70 (1970).

(73) A. Beckett and G. Porter, Trans. Faraday Soc., 59, 2051 (1963).

(74) G. S. Hammond, N. J. Turro and P. A. Leermakers, J. Phys. Chem., 66, 1144 (1962).

(75) J. R. Merrill, J. Phys. Chem., 65, 2023 (1961).

(76) N. M. Cullinane, R. A. Woolhouse, and V. V. Bailey-Wood, Rec. Trav. Chim., 80, 116 (1961).

(77) N. Fuson, M.-L. Josien, and E. M. Shelton, J. Am. Chem. Soc., 76, 2526 (1954).

(78) D. C. Nonhebel, Tetrahedron, 24, 1869 (1968)

(79) W. L. Dilling, J. Org. Chem., 31, 1045 (1966).

(80) A. Heller and E. Wasserman, J. Chem. Phys., 42, 949 (1965).

(81) O. Kysel', R. Zahradník, and B. Pakula, Coll. Czech. Chem. Commun., 35, 3020 (1970).

(82) J. H. Chaudet, G. C. Newland, H. W. Patton, and J. W. Tamblyn, SPE Transactions, 1961, 26.

(83) P. Hrdlovič, D. Belluš, and M. Lazár, Coll. Czech. Chem. Commun., 33 59 (1968).

(84) D. N. Shigorin, N. A. Shcheglova, and Y. I. Kozlov, in "Proc. Int. Conf. Luminescence, Budapest", (G. Szigeti, Ed.), Vol. 1, Akad. Kiadó, Budapest, 1968, p. 528.

(85) O. Kysel' and R. Zahradník, Coll. Czech. Chem. Commun., 35, 3030 (1970).

(86) O. Kysel', R. Zahradník, D. Belluš, and T. Sticzay, Coll. Czech. Chem. Commun., 35, 3191 (1970).

(87) J. Voigt, "Die Stabilisierung der Kunststoffe gegen Licht und Wärme," Springer, Berlin, 1966.

(88) B. K. Manukian, P. Niklaus, and H. Ehrsam, Helv. Chim. Acta, 52, 1259 (1969).

(89) G. M. Wyman and B. M. Zarnegar, J. Phys. Chem., 77, 1204 (1973).

(90) M. Klessinger and W. Lüttke, Chem. Ber., 99, 2136 (1966).

(91) T. G. Bonner and J. Phillips, J. Chem. Soc. B, 1966, 650.

(92) M. I. Knyazhanskii, M. B. Strynkov and V. I. Minkin, Opt. Spectros., (Engl transl.), 33, 484 (1972).

(93) H. J. Heller, 11th IUPAC Microsymposium on Stabilisation of Polymers, Prague, 1972.

(94) J.-E. A. Otterstedt, J. Chem. Phys. 58, 5776 (1973).

Excited State Behavior of Some Bichromophoric Systems

F. C. DE SCHRYVER AND N. BOENS
Department of Chemistry, University of Leuven,
Celestijnenlaan 200 F, B-3030 Heverlee, Belgium

J. PUT
Limburgs Universitair Centrum, Universitaire Campus,
B-3610 Diepenbeek, Belgium

CONTENTS

*The review is restricted to systems that at the most show weak coupling between the two chromophores. It furthermore does not include a discussion on the di-π-methane type rearrangements.

I. INTRODUCTION

"Much information can probably be obtained from bichromophoric molecules in which the two chromophores interact more strongly in the excited state than in the ground state of the molecules," wrote G.S. Hammond in Advances in Photochemistry 7, 389 (1969).

Indeed, incorporation of two chromophores in one molecule separated by a certain number of atoms in such a way that there are no important interactions in the ground state can substantially change the photochemical properties of the respective chromophores. Several processes from very short-lived excited states, which otherwise might have remained uncovered, are observed in such systems. On excitation of one of the chromophores (C) of a nonconjugated bichromophoric system several processes can occur (Scheme 1).

Scheme I

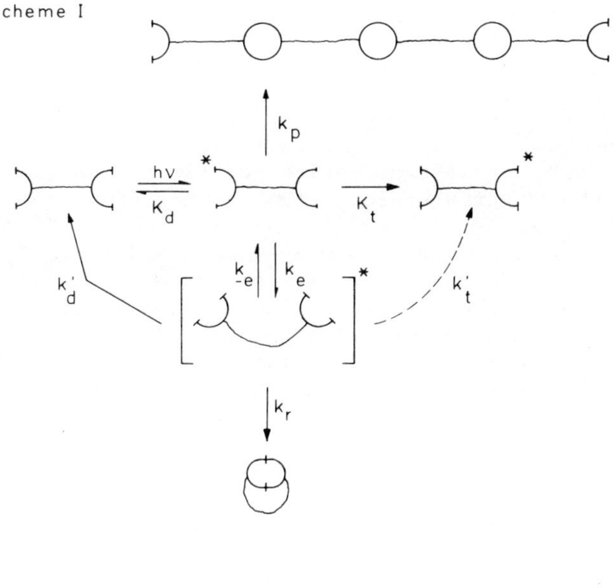

The excited chromophore can deactivate, k_d, by a radiative or nonradiative process; it can form an excited state "complex," k_e, ranging from an intermediate with well-defined properties to a collision-type complex. This excited state complex can either fall apart, k_{-e}, deactivate with or without radiation, k_d', or in case the chromophore is apt to it, it can in concerted or non-concerted fashion react intramolecularly, k_r. The excited chromophore can react intermolecularly, k_p, to form high molecular weight materials. If the two chromophores are not identical and the respective energy levels are spaced adequately, energy transfer can occur either directly, k_t, or may in some specific cases occur through the intermediacy of complex k_t'. Several of these processes might take place in a particular bichromophoric system. We will first discuss systems in which complex formation and energy transfer were studied and stress then some recent developments in the photocyclo- and photopolymerization.

II. EXCITED STATE COMPLEX FORMATION

A. Excimers and Exciplexes

The formation of a complex between an organic molecule in the excited state and one in the ground state leading to a new spectroscopic state, was first observed by Förster and Kasper (1) for pyrene. Such complexes were called excimers (2-4) if the two molecules are identical and heteroexcimers or exciplexes (5,6) if they are different. These complexes only exist in the excited state and they dissociate into monomers upon deactivation. While the existence of singlet excimers and exciplexes has been proved by many investigators, only a few reports deal with triplet complexes.

The most significant indication for the existence of excimers is a broad and structureless fluorescence band shifted 5000 to 6000 cm^{-1} to lower frequencies relative to the monomer fluorescence. This fluorescence has been observed as well in solutions as in the liquid and in the crystalline state. The typical emission may however be absent, because of rapid dissociation or efficient radiationless deactivation (7).

The nature of the bonding in these excited state complexes has been the subject of several theoretical studies recently reviewed by Birks (4). The first theoretical approaches, based on charge resonance and excitation resonance interaction, were not completely successful. Konijnenberg (8) proposed a combined model of configuration interaction between charge resonance and excitation resonance states. In heteroexcimers or exciplexes, excitation resonance is impossible because of the difference in energy levels, and bonding is provided by charge transfer only, the difference between the donor ionization potential and the acceptor electron affinity being the determining factor. Because of the nature of the bonding, exciplex fluorescence is more sensitive to solvent polarity.

To observe complex formation in monofunctional systems, rather high concentrations may be needed, while in the bichromophoric systems under investigation, two chromophores are near each other even in very dilute solutions. This makes it possible to investigate the excited state behavior of such complexes at high dilution, although the chain, linking the chromophores, may impose new restrictions on intramolecular processes. An extensive review on intramolecular excimers, including excimer formation in polymers, has been made recently by Klöpffer (9). Recent developments in this area have broadened substantially the scope of intramolecular complex formation.

B. Structure of Intramolecular Excited Complexes

1. Intramolecular excimers. Stevens (10) showed that excimers are formed by aromatic molecules with parallel planes in the crystal lattice. Thus, it is to be expected that, to form an intramolecular excimer, the two chromophores must have a sandwich configuration with parallel planes. As the chain, linking the two chromophores, will prefer other conformations in the ground state, this sandwich arrangement has to be reached during the lifetime of the excited state.

On the basis of a study of the fluorescence spectra of a series of α,ω-difenylalkanes, Hirayama put forward the so-called n=3 rule, which states that excimers can only be formed if the two chromophores are separated by four saturated bonds, (3 methylene groups) as in 1,3-difenylpropane (11). This n=3 rule was based on the fact that a sandwich configuration has to be reached, with an approximate distance between the chromophores of 3 Å, while the trimethylene chains take a gauche-gauche conformation (Fig. 1).

Only the trimethylene chain can take this rather favorable conformation in an excimerlike structure. Together with the fact that for longer chains the probability to reach a favorable spatial arrangement within the lifetime of the excited state is drastically diminished, this destabilization explained the absence of excimer fluorescence in higher α,ω-diphenylalkanes. The n=3 rule was confirmed by the study of several other systems: 1,3-biscarbazolypropane (12-14), meso and d,1-2,4-diphenylpentane (15), 1,3-bis-(α-naphthyl)-propane (16), 1,3-bis-(β-naphtyl) propane (16), and 9,9'-trimethylenebisadenines (17). Although

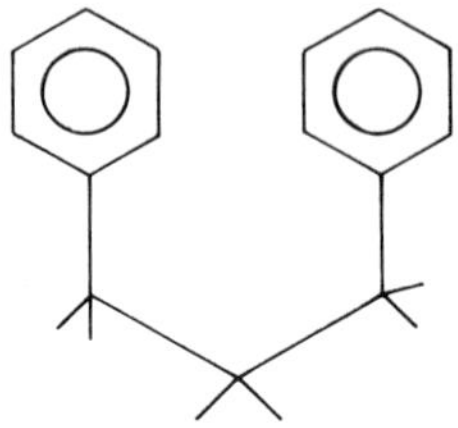

FIGURE 1. Excimer conformation of 1,3 diphenyl-propane.

Chandross and Dempster (16) provided further proof for the necessity of the symmetric sandwich configuration, since 1-(α-naphtyl)-3-(β-naphtyl)-propane did not show excimer fluorescence, it should be noted that they observed a weak excimer emission in 1,4-bis (α-naphtyl)-butane, as a first exception to the Hirayama rule.

Further confirmation for the necessity of a sandwich configuration was obtained in the paracyclophane series (18-19) in which the 4,4-paracyclophane showed an excimer fluorescence while the 4,5-paracyclophane did not. In the 2,2- and the 3,3-paracyclophane, appreciable interaction between the chromophores in the ground state is observed in the ultraviolet spectra. Therefore in these systems, the two chromophores are held in a sandwichlike configuration already in the ground state, they are therefore irrelevant to check Hirayama's rule.

A series of dinucleotide analogs was studied by Leonard and co-workers (17). It has been observed by optical rotary dispersion (ORD) that dinucleotides have a stacked structure, due to attractive forces between the bases. This stacked structure is also considered responsible for the hypochromism observed in solutions of these substances. In those systems in which the strongest ground state interaction existed, only fluorescence from its excited state complex could be observed. Because of the stacking, those systems have no direct bearing on the n=3 rule.

Recently De Schryver and co-workers supposed the intermediacy of excimers in photochemical reactions of bichromophoric systems in which the two chromophores were linked by a chain with much more units. In the case of the bismaleimides and the biscoumarins, because of the very efficient collapse to photoproduct, no emission from an excited state complex is observed. In a series of bisanthracenes they could show as well excimer emission (Fig. 2) as photoreaction (vide infra).

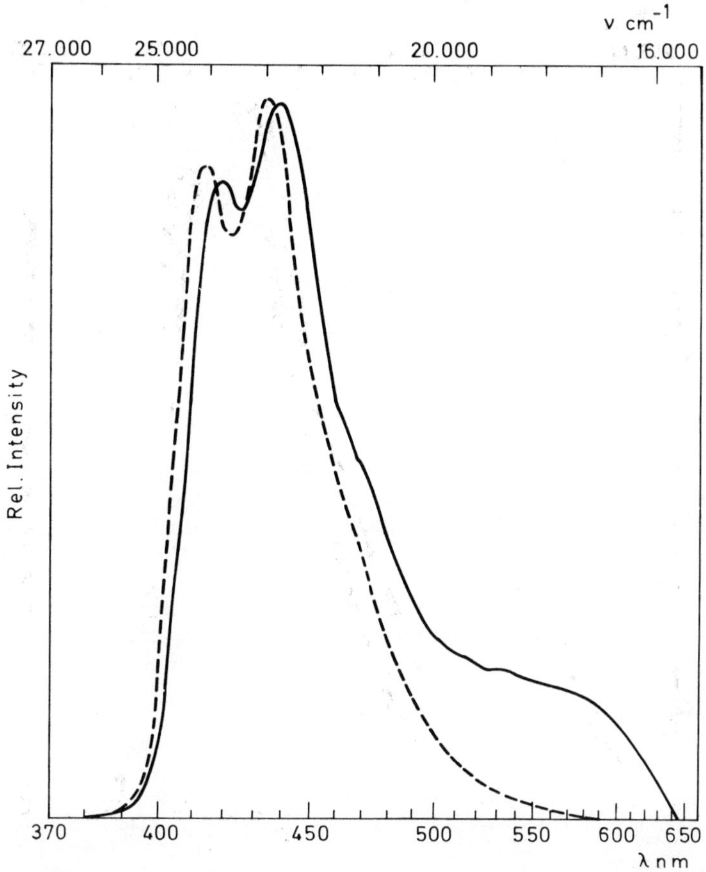

FIGURE 2. Fluorescence of (—) bisanthracene 37a and of n-hexyl-2-anthroate (---) in CH_2Cl_2 at 25°C.

On the basis of these results De Schryver and co-workers reformulated Hirayama's rule as follows: the possibility of intramolecular complex formation is limited by the probability to reach, within the lifetime of the excited state involved, a favorable conformation and by the extent of stabilization in the complex, covalent bond formations being the extreme.

This proposal was confirmed by Zachariasse (21) in an elegant study of a series of α,ω-di-(3-pyrenyl)alkanes, Py-$(CH_2)_n$-Py [1].

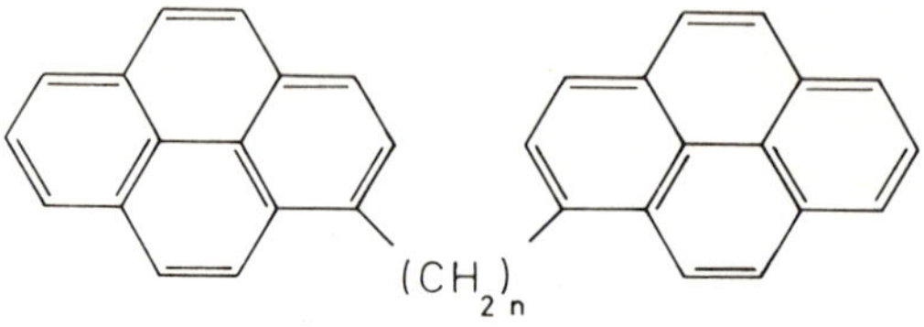

1 n = 3,4,5,6,8,9,10,11.

Excimer fluorescence in these bichromophores was observed for compounds up to 11 methylene units in the linking chain. He concluded that intramolecular excimer formation was possible in this system, due to the large excimer stabilization energy of pyrene, which would compensate the steric and statistical effects. The wavelengths of the maximum of the excimer emission band, however, varies with the chain length, which is explained by assuming changes in excimer geometry as a function of the chain length (Fig. 3).

Table 1, part 1 lists the bichromophoric systems in which excimer fluorescence has been observed until now; the frequency of the maximum and the quantum yield of fluorescence (if reported) are given.

2. Intramolecular exciplexes. The geometrical requirements in exciplex formation seem to be less important than in excimer formation. Chandross and Thomas (24) studied intramolecular exciplex formation in naphtylalkylamines [2] in which n varied from 1 to 4.

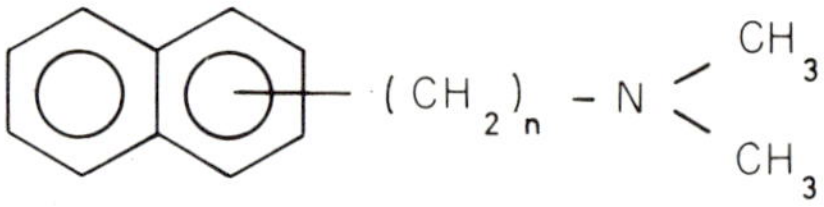

2 n = 1,2,3,4

They observed a maximal interaction for [2] n = 3 and they found the exciplexes with the chain in α to be more stable than those with the chain in β. The preferred configuration is the one with the N atom in the neighborhood of C_1. Chandross and Thomas further studied the formation of a complex between those exciplexes and the solvent. They concluded that this interaction is

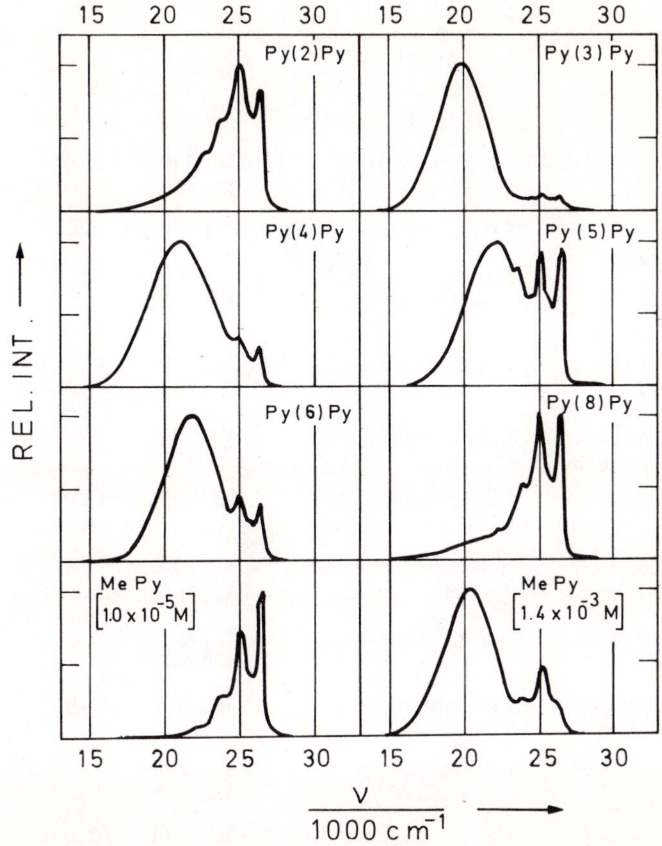

FIGURE 3. Fluorescence spectra of α,ω-di 3-pyrenyl alkanes in Me cyclohexane at 22°C.

of the dipole-dipole type and not charge transfer like (25).

Brimage and Davidson (26) also reported evidence for excited state complex formation in N,N-diethyl-α-naphtylalkylamines [3] and in N,N-dialkyl-9-anthrylalkylamines [4].

TABLE 1. Bichromophoric Systems Showing Emission from Intramolecular Excited Complexes.

Substance	Solvent	$\nu_{max}(cm^{-1})$	Φ	Ref.
	Part 1: Excimers			
1,3-Diphenylpropane	Cyclohexane	30,000	0.018	(11)
	Dioxane	30,300	0.016	(19)
1,3-Diphenyl-2-methyl-propane	Cyclohexane	29,900	0.018	(11)
1,3-Diphenyl-1-methylpropane	Cyclohexane	30,200	0.018	(11)
1,3-Diphenyl-2-ethylpropane	Cyclohexane	29,900	0.021	(11)
1,3-Diphenyl-1,1-dimethylpropane	Cyclohexane	30,000	0.015	(11)
1,3,5-Triphenylpentane	Cyclohexane	30,000	0.016	(11)
Meso-2,4-diphenylpentane	3-Methylpentane	33,700 +32,600	-	(15)
d,l-2,4-Diphenylpentane	3-Methylpentane	33,700 +32,200	-	(15)
1-(3-Methylphenyl)-3-phenylpropane	Cyclohexane	30,300	0.021	(11)
1-(2-Methylphenyl)-3-Phenylpropane	Cyclohexane	30,300		
1,3-Bis(2-methylphenyl)-1-methyl propane	Cyclohexane	30,300	0.028	(11)
1,3-Bis(N-carbozolyl)propane	Cyclohexane	23,500	0.05	(12)
	Methanol	~23,500	0.04	(12)
1,3-Di(α-naphtyl)propane	Methylcyclo-Hexane/isopentane	~25,000		(16)
1,3-Di-(β-naphtyl)propane	Methylcyclo-Hexane/isopentane	25,000		(16)
1,4-Di-(α-naphtyl)butane	Methylcyclo-Hexane/isopentane	~25,000[a]		(16)
2,2-Paracyclophane	EPA	28,600		(19)
	3-Methylpentane	27,200	0.002	(15)

TABLE 1 (cont'd)

Substance	Solvent	$\nu_{max}(cm^{-1})$	Φ	Ref.
4,4-Paracyclophane	1,2-Dichloroethane	29,400		(19)
Syn-(2,2)-paracyclonaphtane	EPA	21,700		(22)
Anti-(2,2)-paracyclonaphtane	EPA	24,800		(22)
Ethylene di-2-anthroate	CH_2Cl_2	18,000	0.042	(108)
Trimethylene di-2-anthroate	CH_2Cl_2	18,000	0.030	(108)
Tetramethylene di-2-anthroate	CH_2Cl_2	18,000	0.018	(108)
Pentamethylene di-2-anthroate	CH_2CL_2	18,000	0.005	(108)
α,ω-Di-3(pyrenyl) ethane	Methylcyclohexane			(21)
α,ω-Di-3(pyrenyl) propane	Methylcyclohexane	20,000		(21)
α,ω-Di-3(pyrenyl) butane	Methylcyclohexane	21,000		(21)
α,ω-Di-3(pyrenyl) pentane	Methylcyclohexane	22,000		(21)
α,ω-Di-3(pyrenyl) hexane	Methylcyclohexane	21,500		(21)
α,ω-Di-3(pyrenyl) oxtane	Methylcyclohexane	20,000		(21)
	Part 2: Exciplexes			
(a)				
1-α-Naphtyl-2-dimethylamino ethane	Methylcyclohexane	>25,000		(25)
1- -Naphtyl-3-dimethylamino propane	Isopentane Methylcyclohexane	24,000 24,500 24,100		(25)

TABLE 1 (cont'd)

Substance	Solvent	ν_{max} (cm^{-1})	Φ	Ref.
1-α-Naphtyl-4-dimethylaminobutane	Methylcyclohexane	24,500 24,300		(25)
1-β-Naphtyl-2-dimethylaminoethane	Methylcyclohexane	No exciplex fluorescence		(25)
1-β-Naphtyl-3-dimethylaminopropane	Methylcyclohexane	24,500 24,400		(25)
1-β-Naphtyl-4-dimethylaminobutane	Methylcyclohexane	No exciplex fluorescence ∿24,700		(25)
3 n=2	Benzene	23,800		(26)
n=3	Methylcyclohexane	24,500		(26)
	Benzene	23,000		(26)
4 n=3	Acetonitrile	19,400		(26)
5		20,800		(27)
6 n=0	Isobutylalcohol	19,500		(28)
n=2	Isobutylalcohol	17,000		(28)
n=3	Isobutylalcohol	17,500		(28)
7a	Cyclohexane	19,900		(29)
	Ether	18,800		(29)
	Tetrahydrofuran	19,600		(29)
8 n=3	Carbontetrachloride	19,300		(30)
n=4	Carbontetrachloride	19,900		(30)
9a	3-Methylpentane	18,800 (exc) 18,700 (CT)		(33,34)
9b	3-Methylpentane	20,000 (exc) 18,700 (CT)		(33,34)
(b) Dinucleotides and Analogs[a]				
ApC		27,900 27,800	$6.2\ 10^{-2}$	(20) (23)
Ap(2'-5')		30,000 28,600	$4.2\ 10^{-2}$	(20) (23)
CpA		26,300 27,400	10^{-2}	(20) (23)

TABLE 1 (cont'd)

Substance	Solvent	ν_{max} (cm^{-1})	Φ	Ref.
ApG		29,900	$9.5\ 10^{-2}$	(20)
		28,400		(23)
GpA		27,800	$3.2\ 10^{-2}$	(20,23)
GpA(2'-5')		29,800	$4.7\ 10^{-2}$	(20)
ApV		27,600	$2.8\ 10^{-2}$	(20)
		27,400		(23)
UpA		27,800	10^{-2}	(20,23)
GpV		24,700	9.10^{-3}	(20)
CpU		29,300	$1.9\ 10^{-2}$	(20)
ApT		28,200	$5.8\ 10^{-2}$	(20)
TpA		30,300	$5.2\ 10^{-2}$	(20)
TpT			0.14	(17)
AdC_3Ad		28,248	0.045	(17)
AdC_3Gu		27,624	0.12	(17)
GuC_3Gu		27,777		(17)
GuC_3Cy		27,021		(17)
GuC_3Th		27,021		(17)

a) For nomenclature in this series of compounds see Ref. 17.

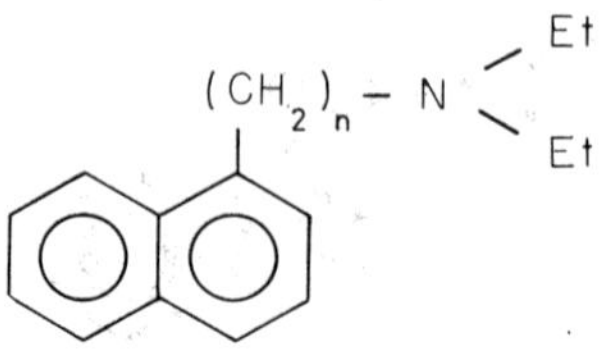

3 n = 2,3,

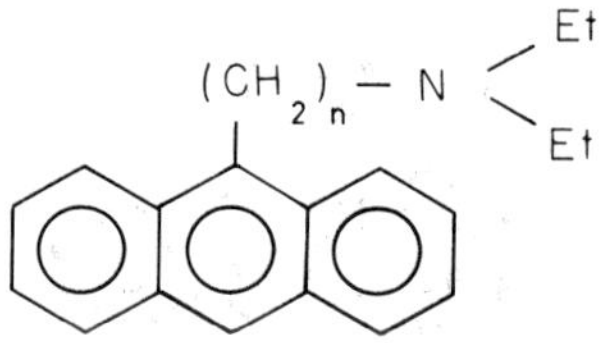

4 n = 2, 3,

Compounds [3] and [4] did not show exciplex fluorescence when n > 3, but Davidson found formation of intramolecular complexes in compound [5], although more than three carbon atoms link the two functions in this case (27).

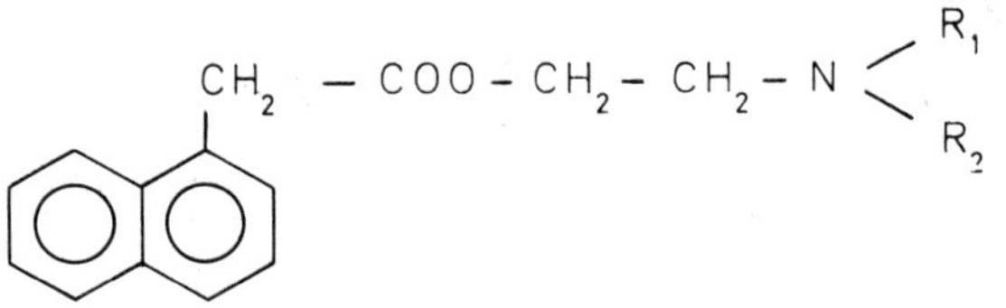

5 a : $R_1 = R_2 =$ Me
b : $R_1 =$ Me ; $R_2 =$ Phenyl

This again shows that repulsion in the linking chain plays an important role in the formation of intramolecular excited state complexes.

Mataga and others (28) have made a very interesting study of intramolecular complex formation in compounds [6]. These results indicate that within the excited state lifetime an equilibrium (A^*- D) $\rightleftarrows$ A^- - D^+ occurs. The wavelength of the emission maximum of the exciplex is strongly dependent on solvent polarity.

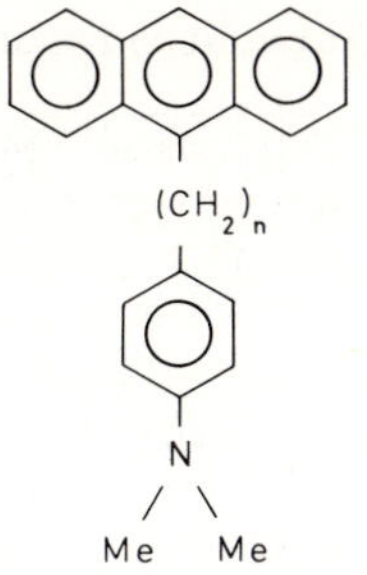

6 n = 0, 1, 2, 3.

Mataga showed that the geometric requirements for intramolecular interaction are not very strict. The two rings do not have to cover each other completely or do not have to be parallel. Analogous conclusions were drawn from a study of compounds [7] (29).

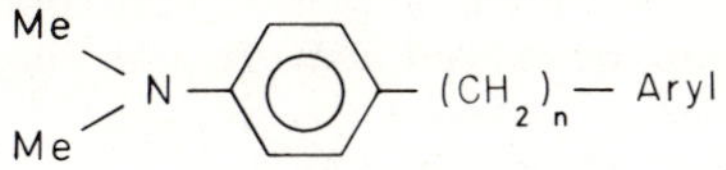

7 a: Aryl = 2 Anthryl; n = 3
b: Aryl = 1 Pyrenyl; n = 1, 2, 3.

The absorption spectra show only for 7 (n=0) an interaction in the ground state.

De Boer and co-workers (30) have studied the fluorescence resulting from intermolecular interactions in compounds [8].

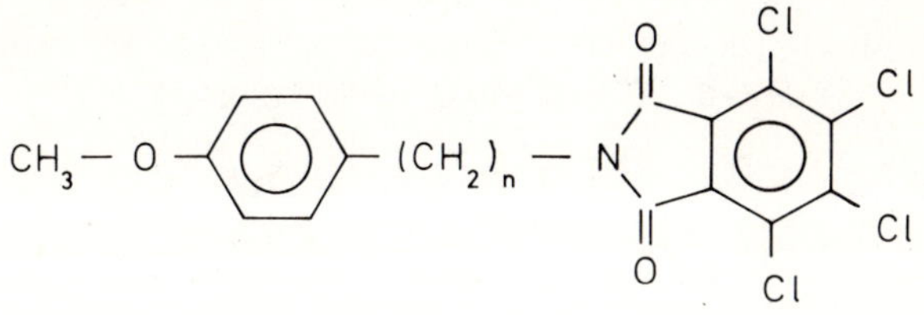

8 n = 1, 2, 3, 4.

These compounds, with n varying from 1 to 4, show an absorption band attributed to a donor-acceptor complex in the ground state (31,32). For n=1 and 2, the flexibility of the chain is insufficient to allow intramolecular interaction, while for n=3 and 4 a fluorescence band identical to that of an intermolecular complex between N-ethyltetrachlorophtalimide and p-methylanisol is observed.

Itoh and co-workers (33) studied the intramolecular electronic interaction between the donor and acceptor moieties in compounds [9], both in the ground and excited states.

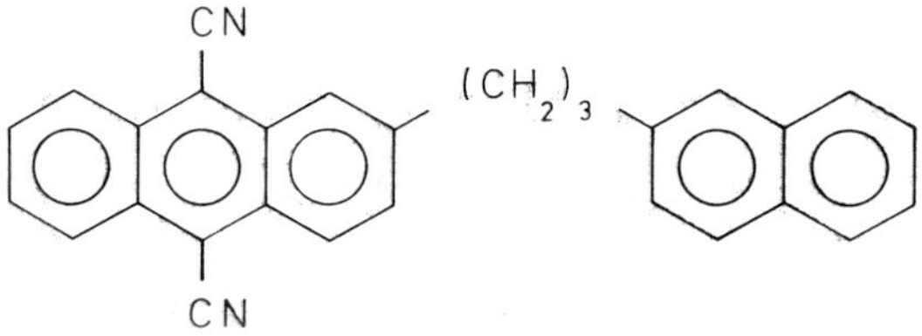

9 a : 1 Naphtyl
b : 2 Naphtyl

They observed the fluorescence in polar and nonpolar solvents. In nonpolar solvents below 150°K another fluorescence than the exciplex fluorescence was observed and it was ascribed to intramolecular charge-transfer fluorescence between the dicyanoanthracene and naphtalene moieties (34).

Table 1, part 2 lists the systems in which fluorescence from an intramolecular exciplex is observed.

C. Kinetics of Intramolecular Excimer Formation

A kinetic scheme for intramolecular excimer formation was proposed by Hirayama (10) in analogy with one developed by Förster (35) and by Hauser (36) for intermolecular excimer formation.

$$C\text{-}R\text{-}C \xrightarrow{I_{abs}} C^*\text{-}R\text{-}C \underset{k_d}{\overset{k_e}{\rightleftharpoons}} [C\text{-}R\text{-}C]^*$$

$$C^*\text{-}R\text{-}C \xrightarrow{k_s} C\text{-}R\text{-}C \qquad C^*\text{-}R\text{-}C \xrightarrow{k_f} C\text{-}R\text{-}C + h\nu \qquad [C\text{-}R\text{-}C]^* \xrightarrow{k_s'} C\text{-}R\text{-}C \qquad [C\text{-}R\text{-}C]^* \xrightarrow{k_f'} C\text{-}R\text{-}C + h\nu'$$

(k_s and k_s' include all nonradiative deactivation processes).

Klöpfer and Liptay developed the steady state kinetics (14a) assuming that the rate constants for fluorescence (k_f) and non-radiative deactivation (k_s), are identical in mono- and bifunctional systems. The equations are repeated here briefly. If C_2 is the concentration of C-R-C* and C_3 that of the excimer [C-R-C]*, their respective concentrations in steady state conditions equal:

$$(C_2)_{stat} = \frac{I_{abs}(k_s' + k_f' + k_d)}{(k_s' + k_f' + k_d)(k_s + k_f) + k_e(k_s' + k_f')} \quad (1)$$

$$(C_3)_{stat} = \frac{I_{abs}\, k_e}{(k_s' + k_f' + k_d)(k_s + k_f) + k_e(k_s' + k_f')} \quad (2)$$

If Φ_M is the quantum yield of monomer fluorescence and I_M the intensity, Φ_E and I_E the respective values of the excimer fluorescence and Φ_0 and I_0 those of the monofunctional derivative the following equations can be derived:

$$\Phi_M = \frac{k_f(C_2)_{stat}}{I_{abs}} \qquad \Phi_E = \frac{k_f'(C_3)_{stat}}{I_{abs}} \qquad \Phi_0 = \frac{k_f(C_0)_{stat}}{I_{abs}}$$

$$\frac{\Phi_0}{\Phi_M} = 1 + \frac{(k_s' + k_f')k_e}{(k_s' + k_f' + k_d)(k_s + k_f)} = \frac{I_0}{I_M} \quad (3)$$

$$\frac{\Phi_0}{\Phi_E} = \frac{[(k_s' + k_f' + k_d)(k_s + k_f) + (k_s' + k_f')k_e]k_f}{(k_s + k_f)k_f'k_e} = \frac{I_0}{I_E} \quad (4)$$

$$\frac{\Phi_M}{\Phi_E} = \frac{(k_s' + k_f' + k_d)k_f}{k_f'k_e} = \frac{I_M}{I_E} \quad (4)$$

$$\Phi_0 = \frac{k_f}{k_f + k_s} \quad (5)$$

In presence of quencher this quantum yield of fluorescence, Φ_q, equals

$$\Phi_q = \frac{k_f}{k_f + k_s + k_q[Q]} \quad (6)$$

The lifetime of the singlet excited state in absence of quencher equals

$$\tau_o = \frac{1}{k_s + k_f}$$

Using these equations the following rate constants in cyclohexane solution could be calculated (s^{-1})

	k_e	k_d	k_s	k_f	k_f'	k_s'
1,3 diphenyl-propane	11 10^8	1.5 10^7	2.6 10^7	5.8 10^6	6.3 10^5	3.4 10^7
1,3 biscarbo-zolylpropane	2.8 10^8	3.8 10^7	3.8 10^7	2.7 10^7	1.9 10^6	2 10^7

Since the rate constants for radiative and nonradiative deactivation of the singlet state, k_f and k_s, are smaller than k_e, excimer formation is the main deactivation process for this state. The rate constants for deactivation of the excimer k_f' and k_s' are in these systems of the same order as the dissociation rate constant, k_d, indicating the existence of an equilibrium.

Chandross and Depster (16) studied the temperature dependence of the rate constant of excimer formation for 1,3-bis(α-naphtyl)-propane and for 1,3-bis(β-naphtyl)-propane. They found an activation energy of 4 kcal/mol, comparable to the energies related to the rotation barrier in a methylene chain. Recently Avouris, Kordas, and El-Bayoumi studied the time dependence of the intramolecular excimer fluorescence of 1,3-dinaphtylpropane. They determined rate constants k_e and k_d at different viscosities (37).

Using the value of the activation energy, found by Chandross and Dempster, they calculated values for activation enthalpy and entropy:

$$\Delta H^{\neq} = 3.5 \text{ kcal/mol}$$

$$\Delta S^{\neq} = -10 \text{ cal mol}^{-1}\text{deg}^{-1}$$

D. Thermodynamics of Excimer Formation

At high temperatures, where excimer formation and dissociation are rapid, compared to the deactivation processes: $k_e \gg k_f+k_s$ and $k_d \gg k_f'+k_s'$, an equilibrium is established. In this region, both monomer and excimer fluorescence will decay exponentially, with a common rate constant (38).

At low temperatures dissociation can be neglected and two exponential decays are observed. In the transition range the decay is nonexponential. This transition temperature depends on the thermodynamic parameters free energy ΔG, enthalpy ΔH, and entropy ΔS (9). The equilibrium constant for excimer formation can be obtained from the concentration under stationary conditions (5).

$$K_E = \frac{(C_3)_{stat}}{(C_2)_{stat}} = \frac{k_e}{k_s'+k_f'+k_d'}$$

True equilibrium is reached only if $k_d >> k_s'+k_f'$, so that the equation simplifies to:

$$K_E = \frac{k_e}{k_d}$$

and this K_E is related to the thermodynamical quantities:

$$\Delta G = -RT \ln K_E = \Delta H - T\Delta S$$

The thermodynamic quantities could be calculated from measurements of the equilibrium constant as a function of temperature. Klöpffer (9) estimated a value of -940 cal/mol for ΔG in biscarbazolylpropane. Recently, Johnson (14b) made a very elegant and detailed study of 1,3-biscarbozolylpropane in 2 Me THF using steady state and dynamic measurements. They found an activation energy for excimer formation of 4.24 kcal/mole. The enthalpy of formation was observed to be -2.75 kcal/mole and the entropy of formation equals -7.2 cal $mole^{-1}$ deg^{-1}. The entropy loss in the intramolecular excimer formation is smaller than in the intermolecular case (16). Therefore intramolecular excimer formation is possible with chromophores that do not show intermolecular excimer formation in the monofunctional models, as is the case in biscarbazolylpropane relative to N-isopropylcarbazole.

The importance of the entropy factor is also illustrated by the fact that no excimer is found in 1,4-diphenylbutane while in 4,4-paracyclophane it is found, although in the latter case chain interactions will be more important while excimer stabilization will be the same. The entropy loss will be certainly smaller in the paracyclophane than in the open chain derivative.

E. Intramolecular Triplet Excimers

As both the excitation resonance and charge transfer should be weaker in the triplet state than in the singlet, the binding energy in triplet excimers is appreciably smaller. As the entropy loss is supposed to be smaller in intramolecular excimer formation, triplet excimers are more likely to be observed in bichromophoric systems than in monofunctional ones (9). A further difficulty that arises is that the study of phosphorescence in liquid solutions is very difficult. Molecules with a ground state sandwich configuration in frozen solutions are, therefore, most suitable for the study of triplet excimers.

In the paracyclophane series, Hillier, Glass, and Rice (39) found no evidence for triplet excimer formation in the (4,4)-derivative, while the (2,2)-derivative showed a phosphorescence that could be attributed to triplet excimer. However, as already mentioned, the 2,2-paracyclophane shows a strong ground state interaction.

Chandross and Dempster performed a photochemical cleavage in frozen solution of the intramolecular photocycloaddition product of 1,3-di(α-naphtyl) propane, so that they obtained the bichromophore in a sandwich configuration (40,41). Only excimer fluorescence is observed in such a sample, but phosphorescence is of the monomer type with a small bathochromic shift. In dinucleotides and their analogs (17), where singlet excimer formation is observed, the phosphorescence always originates from the base which has the lowest triplet state. Recently Avouris and El Bayoumi (42) presented evidence for the formation of an intramolecular triplet excimer in the 1,3-diphenylpropane during the softening of an isopentane degassed glass matrix. At $77^{\circ}K$ they observed a monomer phosphorescence but on warming up to about $115^{\circ}K$, a red-shifted band at 420 nm appeared and was interpreted as excimer phosphorescence.

It is to be expected that triplet excited state complex formation will be appreciable, only when charge transfer interaction is great, as the contribution of excitation resonance is small. In these systems however, where charge transfer interaction is great, ground state interaction often is observed.

F. Conclusion

Intramolecular excited state complex formation can be an important deactivation process in bichromophoric systems. The governing factors are the lifetime of the excited state involved and the free energy change accompanying the reaction. Three possibilities exist:

1. If very rapid deactivation processes exicst for the monomer excited state (very short lifetime) little or no complex formation will be observed.
2. Complex formation can be fast compared with other deactivation processes of the monomer excited state, but dissociation can be negligible. The dissociation can be inefficient for two reasons:
   - the rate of dissociation can be small compared with other deactivation processes of the complex, for example, photochemical reaction;

- at low temperature the dissociation is slow because of the small but definite activation energy.

3. Complex formation as well as dissociation are fast compared with other deactivation processes. In this case, true equilibrium between the monomer and the complex exists. This equilibrium can be shifted to the monomer or the complex side according to the free energy change of the process. Mainly three factors influence this free energy change: the stabilization of the complex, the destabilization as a result of interactions in the linking chain, and the entropy loss in the process.

The stabilization is largest in the singlet (homo) excimers because both excitation resonance and charge transfer contribute to the binding. In such systems therefore, intramolecular excimer formation can be observed between chromophores linked by long chains e.g. bispyrenes. In exciplexes and triplet excimers, the stabilization is much less because only charge transfer can contribute, but the geometrical restrictions are less stringent. The stabilization will increase in these complexes as the difference between the donor ionization potential and the acceptor electron affinity increases.

Intramolecular excimer and exciplex formation has been used to study the structure and mobility of chain segments in polychromic systems. A review of the results of these studies can be found in ref. 9.

III. INTRAMOLECULAR EXCITATION TRANSFER

If there is a difference in the respective energy levels of the excited states of both chromophores in a bichromophoric system, excitation transfer can occur from one chromophore to the other. This energy transfer can occur as well before as after intersystem crossing of the originally excited chromophore. The systems studied are divided into two main classes, according to the method used in the detection. The excitation transfer can be monitored either by spectroscopic methods or by reaction of the acceptor chromophore.

Besides the respective energy levels, two more factors are of importance: the distance between the two chromophores and their relative orientation. Each main class can, therefore, be subdivided into a subclass in which the two chromophores are joined by a nonrigid chain, in which a great variation in distance and orientation can occur, and into a subclass in which the chromophores are coupled by a rigid link imposing more restrictions on distance or/and orientation.

A. Spectroscopic Studies

1. Chromophores With a Flexible Link. Bichromophores containing at one side a naphtyl group and at the other side a series of chromophores, linked by a polymethylene chain [10] have been studied by means of emission spectroscopy.

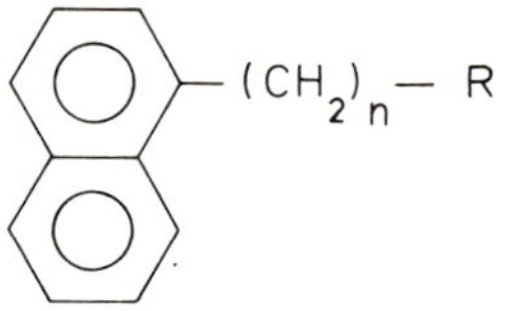

	R = a : anthryl	n = 1,2,3	(43)
	b : benzophenone	n = 1,2,3	(44,45)
10	c : Phtalimidyl	n = 0,1,2,3	(46)
	d : N-carbazolyl	n = 1	(46)

Schnepp and Levy (43) found that the efficiency of singlet energy transfer in systems [10a, 1,2,3] did not change with chain length and occurred as well at room temperature as in a glass at 77°K. Since the quantum yield of excitation transfer equals the quantum yield of fluorescence of the isolated donor, they concluded that the rate of energy transfer is situated between the rate of fluorescence of naphtalene $10^7 s^{-1}$ and the rate of radiationless decay $10^9 s^{-1}$ (47).

A Förster-type transfer mechanism (48) was proposed to occur. Batochromic shifts and hyperchromicity were observed in the absorption spectra of systems [10b,1,3] and to the largest extent in [10b,2] (44,45). On specific excitation of the benzophenone chromophore, triplet transfer to the naphtalene moiety occurs with 100 percent efficiency. The rate of transfer is greater than $10^{10} s^{-1}$ and not measurably dependent on chain length. Since the ratio of intersystem crossing in benzophenone is of the order of $10^{12} s^{-1}$ (49), energy transfer will occur, after intersystem crossing, by an exchange mechanism in which the possibility of complex formation cannot be excluded. On excitation of the naphtalene chromophore, singlet energy is transferred in [10b,1,2,3] by two mechanisms, one of them being the trivial process. The latter dominates in [10b,2]. The rate of singlet energy transfer is of the order of 10^7 to $10^8 s^{-1}$.

Fluorescence, phosphorescence and absorption spectra of systems [10c,0,1,2,3] and of [10d,1] at 77°K were studied by

Breen and Keller (46). Additional absorption bands, relative to the spectra of the isolated chromophores were explained as charge transfer bands from naphtalene to phtalimide in [10c] and could account for the very efficient triplet energy transfer from the phtalimide to the naphtalene group. A dipole-dipole interaction mechanism is proposed for the singlet energy transfer from naphalene to the other chromophores in [10d] and [10d].

A series of compounds in which para-fluorobutyrophenone is linked to another chromophore, for example, [11], was studied by Thierry (50).

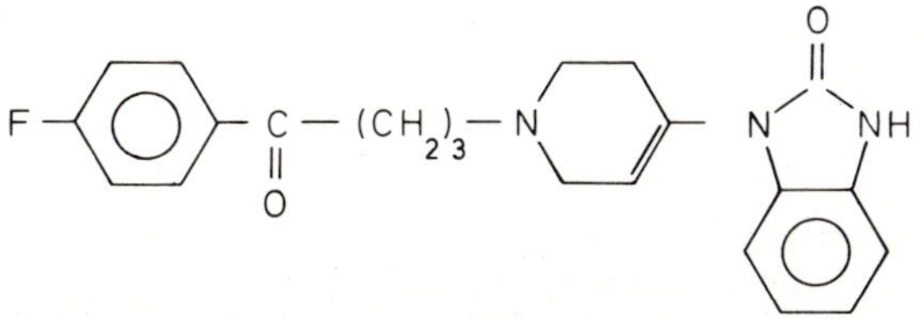

11

The fluorescence of the other chromophore was efficiently but not totally quenched in the bifunctional systems but not at all in equimolar mixtures of the monofunctional models. All phosphorescence, however, occurred from the parabutyrophenone part, indicating very efficient triplet transfer. Bathochromic shifts and a hyperchromic effect in the longest wavelength band of the absorption spectra are an indication for ground state interaction, which might lead to nonplanar conformations of the molecule playing a role in the excitation transfer.

From phosphoresence spectra in a series of dinucleotide analoges (20) it was found that with exception of the bichromophore containing adenine and thymine, the emission occurred from the chromophores with the lowest triplet level. Analogous results were found in dinucleotides (51) and in DNA (20). Since thymine has the lowest triplet level, emission of thymine is found to be more intense than could be expected on the basis of its intersystem crossing efficiency. The possibility that energy transfer could occur via exciplex formation was put forward by Lamola (51).

Energy transfer has been used to calculate distances between chromophores in biomolecules (51), not only polynucleotides but also in proteines. Conrad and Brand (52) studied bichromophoric systems [12] in order to explore the potential of energy transfer measurement as a means of calculating distances between fluorescent residues on polypeptides and proteins in solution.

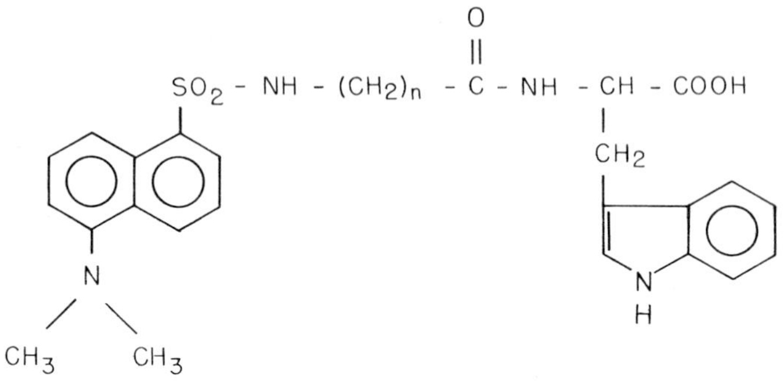

12

They found that Förster's theory of resonance energy transfer was applicable to such systems if the interchromophoric distance and orientation, the fluorescence efficiency of the donor, the extinction of the acceptor, and the overlap between the emission of the donor and the absorption of the acceptor are of magnitudes which produce a transfer rate constant of less than $10^{11}s^{-1}$ and a transfer efficiency which is not too close to 0 or 1.

Such a method has been used, for example, by Epand (54) in a glucagon while Galley and Stryer (54b) studied triplet-triplet energy transfer in proteins as a criterion of proximity.

Another bichromophoric system, studied as a model for biomolecules, consists of two covalently linked metal porphyrins [13] (55).

In rigid solutions of 2-methyltetrahydrofuran at $77^{o}K$, intramolecular energy transfer from the zinc triplet state to that of the copper tripdoublet was manifested in a shortening of the zinc porphyrin triplet state lifetime from 46±1 to 34±1 ms in the ethylene bridged systems. In the cobalt-zinc ethylene bridged systems triplet energy transfer from the zinc chromophore was more efficient as manifested in the complete quenching of the zinc phosphorescence. Calculations suggested that in the copper-zinc case, energy transfer proceeded by a triplet-triplet exchange mechanism, whereas in the cobalt-zinc case, quenching of the zinc triplet proceeded by the formation of a charge transfer complex. There was no evidence for triplet-triplet energy transfer in the phenylene bridged systems. Singlet-singlet energy transfer was not observed in any of the systems.

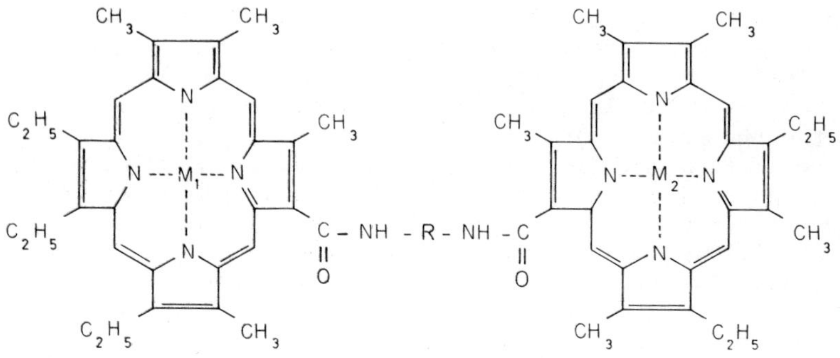

13

a : $R = -(CH)_2-$ $M_1 = Zn$; $M_2 = Cu$ or Co.

b : R = 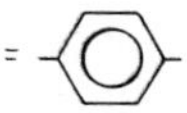 $M_1 = Zn$; $M_2 = Cu$ or Co.

The intramolecular singlet energy transfer from the phenol to the indole chromophore in system [14] has been studied by Tamaki (56).

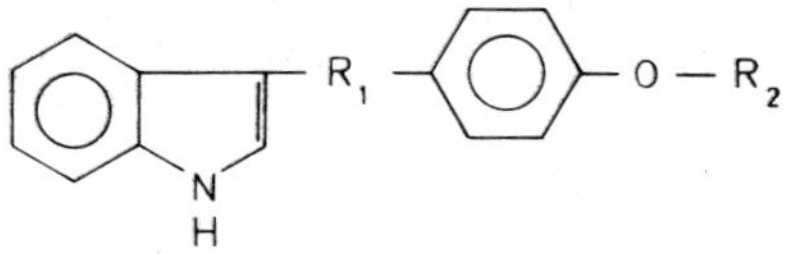

14

a : $R_1 = (CH_2)_2$; $R_2 = CH_3$

b : $R_1 = (CH_2)_3$; $R_2 = CH_3$

c : $R_1 = (CH_2)_3$; $R_2 = H$

d : $R_1 = CH_2-C(=O)-NH-CH_2$; $R_2 = CH_3$

e : $R_1 = CH_2-C(=O)-NH-CH_2$; $R_2 = H$

f : $R_1 = (CH_2)_2-C(=O)-NH-(CH_2)_2$; $R_2 = CH_3$

He found highly efficiently singlet energy transfer in the compounds containing those chromophores joined by the methylene and the amine groups.

The total emission spectra of these compounds consisted only of those from the indole chromophore; neither the fluorescence, nor the phosphorescence of the phenol chromophore contributed to them. From the distortion of the absorption spectra can be concluded that a weak intramolecular hydrogen bonding of the indole chromophore to the phenolic oxygen or the π orbitals of the phenol ring exists. The low fluorescence yields of compounds [14b] and [14c] were attributed to the formation of an intramolecular complex between the indole group in the excited state and the phenol group in the ground state.

Recently, in an elegant study, Wagner (57) used a bichromophoric system to explain n,π^*-like chemical reactivity in ketones with π,π^* lowest triplets. 1-benzoyl-4-anisoylbutane [15] undergoes normal type II photoelimination from each end.

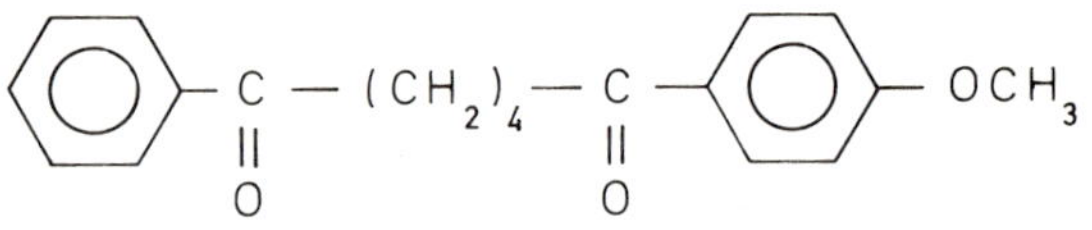

15

From his results he concluded that ketones such as the p-methoxybenzoyl chromophore react from equilibrium concentrations of their n,π^* triplets. Support for this view is found in the fact that reversible internal conversion of the two triplets is much faster than any irreversible decay modes of either individual triplet. Furthermore, the observation that interchromophore energy transfer is sufficiently rapid to allow almost complete equilibration before any chemical reaction, certainly lends support to the supposition of rapid intrachromophore energy transfer.

In these types of bichromophoric systems, linked by a short flexible chain, singlet transfer does occur efficiently in most cases, mostly through a dipole-dipole mechanism, while triplet transfer occurs with very high efficiency through an exchange mechanism. Due to very fast and almost total transfer, rates of transfer are difficult to measure. In several cases, there are hints at the possibility of complex formation being involved in excitation transfer.

2. Chains with Restricted Mobility. Blout and co-workers (58) studied energy transfer between chromophores linked by a bissteroid bridge [16].

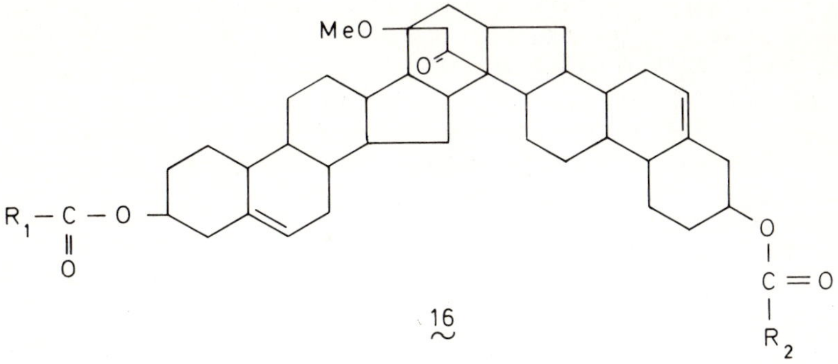

16

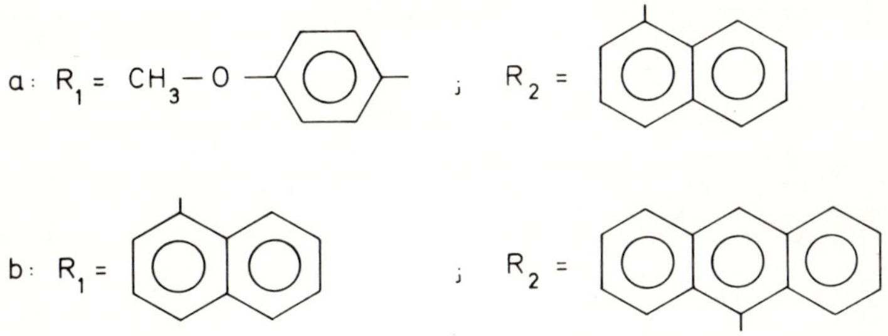

The distance between the chromophores is approximately 21 Å as measured on molecular models and no ground state interaction could be observed. In dioxane, the singlet energy transfer occurs by a Förster-type mechanism. On the basis of this mechanism, the distance can be calculated from the transfer efficiency. For [16a] a value of 21.2±1.6 Å was found, while for [16b] one obtained a value of 16.7±1.4 Å. This latter discrepancy was explained on the basis of preferential orientation of the chromophores. That this phenomenon indeed plays an important role, was proved by Leermakers (59,60) in a series of indole alkaloids. The same author (59) also illustrated the absence of triplet-triplet transfer at low temperature in compound [16b].

Keller and Dolby (61,62) found that in the bichromophoric systems [17a] and [17b] efficient singlet energy transfer occurred according to a Förster mechanism. A distance of 15 Å for [17b] and 14 Å for [17a] was found on the basis of transfer efficiency, while molecular models lead to a value for separation of the chromophores of 15 Å. Triplet energy transfer was found to be much less efficient (35% for [17a], 30% for [17b]) as compared to triplet transfer between the same chromophores when linked to each other with a flexible chain [10b, 10d], sustaining the hypothesis that triplet transfer occurs by a contact mechanism and over much shorter distances.

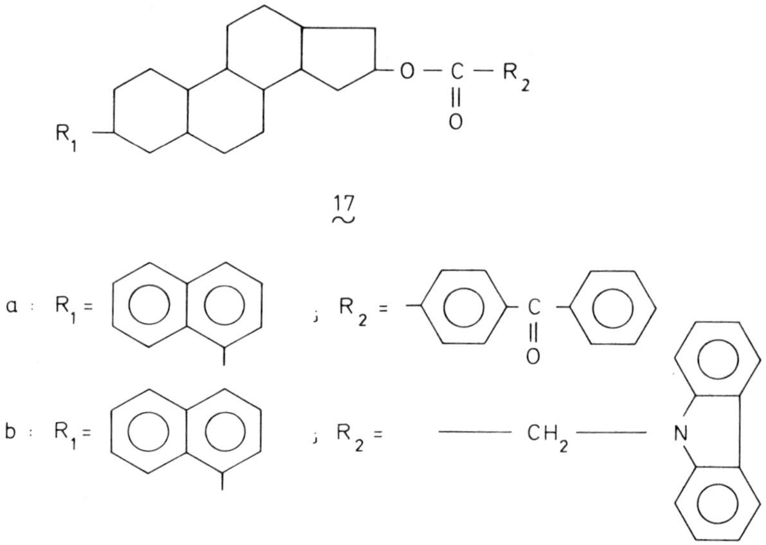

17

The presence of an orientation factor in the equation for Förster-type energy transfer has led to the study (63-67) of nonconjugated bichromophores held together by a rigid link, in such a way as to fix the orientation of one chromophore with regard to the other one, for example, [18] (63).

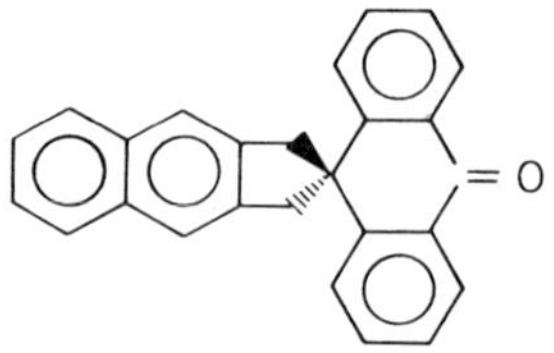

18

Several discrepancies (64,65) were found in these studies, leading to no clear picture of the importance of the orientation factor. Important problems are inherent to these studies, as has been pointed out by Lamola (65). Indeed, the quantitative relation, worked out by Förster, is only valid for distances between the chromophores for which the coulombic interactions between donor and acceptor oscillators prevail and for which the point-dipole approximation is valid. This is certainly not the case in the compounds studied. Furthermore, most transitions have polarizations along more than one molecular axis, resulting in small dipole-dipole interactions, even if the chromophores are held perpendicular to each other.

B. Photochemical Reaction of the Acceptor

1. Flexible Bichromophoric Systems. Chemical changes in the acceptor moiety have been used as detection method for excitation transfer of the bichromophoric systems with a flexible link mentioned in Table 2.

On specific irradiation of the carbonyl function in 19,I n=1 trans, 19,I n=1 cis is formed together with other products. Morrison (68,69) proposed that this isomerization was a result of triplet energy transfer from the carbonyl function to the olefinic part of the molecule. This hypothesis could be further verified for the systems 20-22 of Table 2. In all cis-trans isomerizations studied, a scheme is put forward in which energy transfer from the phenyl rings or from the carbonyl group in its triplet manifold to the either cis or trans double bond occurs, resulting in a common related olefin triplet. On the basis of this scheme, rate constants k_{et} could be determined by quenching experiments. The rate constants obtained in this way are assembled in Table 3.

In the study of [19] n=2 and [20] n=3 Morrison and co-workers observed as the most important process, intramolecular cycloaddition (vide infra) occurring from the singlet excited state manifold of, respectively, the carbonyl and the phenyl group through the intermediacy of an excited state complex. They also found that compounds [26] and [27], which do not form adducts, show strongly decreased fluorescence and, consequently intersystem crossing efficiencies (70).

Compound [20] n=2 on the contrary shows no important change in fluorescence quantum yield as compared to hexylbenzene, indicating that complex formation from the singlet excited state is not a competing process in this series unless three methylene groups separate the two chromophores.

TABLE 2.

	n	System	Nature of Change	Ref.
[19]	1	$CH_3-C(=O)-(CH_2)_n-CH=CH-CH_3$	*cis*-*trans* isomerization	(67)
	2			(69,70)
[20]	1	$\Phi-(CH_2)_n-CH=CH-CH_3$	*cis*-*trans* isomerization	(71,72)
	2			(73)
	3			(74)
[21]	1	$\Phi-C(=O)-(CH_2)_n-CH=CH-$	*cis*-*trans* isomerization	(75,76)
	2			(75,76)
	3			(75,76)
	4			(75,76)
[22]		$\Phi-C(=O)-C(CH_3)_2-CH=CH-CH=CH_2$	*cis*-*trans* isomerization	(77)
[23]		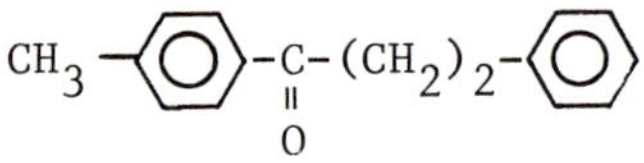	Racemization	(78)
[24]	0	$\Phi-(CH_2)_n-C(=O)-NH-CH(CH_3)-COOH$	Hydrogen abstraction	(78)
	1			(79)
	2			(79)
	3			(79)
	4			(79)

TABLE 2 (cont'd)

	n	Systems	Nature of Change	Ref.
[25]	0	$\Phi-(CH_2)_n\underset{\underset{O}{\|}}{C}-CH_2-CH_2-O-CH_2-CH_3$	Hydrogen abstraction	(80)
	1			
	2			
	3			
	4			
[26]		$CH_3-\underset{\underset{O}{\|}}{C}-(CH_2)_3-CH=CH_2$		(70)
[27]		$CH_3-\underset{\underset{O}{\|}}{C}-\underset{CH_3}{CH}-CH_2-CH=CH-CH_3$ (<u>trans</u>)		(70)
[28]		$CH_3-\underset{\underset{O}{\|}}{C}-(CH_2)_2-C{=}C-CH_3$		(70)
[29]		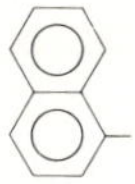 $-(CH_2)-\underset{\underset{O}{\|}}{C}-CH_2-CH_2-O-CH_2-CH_3$		(80)

The importance of orbital overlaps in triplet energy transfer in bichromophoric systems is clearly illustrated by the variation of the rate of transfer with chain length in the series [21] n=2 to n=4. Since intersystem crossing is fast in these compounds, no doubt exists that energy transfer occurs from the triplet state.

The importance of complex formation in singlet energy transfer was pointed out earlier by Hammond and Cooke (78) in the racemization of the sulfoxide group in compound [23]. They proposed a mechanism involving two excited states: a singlet excited state with excitation localized on the naphtalene ring and an excited state complex with delocalized excitation. Only the former could be quenched and no emission from the second could be observed. The results led to the conclusion that both states were in equilibrium with the first one as the principal species present in this

TABLE 3. Rate Constants of Triplet Energy Transfer.

	Compound	k_{et} (s^{-1})	Ref.
[19]	n=2	$<4.10^9$	(70)
[20]	n=1	2.10^8	(71)
		$<4.10^9$	(74)
	2.10	2.10^6	(72)
	n=2	6.10^7	(73)
	n=3	$<9.10^8$	(74)
[21]	n=2	$7.2\ 10^{10}$	(76)
	n=3	1.10^{10}	(76)
	n=4	$3.3\ 10^9$	(76)
[22]		$<10^{10}$	(77)

[a]See ref. 74, footnote 58 for discussion.

[b]Estimated from the fact that no diffusion controlled quanching of benzophenone occurs.

equilibrium. They also considered the possibility of irreversible formation of the complex, while quenching of only the state with localized excitation is possible. On the basis of this assumption, they calculated a rate constant for complex formation of 9.10^7 s^{-1} which they considered to be too slow.

In a series of amides [24], maximum efficiency of hydrogen abstraction was found for [24] n=3, but no mechanistic details are available. Brain and Morrison (80) on the contrary did not find a maximum efficiency for a chain length of three units in systems [25], but they observed a decrease in the quantum yield of hydrogen abstraction on lengthening of the alkylene chain. They took this as an indication for a specific steric requirement necessary for energy transfer to occur, as in the series studied by Cowan and Baum [21] (76). Since fluorescence quenching occurs at the same rate as quenching of the reaction, energy transfer from the singlet excited state is assumed. This could explain the inactivity of naphtyl analogs since their singlet energy levels would be too low for efficient transfer.

Excited state interaction through species in the singlet and triplet excited state has been invoked to explain the reduced photoreactivity of β-phenyl-substituted ketones [30][31] (81-84).

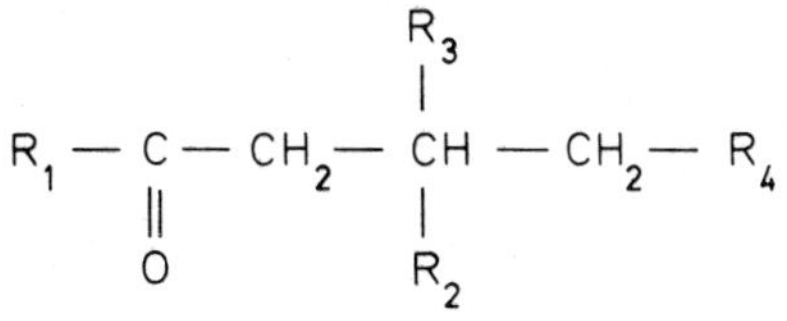

30

a :	$R_1 = CH_3$	$R_2 = \phi$	$R_3 = H$	$R_4 = H$	(81)
b :	$R_1 = CH_3$	$R_2 = \phi$	$R_3 = H$	$R_4 = \phi$	(81)
c :	$R_1 = CH_3$	$R_2 = \phi$	$R_3 = H$	$R_4 = H$	(83)
d :	$R_1 = CH_3-CH_2$	$R_2 = H$	$R_3 = H$	$R_4 = \phi-X$	(83 84)
e :	$R_1 = CH_3$	$R_2 = \phi-X$	$R_3 = CH_3$	$R_4 = H$	(84)

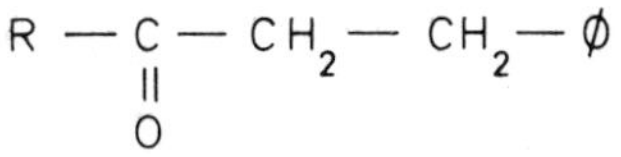 (82)

31

a : $R = CH_3$

b : $R = \phi$

The efficiency of the type II process, which in these systems occurs from the singlet state only, is highly diminished in comparison with carbonyl compounds such as butyrophenone or valerophenone. The rate of the type II process, intersystem crossing, and internal conversion are sensitive to phenyl substitution (84). This observation lends further support to specific excited state interaction between the phenyl group and the excited ketone.

2. Rigid Links. Only a few examples [32, 33] are known of rigidly held nonconjugated bichromophoric systems in which energy transfer is detected by a reaction of the acceptor. In the phototransformation of [32] (85) the authors propose that this reaction is sensitized from the second excited triplet state of the enone.

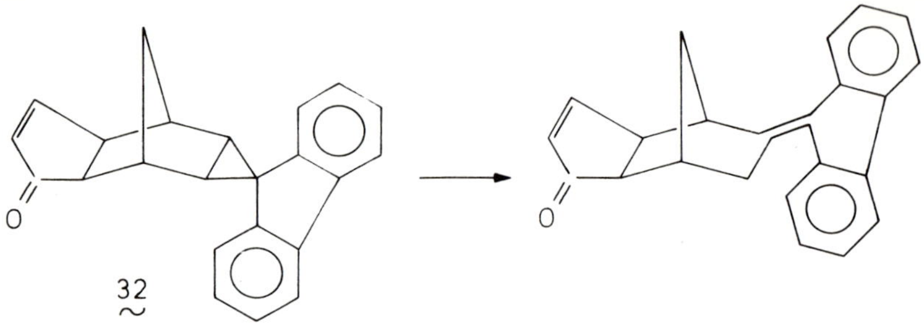

32

Excitation of [33] in the dimethoxybenzene part leads, after very efficient intramolecular triplet energy transfer, to reaction of the norbornene part of the molecule.

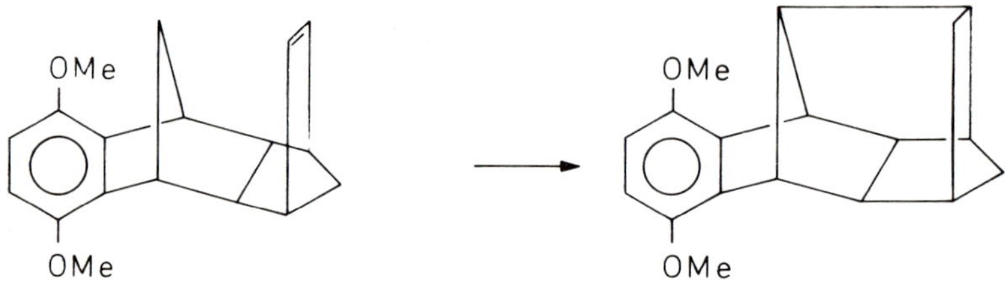

33

Absorption spectroscopy indicated no strong coupling between the chromophores. Sensitization, determination of quantum yield of fluorescence (Φ_F = 0.2), in comparison with the hydrogenated model show that the reaction occurs with high efficiency (Φ_r=0.42) from the first triplet excited state only.

IV. INTRAMOLECULAR REACTION

As pointed out in the introduction, if the two chromophores of the system are suitably chosen, an intramolecular cycloaddition process can occur. The photochemical intramolecular cycloaddition of nonconjugated bisolefines has been the subject of several articles (87-98). Many of these cyclomerizations involve polycyclic systems.

Emphasis will in this survey be put on bichromophoric systems linked by a flexible chain. Symmetrical molecules in which no competition with energy transfer processes exists, will be reported first followed by a discussion of some nonsymmetrical systems.

A. Systems With Two Identical Chromophores

1. Bisanthracenes. A series of 9,9'-linked bisanthracene derivatives (Table 4) are cyclomerized on irradiation according to the following reaction scheme [34 → 35].

TABLE 4. Bisanthracenes.

[34]	R	Ref.
a	-C(=O)-O-C(=O)-	(99)
b	-CHOH-	(100)
c	O-C(=O)-O	(101)
d	-NH-C(=O)-	(101)
e	-N=N-	(101)
f	$-(CH_2)_2-$	(102,103)

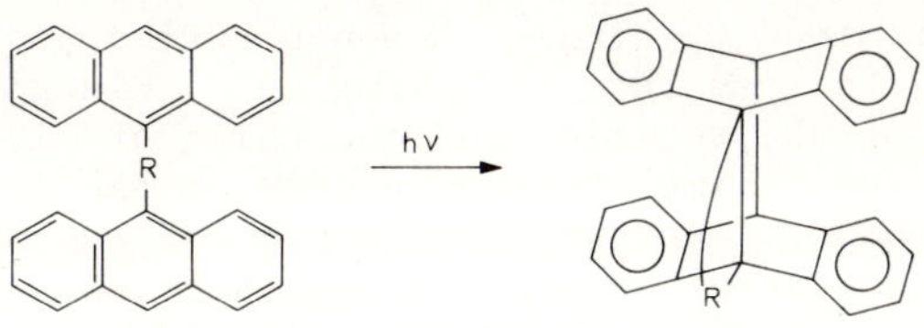

34 35

A bisanthracene with structure [34f] but substituted at the 10, 10'-positions with either bromine or chlorine did not cyclomerize (101), although paracyclophane [36] (104) and 1,2-bis (10'-methyl-9'-anthryl) ethane (105) do give a cycloadduct on irradiation.

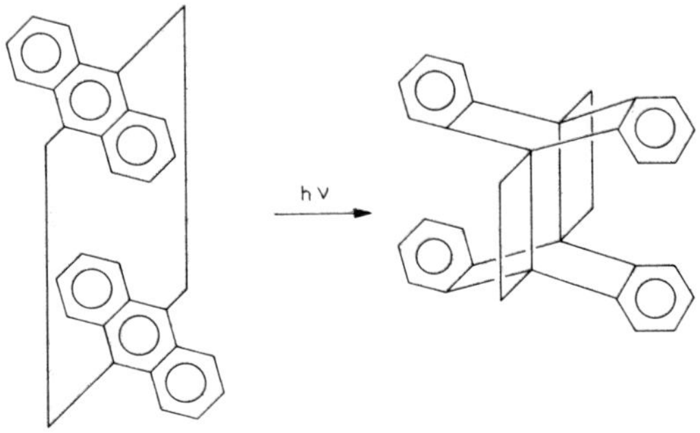

36

A study of Livingston and Wei (102) showed that the dimerization reaction of anthracene is occurring via the singlet excited state although it may be partially due to triplet reaction based on the effect of oxygen on the quantum yield of cyclomerization. Recent results (106), however, indicate that dimerization of 9-methylanthracene occurs via excited state complex formation from the singlet excited state.

Recently intramolecular cycloadditions of bisanthracenes linked at 1,1'- and at 2,2'-positions were observed (107).

Product ratios and quantum yields of cyclomerization are given in Tables 5 and 6. A spectroscopic study of the systems [37a-f] indicated that for $n=2 \rightarrow 5$ a bathochromic emission could be observed (Fig. 2).

The lifetime of the state responsible for this emission could be determined by single photon counting and varied from 46 ns for n=2 to less than 10 ns for n=5. Kinetic analysis by quenching of the reaction and quenching of fluorescence indicated that the excimer formation lies on the reaction coordinate leading to product. An analogous statement was made by Aladekomo in the dimerization of 9-methylanthracene (106). An interesting point is the observation of excimer emission in a photochemical reactive system in which the chain contains more than four

TABLE 5. Cyclomerization of [37a-f](108) in CH_2CL_2

[37]	Φ_C	Product [38]	Ratio [39]
a	$0.5 \cdot 10^{-1}$	90	10
b	$0.9 \cdot 10^{-1}$	85	15
c	$0.8 \cdot 10^{-1}$	70	30
d	$1.1 \cdot 10^{-1}$	65	35
e	$0.6 \cdot 10^{-1}$	60	40
f	$0.5 \cdot 10^{-1}$	50	50

TABLE 6. Cyclomerization of [40a-d] (109) in CH_3CN

[40]	[41]	[42]
a	100	-
b	65	35
c	50	50
d	60	40

saturated bonds. This forms a link between the nonemissive bismaleimides (117) (vide infra) and the recent spectroscopic findings by K. Zachariasse in the bispyrene system (21) (vide supra).

2. Bisnaphthalenes. Chandross and Dempster (16,110) showed that 1,3-bis(α-naphtyl) propane [43], on excitation above $-30^{o}C$, results in the formation of a complex which at this temperature can either fluoresce or collapse into product [44]. Irradiation of [44] at $77^{o}K$ resulted in the reverse reaction and excimer fluorescence, but no excimer phosphorescence of this cleaved dimer could be observed (111). This clearly indicated that on the way to product collapse an excited state complex is formed.

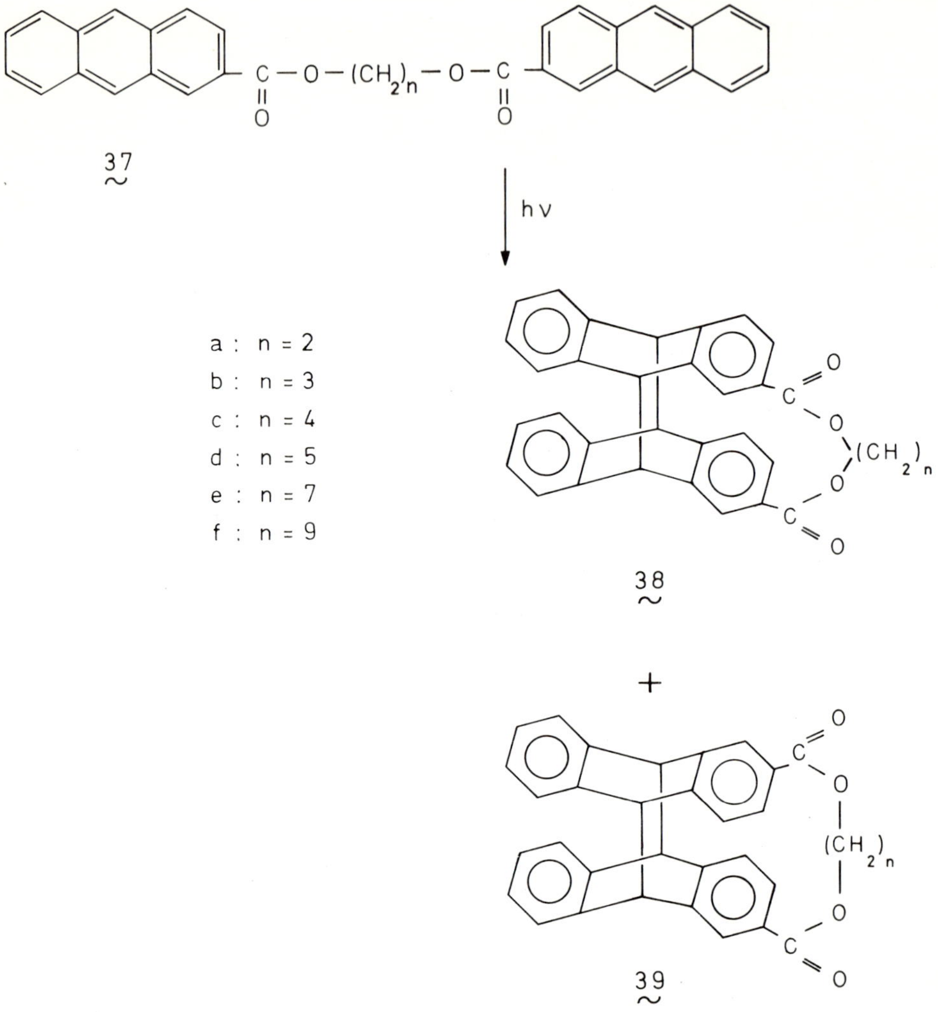

No intramolecular product formation was observed for 1,2-bis (α-naphtyl)-ethane or 1,4-bis(α-napthyl)-butane, while a β,β-linked 1,3-bis(naphtyl)-propane did show excimer fluorescence but no product formation. Excimer fluorescence quenching is observed which according to the authors (16) could be due to formation of an unstable photoproduct that reverses very fast to the open form.

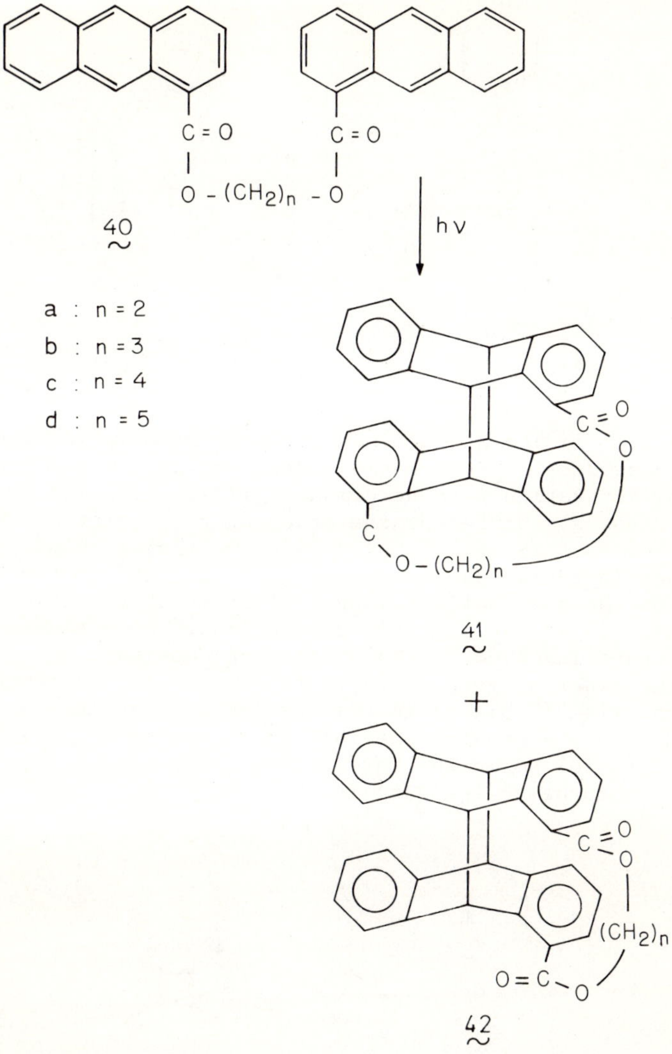

3. Bisthymines. Intramolecular cyclization has been observed in a whole series of bisthymines. An extensive and elegant study was made by Leonard (112-114) and co-workers on the effect of the chain length and the point of attachment on the nature of the reaction products obtained.

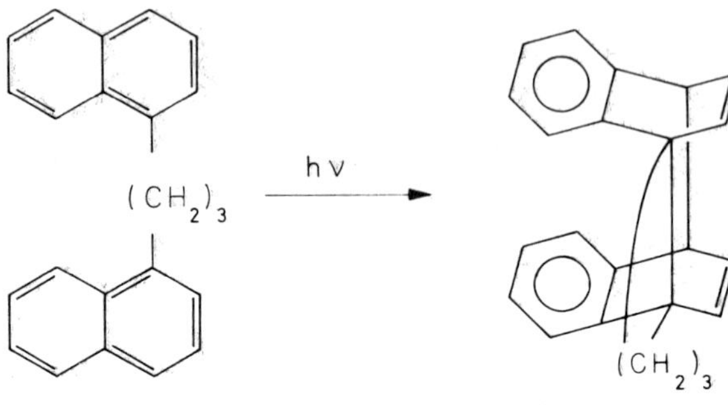

43

44

The 1,1' linked bisthymines yield on irradiation exclusively a cis-syn product. The relative rates for direct and sensitized photocyclomerization were determined (113). Although the ratios in both cases are different the order of reactivity can be established as follows: [45b] > [45c] > [45a] > [45d]. The difference in ratios and increased rate on sensitization is explained on the basis of the longer lifetime of the triplet excited state involved in the sensitized reactions. No evaluation of the excited state involved in the direct cyclomerization is, however, made. The order of reactivity is found back in the percentage of hypochromism (Table 7) in this series indicating that the compound with the greater possibility to assume the planar rings in near parallelism in the interacting conformation does show greatest photochemical reactivity.

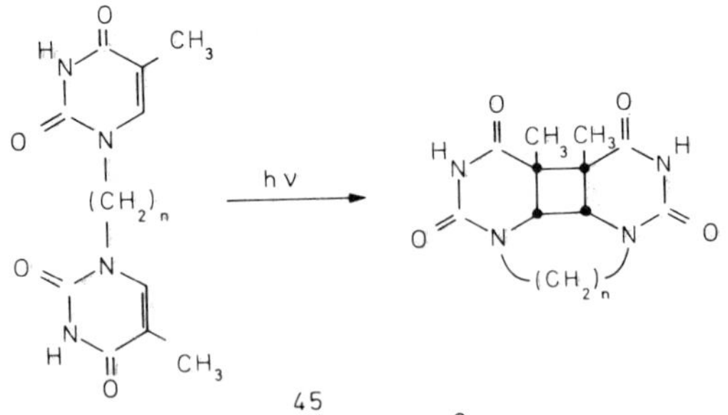

45

a : n = 2

b : n = 3

c : n = 4

d : n = 6

The high reactivity of [45b] is found back in the 5'-deoxy-5'-(1 thyminyl)thymidine which with equal ease as [45b] on irradiation results in a cis-syn dimer with the carbonyl groups cis to the furane oxygen (115).

If the polymethylene chain is attached to the two thymines at N-3, [46], the cis-syn cyclomer is formed exclusively and the trimethylene derivative is found to be the most reactive on sensitization.

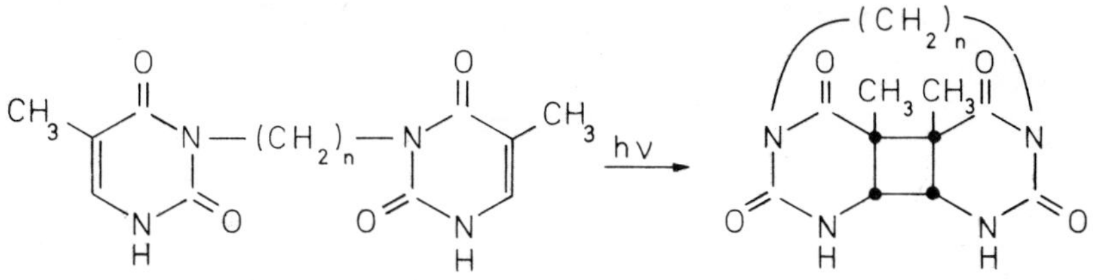

46 a : n = 2
b : n = 3
c : n = 4

If two reactive groups are attached at 1,5-position [47], intramolecular photocycloaddition is restricted to a trans head-to-tail (anti) addition. Furthermore, the chain is now linked to a carbon which is involved in the reaction, [47]n=2 reacts faster than [47]n=3.

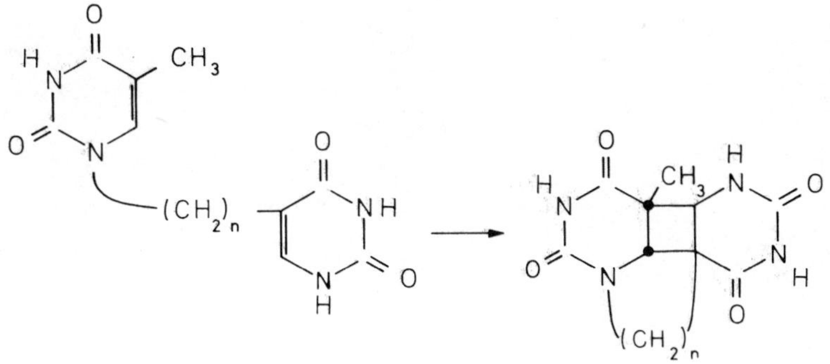

47 a : n = 2
b : n = 3
c : n = 4

On irradiation of 1,3-linked polymethylene bisthymines [48] only the <u>cis</u> head-to-tail cyclomer is formed and no evidence for the formation of oxetane-type adducts was obtained.

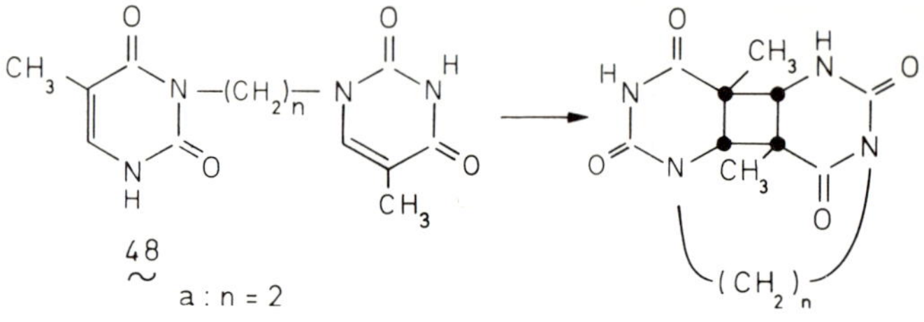

From Table 7 it can be seen that, except for the 1,3 series hypochromism seems to be related to reactivity. In view of the recent results on biscoumarins (vide infra) one can therefore expect that in this series the existence of favorable prearranged ground state conformations must play a very important role.

TABLE 7. Relative Rates of Photosensitized Cyclomerization and and Percentage Hypochromism (H).

	H(%)	Rel. Rate
1,1'[45]	-	
2	6.6	1
3	10.5	5
4	4.6	2
5	7.2	
6	3.9	0.25
1,3[48]		
2	4.8	1
3	8.9	0.8
4	6.7	0.6
3,3[46]		
2	3.9	1
3	4.5	7
1,5[47]		
2	7.8	1
3	6.8	0.7
4	5.2	0.7

4. Bismaleimides. N,N'-alkylenebismaleimides [49] cyclomerize on direct irradiation in quantitative yields to the tetracyclomer [50] if n varies from 3 to 7 (116). Mechanistic studies showed (117) that the reaction occurred from the singlet excited manifold while spectroscopic studies gave no indication of important interaction in the ground state.

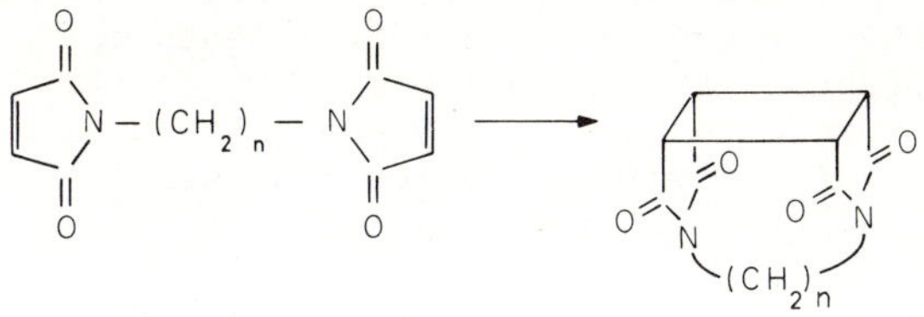

49

50

a : n = 3
b : n = 4
c : n = 5
d : n = 6
e : n = 7

Triplet sensitization did not lead to intramolecular adducts for higher alkanes but resulted in polymer formation (vide infra); a reaction scheme including excited complex formation in a primary step was proposed. The rate constant for excited complex formation, k_e, could be calculated as a function of the number of methylene units and diminished by lengthening of the chain (Table 8). The quantum yield of intersystem crossing was much lower (Table 8) than in the isolated chromophore, N-butylamaleimide (Φ_{isc} = 0.25), and substantially lower than the quantum yield of reaction, Φ_c (Table 8).

TABLE 8. Rate Constants for Intramolecular Excited State Complex Formation, Quantum Yields of Intersystem Crossing, and Reaction of [49] in Dichloromethane.

n	$k_e(s^{-1})$	Φ_{isc}	Φ_c
3	1.5×10^9	0.025 ± 0.005	0.62
4	0.6×10^9	0.045 ± 0.01	0.42
5	0.7×10^9	0.045 ± 0.01	0.54
6	0.5×10^9	0.12 ± 0.02	0.42

It is on the basis of these results that the idea was put forward that intramolecular reaction through an excited state complex is limited by the probability of the formation of a favorable conformation within the lifetime of the exicted state involved and by the complex stabilization, covalent bond formation being the extreme (117). Factors influencing excited state lifetime or complex stability will, therefore, influence the intramolecular reaction. Substitution of the double bond of the maleimide ring with two methyl or two chlorine substituents decreases the cyclomerization probability substantially. White [49] cyclomerizes efficiently even with seven methylene units between the two maleimide rings, no cyclomerization is observed for N,N'-pentamethylene-bisdimethylmaleimide (118).

5. Biscoumarins and Bisfurocoumarins. An extensive study of the 7,7'-polymethylenedioxycoumarins has been made. All cyclomers obtained had a <u>cis</u> (endo-) structure even in those derivatives where an anti adduct could be formed without severe strain in the molecule. The 7,7'-polymethylenedioxycoumarins, except [51]n=2 for strain reasons, form primarily <u>cis</u>-syn adducts (Table 9) and this choice is strongly enhanced by substitution of reactive double bond by a methyl group (119,120).

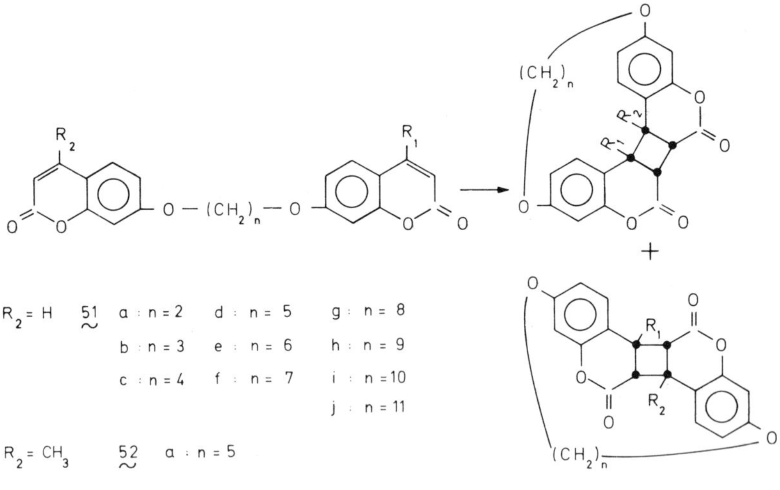

$R_1 = R_2 = H$ **51** a : n = 2 d : n = 5 g : n = 8
b : n = 3 e : n = 6 h : n = 9
c : n = 4 f : n = 7 i : n = 10
j : n = 11

$R_1 = R_2 = CH_3$ **52** a : n = 5

$R_1 = H, R_2 = CH_3$ **53** a : n = 4

TABLE 9. Photocyclomerization of 7,7'-Polymethylenedioxycoumarins

	n	cis head-to-head	cis head-to-tail	$\Phi_{cyclomerization}$
[51]	2	85	15	$0.24\ 10^{-2}$
	3	41	59	$1.25\ 10^{-2}$
	4	42	58	-
	5	40	60	$1.42\ 10^{-2}$
	6	30	70	-
	7	40	60	-
	8	28	72	-
	9	33	67	$0.44\ 10^{-2}$
	10	32	68	-
	11	33	67	-
[52]	5	3	97	
[53]	4	30	70	

A detailed kinetic study (121) of the 7,7'-polymethylenedioxycoumarins indicated that the reaction occurs partially from the singlet excited state manifold and partially from direct excitation of favorable ground state conformation. The latter are extremely important in the total reaction since they collapse to a product with high quantum yield while the reaction from singlet species is much less efficient. The total reaction scheme is presented in Scheme II. The lifetime of the different excited states involved could be estimated and a correlation with recently proposed models of energy surface crossings could be made (122-124).

Indeed the excited head-to-tail conformer E_2, stabilized in the ground state by dipole-dipole interactions, is more destabilized in the excited state and shows a slightly longer lifetime than E_1.

A small change in the nature of the link affects substantially the product distribution of the cyclomers as can be seen in the polymethylene dicarboxylic acid-(7-coumarinyl)-diester [54] and [55] (Table 10).

Irradiation of bisdihydrofurocoumarin [56] results in the formation of cis-syn (80%) and cis-anti(20%)cyclomer while

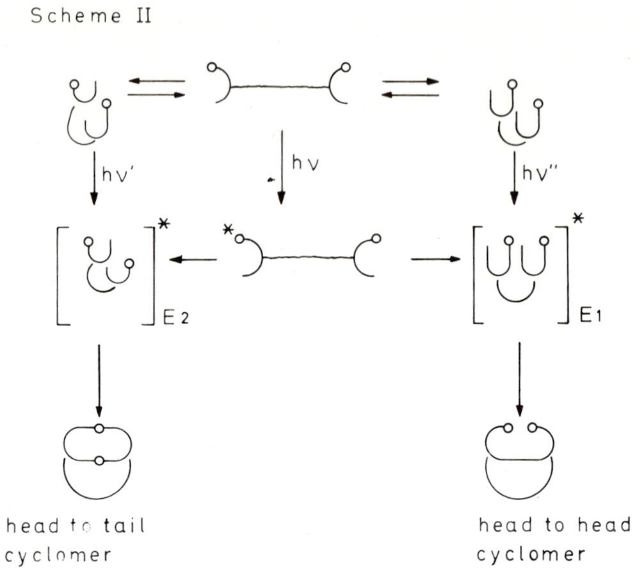

TABLE 10. Photocyclomerization of Polymethylenedicarboxylic Acid-(7-coumarinyl)-diesters.

	n	cis-syn	cis-anti	$\Phi_{cyclomerization}$	Ref.
[54]	3	80	20	$0.62\ 10^{-2}$	(125)
	4	90	10	$0.65\ 10^{-2}$	(125)
	5	80	20	$0.38\ 10^{-2}$	(125)
	6	100	-	$0.21\ 10^{-2}$	(125)
[55]	2	-	-	$0.7\ 10^{-2}$	(125)
	3	-	-	$1.56\ 10^{-2}$	(125)
	4	-	-	$1.32\ 10^{-2}$	(125)
	5	30	70	$1.1\ 10^{-2}$	(125)
	6	-	-	$0.55\ 10^{-2}$	(125)

bisfurocoumarin [57] leads to the formation of three cyclomers, two of which involve addition of the pyrone double bond to the furan side of the molecule (126). It should be noted that no reactivity of the furan side of furocoumarine is observed in the dimerization of the model compounds.

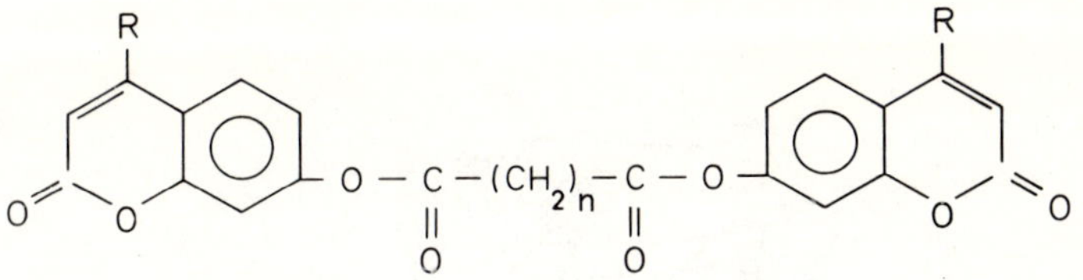

R = H 54 n = 3,4,5,6,7

R = CH_3 55 n = 2,3,4,5,6

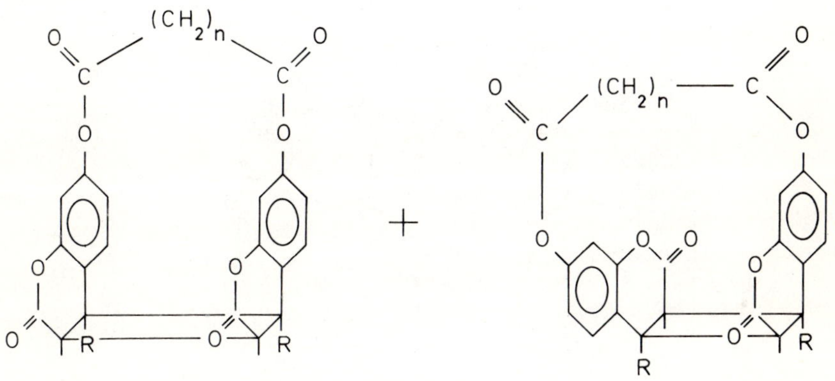

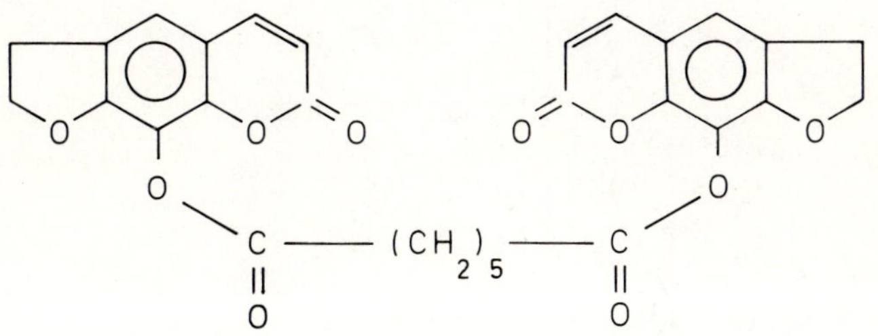

56

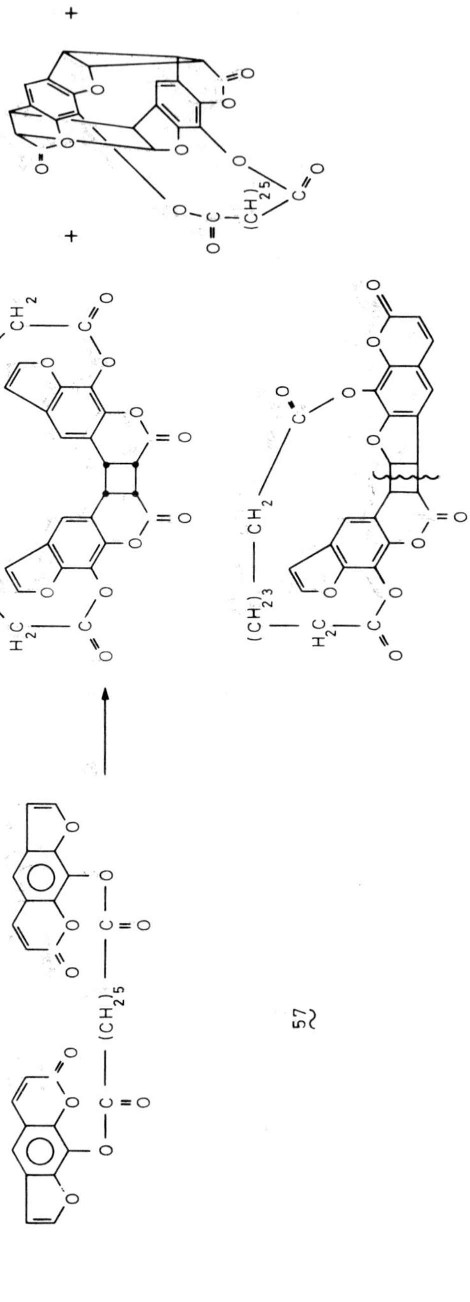

57

6. Bispyrones. More insight into the importance of the number of units in the link between the two chromophores was obtained in the cyclomerization of 6,6'-polymethylene bis(4-methylpyrones). On direct irradiation only the trimethylene derivatives [58] yielded intramolecular cycloadducts (127). The 2+2 adduct is formed by a thermal rearrangement.

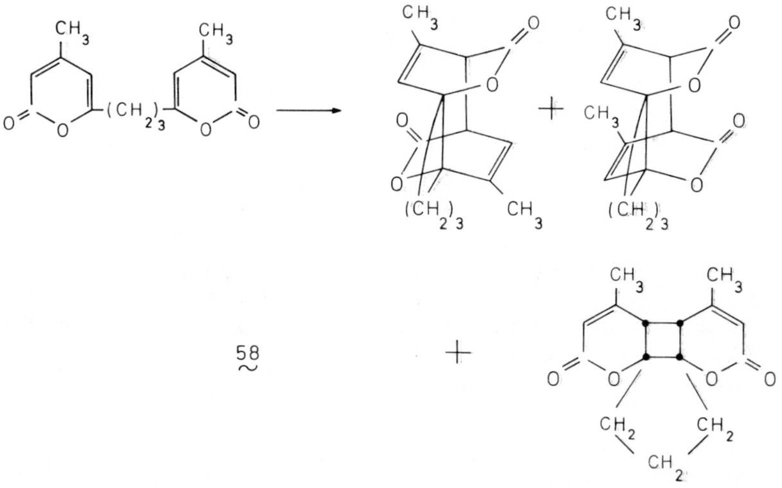

58

On sensitization the trimethylene and tetramethylene derivative give intramolecular 2+4 cycloadducts (128) while in higher homologs 2+2 addition at the double bond α to the methyl takes place. Although no detailed kinetic study of these systems was possible it is tempting to suggest that in the singlet reaction, which might go via an excited state complex, the destabilization is too important for the tetramethylene derivatives to lead to a cyclomer. In the triplet manifold the reactivity is such that the geometric constraints are not as severe and the product formed reflects the stability of intermediates of radical nature rather than that of a delocalized excited state.

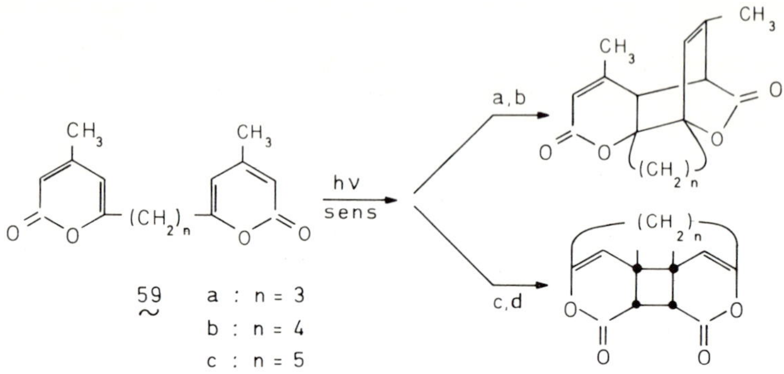

59 a : n = 3
b : n = 4
c : n = 5
d : n = 6

7. Acyclic Bisolefins. Irradiation of α,ω-dicinnamoyloxyalkanes [60] in the presence of benzophenone results in intramolecular cyclomerization as a competitive process for cis-trans isomerization (129-131) even if the two cinnamates are separated by 10 methylene units.

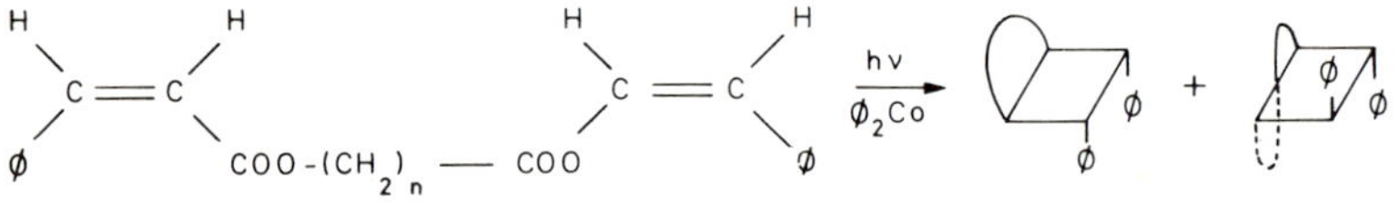

60

Scheffer (132-136) and co-workers showed in an elegant and systematic study that 1,5-, 1,6-, and 1,7-dienes [61-63] did add intramolecularly on sensitization while 1,8- and 1,9-dienes no longer cyclomerize presumably due to the unfavorable thermodynamics involved in the formation of seven- or eight-membered rings.

The authors propose an internal cyclization via stepwise reaction to 1,4 diradical intermediates for which the closure step will be kinematically controlled. The observation that the 1,5 diene undergoes crossed cycloaddition while 1,6- and 1,7-dimers add "straight" can be interpreted in terms of strain, entropy effects, and radical stabilization.

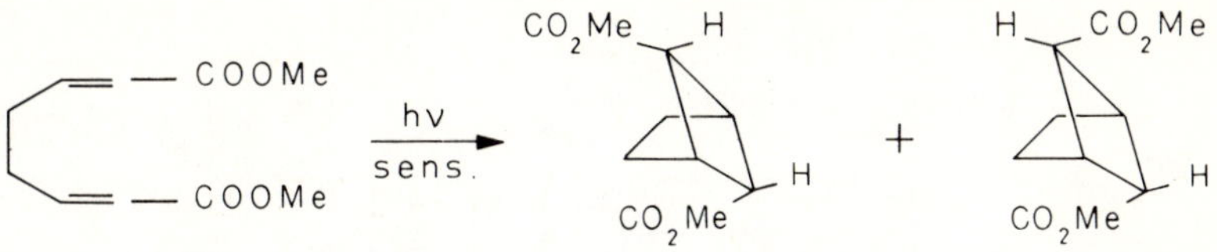

61

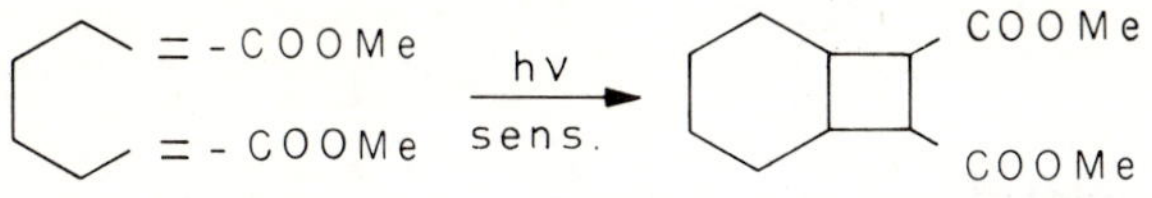

62

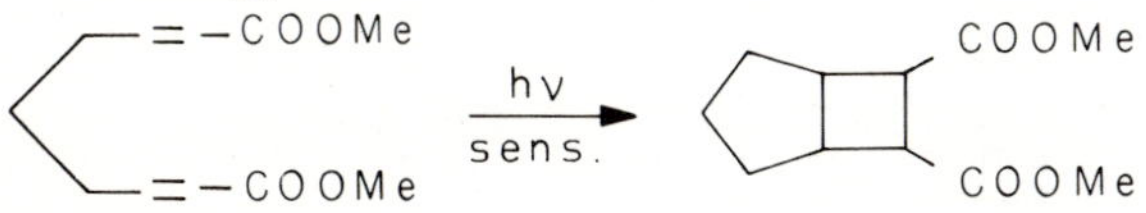

63

Isomeric diene diester [64] has been studied by Bloomfield and Owsley (137). They obtained a crossed addition in the n=2 and straight addition in the n=3 derivatives.

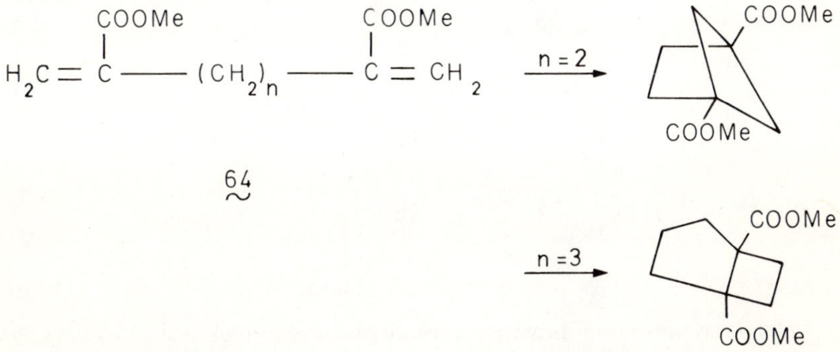

64

Compound [65] on sensitized irradiation leads to two intramolecular cycloadducts (138). The straight 2+2 addition occurs three times more efficiently than the crossed addition.

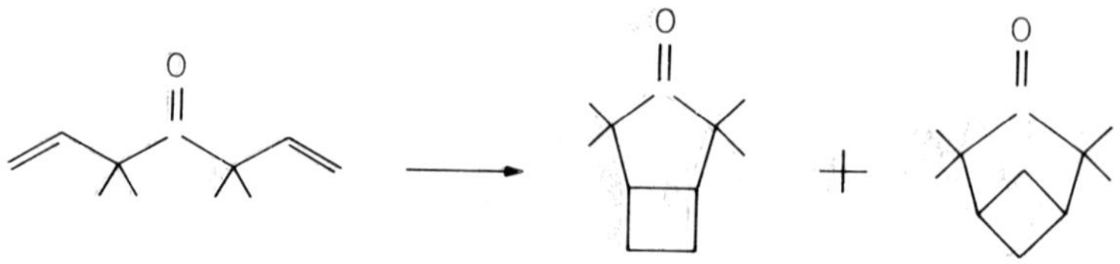

65

Intramolecular cycloadditions were observed in diene [66a-c].

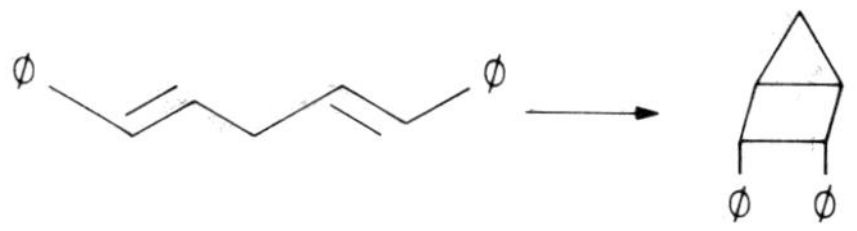

66 a

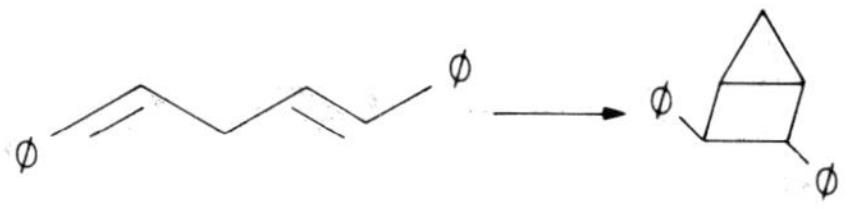

66 b

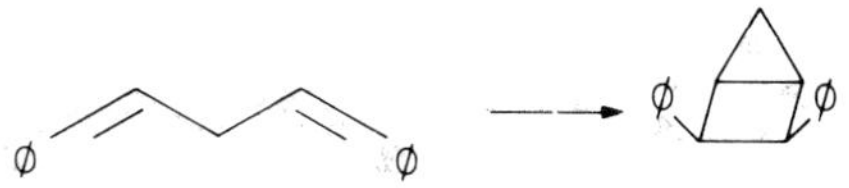

66 c

While in systems having a methyl or phenyl substituent at carbon 3 a di-π-methane rearrangement could be the main process, no evidence is found for vinylcyclopropanes in this series (139).

B. Systems With Two Different Chromophores

If the two chromophores are not identical energy transfer can occur (vide supra II) but in suitably chosen systems intramolecular reaction can be a competitive process. Three main areas of interest can be distinguished: (i) addition of an unsaturated carbon-carbon bond to another one, (ii) addition of a carbonyl to a double bond to form an oxetane, (iii) skeletal rearrangements. Examples of cases (i) and (ii) are reviewed by Dilling (97). This report deals with some of the recent examples without being exhaustive.

1. Addition of an Unsaturated Carbon-Carbon Bond to Another One.

A series of studies have been concerned with the synthetic use of the intramolecular addition of a double bond to an enone. A review of this topic has appeared (140).

Some recent examples in this area are listed below.

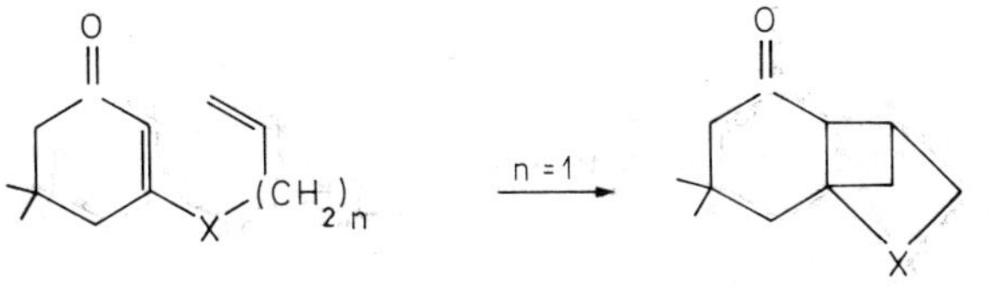

67

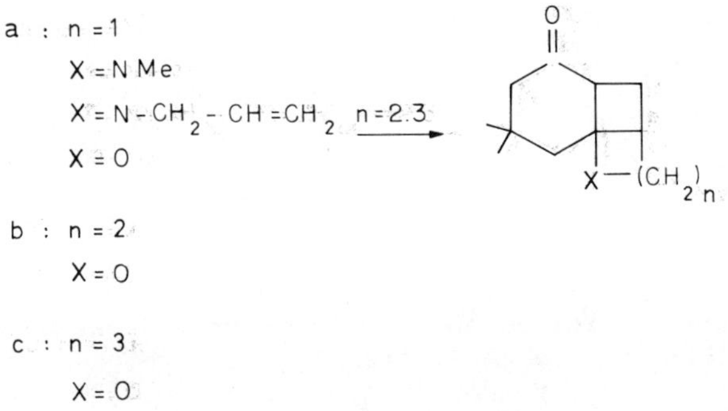

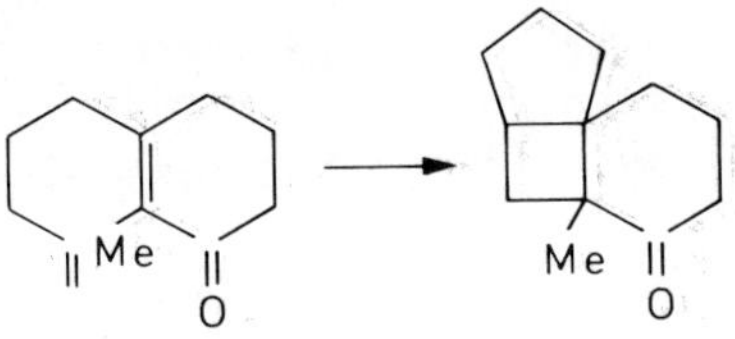

(144)

68

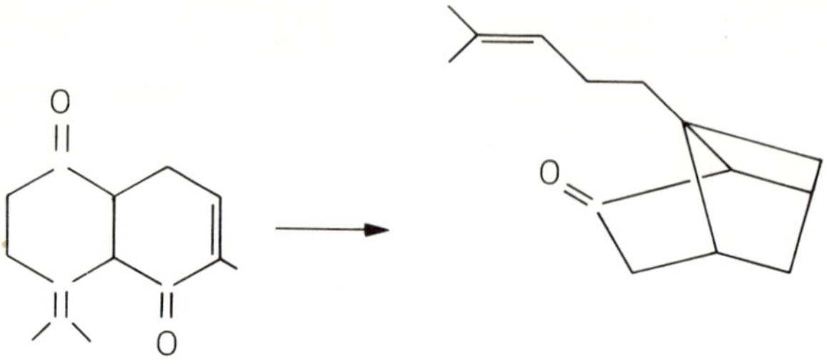

(145)

69

This type of intramolecular reaction has been recently applied in the prostaglandin series (146).

K. Golankiewicz (147) has reported the intramolecular cyclomerization of non-symmetric bispyrimidines.

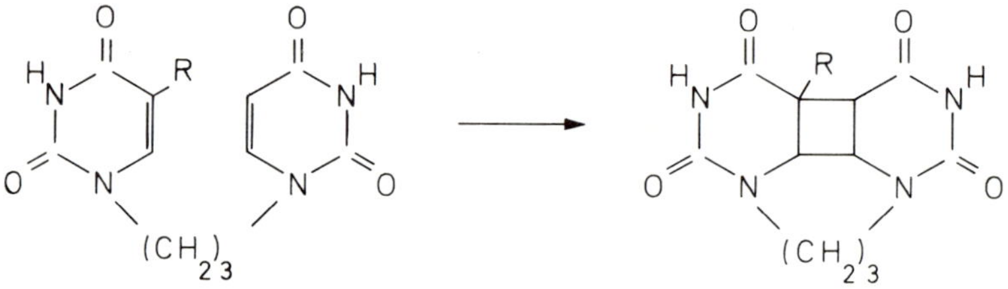

70 a : R = CH_3

b : R = C_2H_5

Acetone-sensitized photolysis of 2,5'-deoxy-5'-(1-thyminyl) uridine [71] leads to two intramolecular cycloadducts (115) which were shown to have, respectively, structures [72] and [73].

Intramolecular additions of a double bond to an aromatic nucleus have been reported. Chandross and Schiebel (148) have observed the intramolecular addition of naphthalene to anthracene in 1-(9-anthryl)-3-(1-naphtyl)-propane. The reaction occurs presumably via an excited state complex between anthracene and naphtalene moeity. Morrison and co-workers studied the intramolecular addition of 6-phenyl-2-hexane [74] (74) and 6-phenyl-2-hexyne [75] (149).

In a very elegant study, Morrison and co-workers could demonstrate that in 6-phenyl-2-hexane [74], besides <u>cis-trans</u> isomerization by energy transfer from the phenyl ring (vide

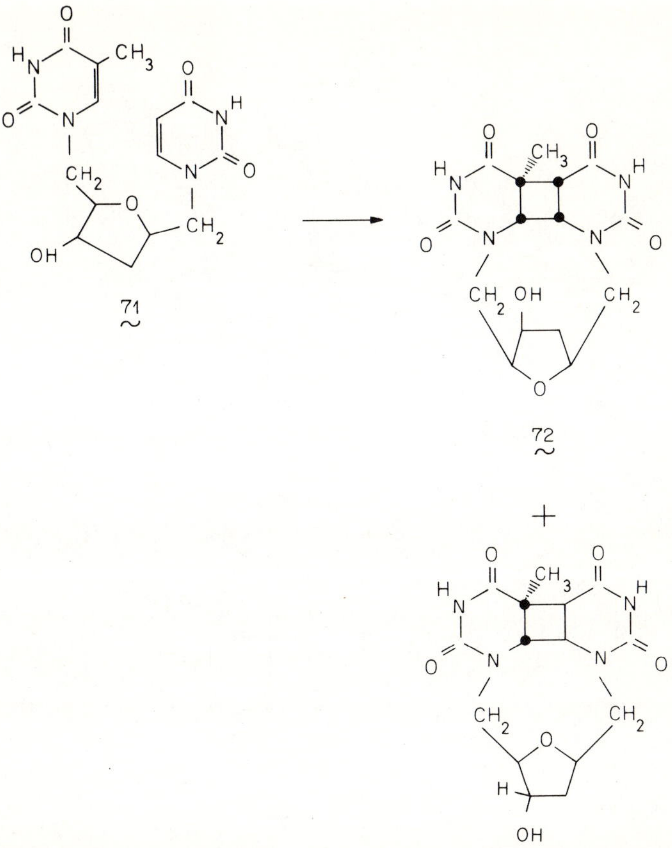

71

72

+

73

supra), efficient (Φ_{cycl}=0.26) cyclomerization occurs (74) from [74] cis. By comparison of the quantum yield of fluorescence in this system with bichromophores where no intramolecular reaction occurred, it could be shown that the reaction proceeds from the singlet excited state.

The formation of an excited state complex as an intermediate is proposed. It should be pointed out that the cyclomerization is restricted to a system with three methylene units between the chromophores.

Quantum yields as well of isomerization as cyclomerization are of the order of 10^{-2} indicating that the major process is the

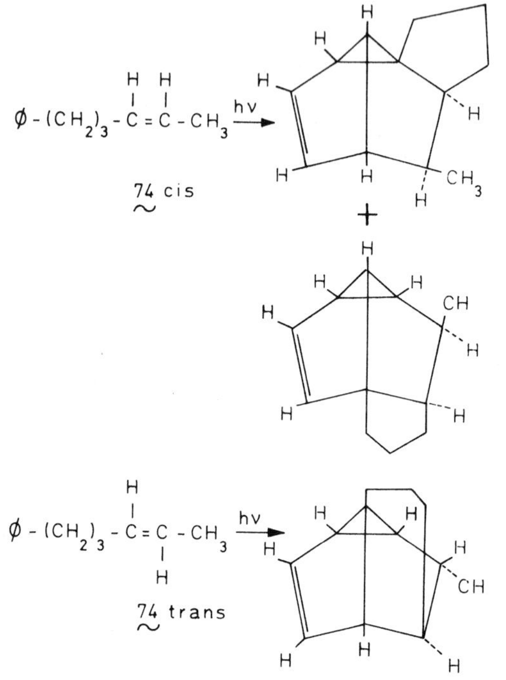

74 cis

74 trans

radiationless decay from the exciplex. The rate constant for the singlet process competing with intersystem crossing or fluorescence was proposed to be larger than 4.10^8 s^{-1}.

The triple bond of 6-phenyl-2-hexane [75] adds intramolecularly to the phenyl ring to a cyclo-octatetraene derivative. The cycloaddition is a hundred times less efficient than in [74] and is found to be a singlet state reaction of the excited phenyl moeity.

75

2. Addition of an Unsaturated Carbon-Carbon Bond to a Carbonyl Function. Intramolecular Paterno-Buchi reactions have been reported in the following systems. It should be pointed out that in most cases <u>cis-trans</u> isomerization, if possible, is a competitive process.

+ Other products (150)

76

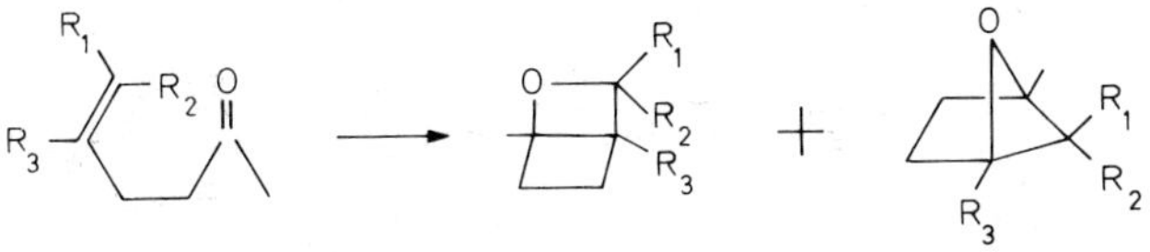

(151)

(152)

(70)

77 78 79

a : $R_1 = R_2 = Me$, $R_3 = H$

b : $R_1 = R_2 = R_3 = H$

c : $R_1 = R_2 = H$, $R_3 = Me$

d : $R_1 = R_3 = H$, $R_2 = Me$

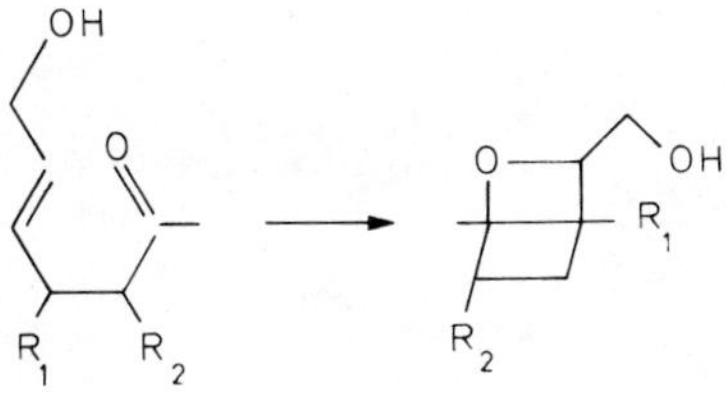

(153)

80

a : $R_1 = R_2 = H$

b : $R_1 = Me$, $R_2 = H$

c : $R_1 = H$, $R_2 = Me$

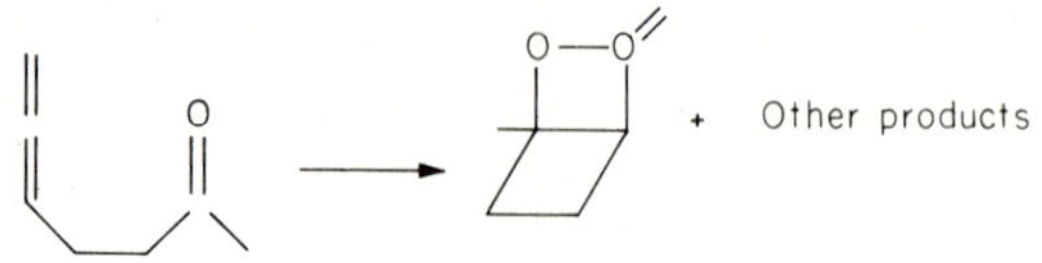

(154)

81

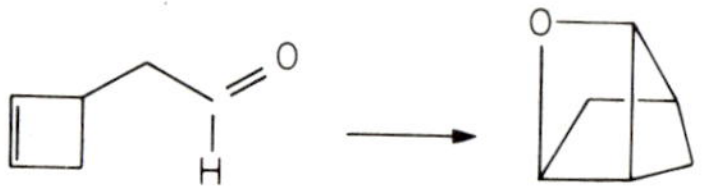

(155)

82

(156)

83

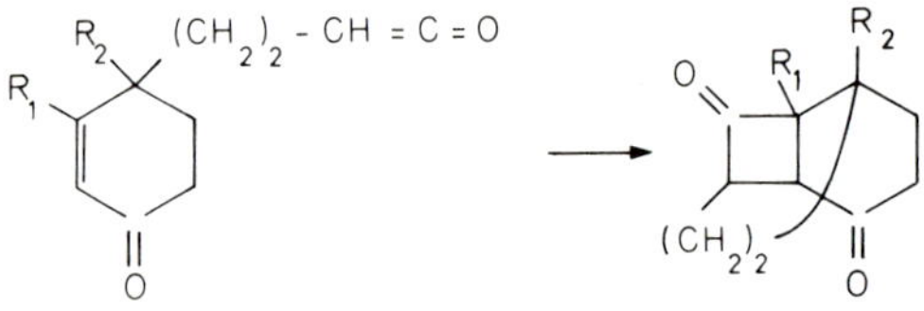

(157)

84

a : R_1, R_2 > Cyclobutyl

b : R_1 = H , R_2 = Me

Morrison has made a detailed mechanistic study (70) of [77c] and [77d]. On irradiation, compound [77d] leads only to [77c] with a quantum yield of 0.029. Compound [77c], however, forms oxetane [78] (Φ =0.015) in competition with isomerization to [77d] (Φ =0.02); no crossed adduct [79] was observed. Triplet quenching did have very little effect on isomerization or oxetane formation, while on sensitization cis-trans isomerization but no oxetane formation occurred. The authors concluded that the reaction occurred via that intramolecular complex formation in the singlet excited state manifold in competition with very efficient radiationless decay to the ground state from the complex.

3. Skeletal Rearrangements. Skeletal rearrangements are observed in the following systems.

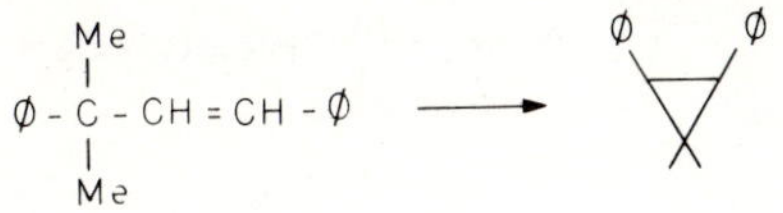

85

(158)

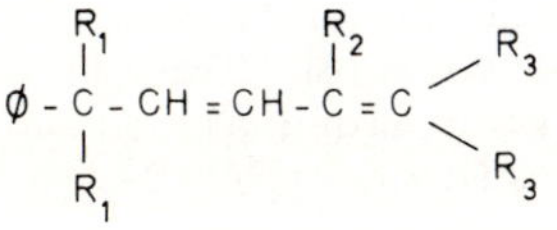

86

a : $R_1 = R_3 = Me$, $R_2 = R$

b : $R_1 = Me$, $R_2 = R_3 = R$

c : $R_1 = R_2 = Me$, $R_3 = R$

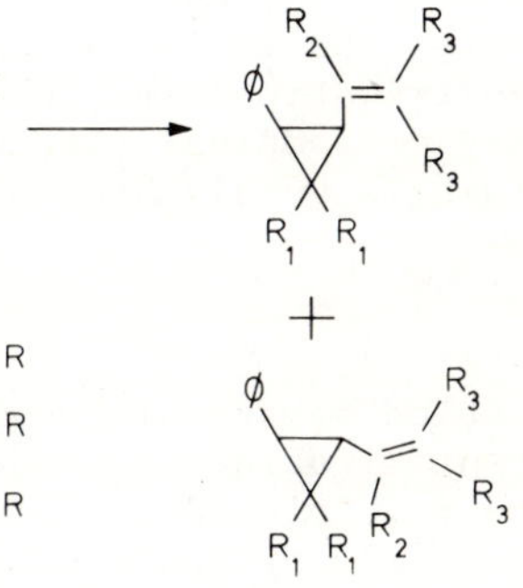

(159)

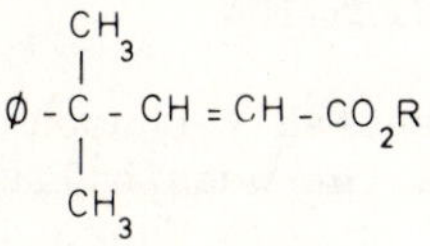

(160)

87

a : $R = C_2H_5$

b : $R = C_2H_4-N(CH_3)_2$

c : $R = C_2H_4-N(CH_3)-C_6H_5$

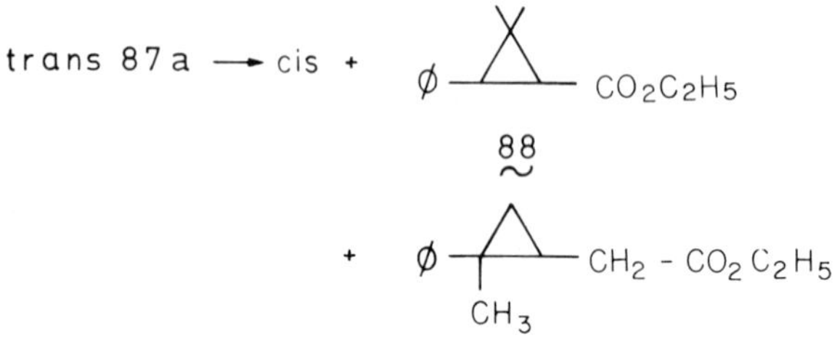

89

The skeletal rearrangment (except formation of [89]) is thought to involve a delocalized excited state. The absence of the rearrangement in [87b], which on excitation leads to a multitude of products, and in [87c], which is photochemically stable, (<u>cis</u>-<u>trans</u> isomerizations are observed from [87a-c]) stresses the possibility of complex interaction among all groups in a system containing several chromophoric units (160).

C. Conclusion

The selection of systems reviewed clearly illustrates the synthetic potential of bichromophoric molecules. It further emphasizes the role played by excitation delocalization and its impact on the choice of deactivation pathways available for the molecule. Systematic studies give an insight into the dynamics of the linking chain, the importance of ground state interaction, and the role played by excited state complexes.

V. PHOTOPOLYMERIZATION

If the bichromophoric system is suitably chosen to eliminate intramolecular processes and to ensure the possibility of intermolecular reaction, a polymer can be formed. Where previously the term photopolymerization was used to indicate photoinitiated chain processes (161), the present discussion of nonconjugated bichromophores is from a different point of view. The term "photopolymerization" as defined here, is a polymerization process in which every chain-propagating step involves the absorption of a photon (162,163). This definition implies that photopolymerization is a stepwise reaction, in contrast to photoinitiated polymerization, which is a chain process. A sensitized photopolymerization is a photopolymerization in which the reacting chromophore is excited by energy transfer from a suitable donor. The term "sensitization" is not correct for radical and ionic chain reactions initiated by a photosensitive compound, unless the decomposition of the initiator is the result of energy transfer. In this case, the term "sensitized photoinitiated polymerization" can be used. Photopolymerizations in solution can, in principle, be divided into two classes according to the nature of the reactive species intervening in the propagation step: In the type a photopolymerization an excited state (singlet and/or triplet) is the propagating species while in type b photopolymerization the propagation step involves a reactive ground state resulting from a photochemical reaction. In solid state photopolymerizations the polymerization behavior is determined by the crystal structure of the monomer.

A. Photopolymerization in Solution

1. Synthesis of Polymers.

a. Type a photopolymerization.

1a1. Singlet photopolymerization.

1a1a. Photopolymerization of bisanthracenes. The photodimerization of anthracene (164) and 9-anthryl carboxylic acid (165) is a well-known reaction and occurs between a singlet excited state and a ground state molecule. This reaction has been extended to bisanthracenes [90].

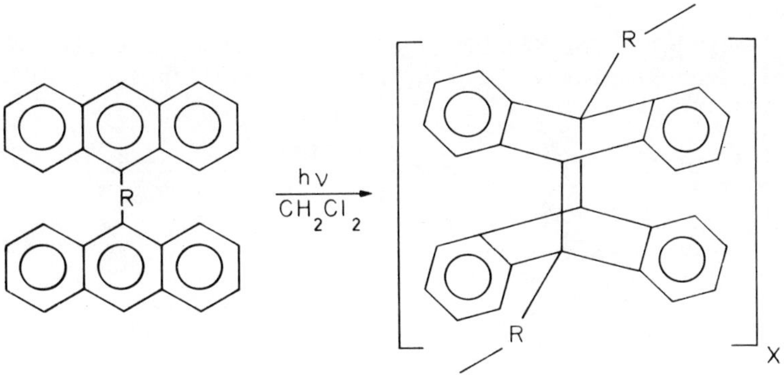

90

Irradiation of the diesters [91] derived from α,ω aliphatic diacid chlorides and 9-hydroxymethylanthracene gave polymers with limited molecular weights (Table 11). Head-to-head and head-to-tail isomers were formed in 1:4 ratio.

TABLE 11. Photopolymerization of Bisanthracenes.

Compound	$\bar{M}_n$ (a)	$[\eta]^{25}_{CHCl_3}$	Ref.
[91] n=7	12,000	0.37	(166)
[91] n=8	11,600	0.25	(166)
[92] n=11	52,000	1.3	(166)
[92] n=12	28,000	0.68	(166)

[a]Number average molecular weight.

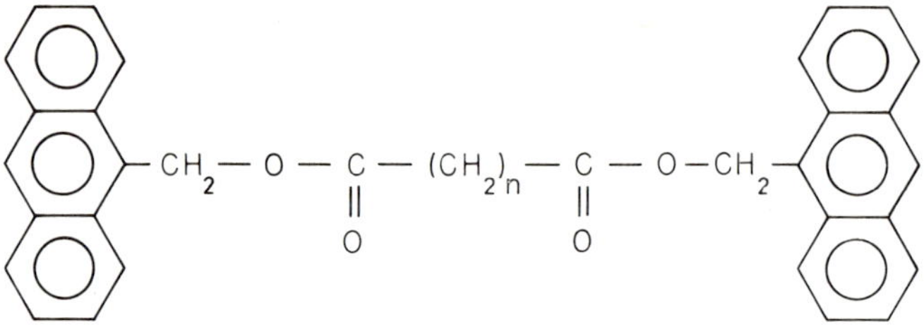

91

The results for the photopolymerization of diesters [92] derived from 9-anthroylchloride and an α,ω aliphatic diol are reported in Table 11.

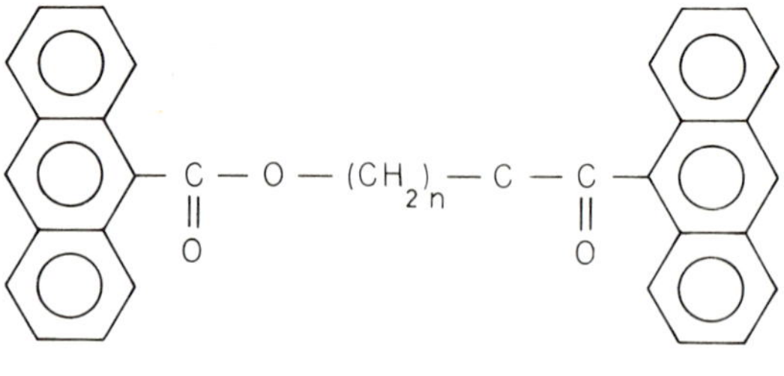

92

Only the head-to-tail isomer was formed in this case. The obtained polymers were film forming and could either photochemically (λ < 300 nm) or thermally (T > 180°C) be degraded back to monomers.

1a1b. Photopolymerization of benzene. The 1,3 cycloaddition of benzene [S_1, $^1B_{2u}$] to olefins is a well-documented photoreaction (167-169). Irradiation of a very dilute solution of cyclooctene in benzene formed the 1:1 adduct [93] which gave the polymer [94] by a stepwise photochemical 1,3 cycloaddition.

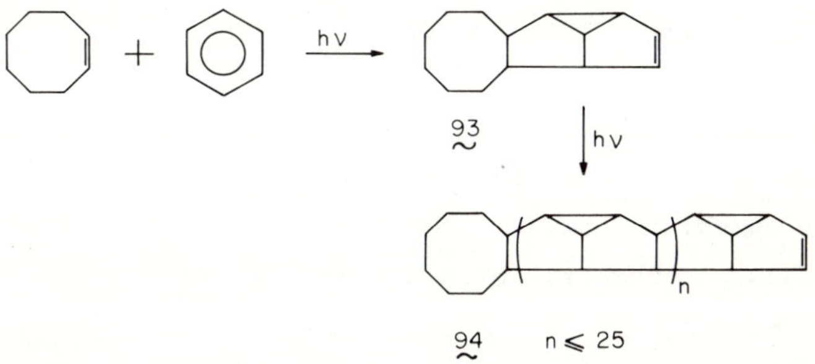

1a2. Triplet photopolymerization.

1a2a. Sensitized photopolymerization of N,N'-alkylenebisdichloromaleimides. The photodimerization of maleic anhydride, maleimide and derivatives has been studied by several research groups (170-172). The benzophenone-sensitized photoreaction of N,N'-alkylenebisdichloromaleimides [95] with a long polymethylene chain between the two chromophores resulted in the formation of macromolecules (173). The data are collected in Table 12.

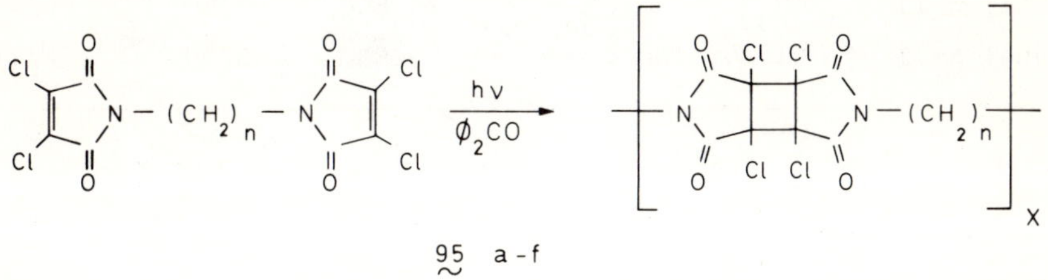

TABLE 12. Sensitized Photopolymerization of [95] in Dichloromethane.

[95]	n	$\bar{M}_n$	$[\eta]^{25°C}_{DMF}$	Ref.
a	7	-	0.25	(173)
b	8	2,000[a]	-	(173)
c	9	30,000[b]	0.45	(173)
d	10	2,700[c]	-	(173)
e	11	-	0.65	(173)
f	12	-	-	(173)

[a]Vapor pressure osmometry in dimethylformamide.

[b]Osmometry in chloroform.

[c]Vapor pressure osmometry in chloroform.

1a2b. Unsensitized photopolymerization of substituted N,N'-alkylenebismaleimides (174,175). Substituted N,N'-polymethylenebismaleimides [96a-d] were also polymerized upon direct irradiation in dichloromethane solution (Table 13).

TABLE 13. Photopolymerization of Substituted Bismaleimides in Dichloromethane.

	R^1	R^2	$\bar{M}_n$	$[\eta]^{25°C}_{CHCl_3}$	Ref.
[96a] n=9	$(CH_2)_2$	$(CH_2)_2$	5,100	0.16	(174)
[96b] n=10	Br	H	9,500	-	(174)
[96c] n=11	Br	H	14,500	0.43	(174)
[96d] n=11	CH_3	CH_3	20,800	0.90	(174)
[96e] n=11	Cl	Cl	3,880	0.09	(174)

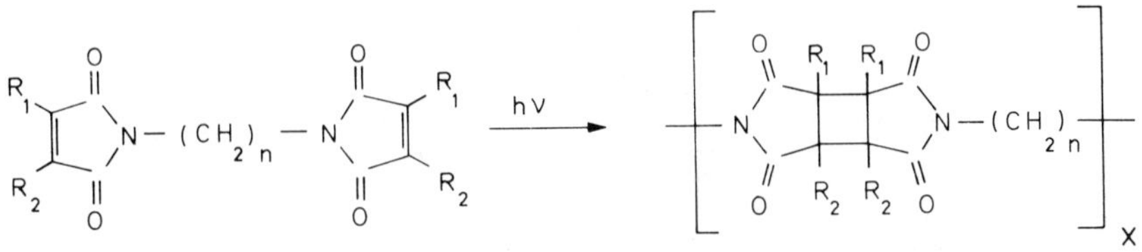

96 a-d

la2c. Sensitized photopolymerization of polymethylenedicarboxylic acid-7-(4-methyl)-coumarinyl diesters (176). The sensitized irradiation of 7-(4-methyl)-acetoxycoumarine in dichloromethane solution leads to a dimer with an exo head-to-head configuration. The sensitized irradiation ($\lambda > 335$ nm) of polymethylenedicarboxylic acid-7-(4-methyl) coumarinyl diesters [97] in dichloromethane solution gave polymers that could photochemically be degraded ($\lambda \le 310$ nm) to monomer (Table 14).

TABLE 14. Photopolymerization of [97] in Dichloromethane.

[97]	n	$\bar{M}_n$	$[\eta]^{25°C}_{CHCl_3}$	Ref.
a	10	6,000[a]	0.29	(176)
b	11	9,400[a]	0.42	(176)
c	12	22,300[b]	0.49	(176)

[a]Vapor pressure osmometry in chloroform.
[b]Osmometry in chloroform.

la2d. Polyoxetane formation. The photochemical addition of carbonyl compounds to unsaturated carbon-carbon bonds results in the formation of oxetanes (Paterno-Buchi reaction). This reaction could be extended to bichromophoric systems, resulting in the formation of polymers with oxetane rings in the main chain.

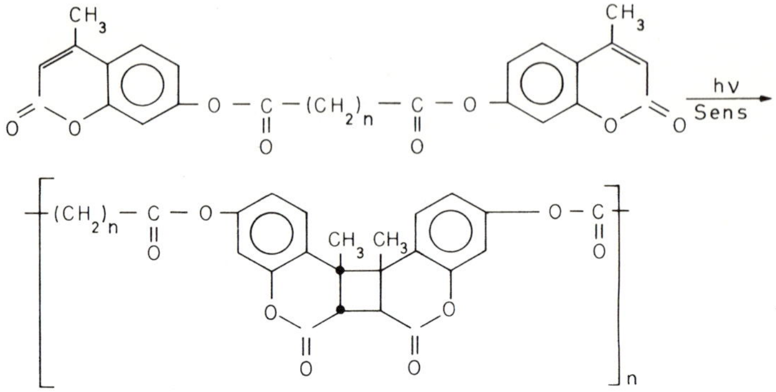

97 a - d

The photocycloaddition of aromatic diketones [98] to furan and tetramethylallene was studied by D.J. Andrews and W.J. Feast (177). The results are given in Table 15.

TABLE 15. Molecular Weights of the Polyoxetanes[a] (177).

[98]	[98] with R	Tetramethylalkene	Furan
a	m-ϕ	6,600	8,900
b	p-ϕ	2,200	1,800
c	p-ϕ-O-ϕ-	6,200	No polymer
d	p-ϕ-ϕ-	No polymer	No polymer
e	p-ϕ-CH_2-ϕ-	13,000	5,400[b]
f	-ϕ-$(CH_2)_2$-ϕ-	2,800	6,100
g	-ϕ-$(CH_2)_6$-ϕ-	6,200	7,300

[a]Perkin-Elmer 115 molecular weight apparatus, benzene solutions.
[b]Polymer precipitated after prolonged irradiation.

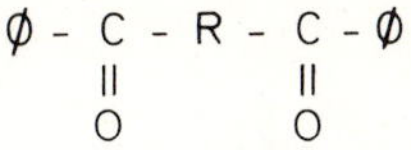

98

Polyoxetane formation was also studied by H. Leblanc (178) (Table 16).

TABLE 16. Polyoxetane Formation.

[98]	with R	$\bar{M}_n$ [a]	$[\eta]^{25^oC}_{CHCl_3}$	Ref.
h	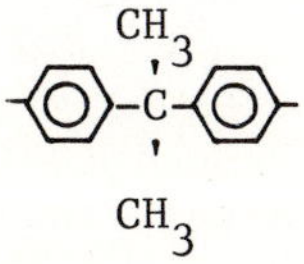	21,300[b]	0.40	(178)
h	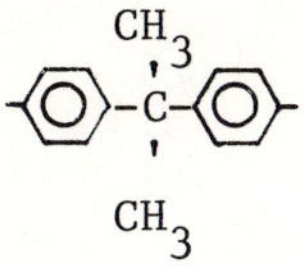	7,800[c]	-	(178)

[a]Vapor pressure osmometry in chloroform.
[b]Diolefine = 2,5-dimethyl-1,5-hexadiene.
[c]Diolefine = 2,4-dimethyl-1,4-pentadiene.

1a2e. Photocopolymerization of benzene and bismaleimides. Upon irradiation in benzene solution maleic anhydride and several maleimides form an 2:1 adduct. The acetophenone sensitized irradiation of bismaleimides [99a-h] in benzene solution resulted in the formation of polyimides with a limited solubility, probably due to crosslinking (179).

Bismaleimides and alkylbenzenes undergo an analogous photopolymerization (180).

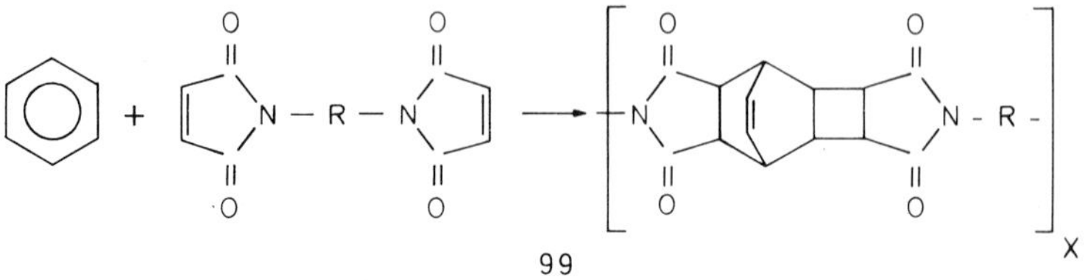

99

a : R = $-(CH_2)_2-$

b : R = $-(CH_2)_3-$

c : R = $-(CH_2)_6-$

d : R = 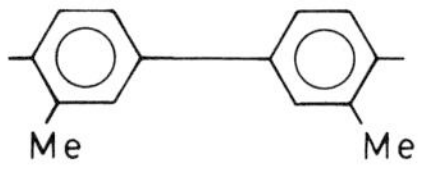

e : R = 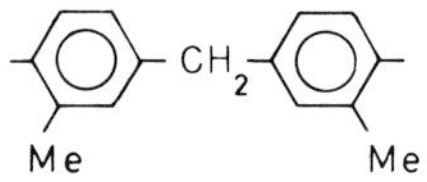

f : R = 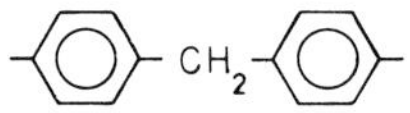

g : R = 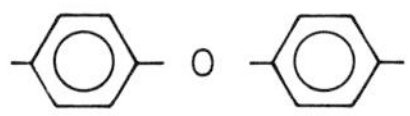

h : R = 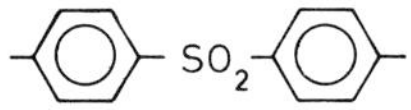

b. Type b photopolymerization.

1b1. Photopolymerization of bisbenzophenones (181-186). The reductive photodimerization of benzophenone has been extended to bisbenzophenones [98]. The results are summarized in Table 17. The polybenzopinacols could be rearranged to polybenzopinacolones in sulfuric acid/dioxane solution (187).

TABLE 17. Photoreductive Polymerization of Bisbenzophenones.

[98]	R	Footnote	$\bar{M}_n$	Ref.
a		a	825	(181)
b		a	435	(181)
c		a	805	(181)
e		b	26,000[c]	(183)
g		b	13,000[d]	(183)
i	$-\phi-(CH_2)_8-\phi-$	b	11,000[d]	(183)
j	$-\phi-(CH_2)_{10}-\phi-$	b	16,000[c]	(183)

[a]Solvent 1:1 benzene-isopropanol or 1:1 benzene-ethanol or 1:1 tetrahydrofuran-isopropanol or isopropanol; irradiation under nitrogen atmosphere; Fischer-Black-Ray long-wavelength UV-lamp: molecular weights are determined by vapor pressure osmometry in benzene.
[b]Solvent 2:1 dichloromethane-isopropanol; irradiation under vacuum with RUL 3500 Å-lamps.
[c]Osmometry in dimethylformamide.
[d]Vapor pressure osmometry in chloroform.

1b2. Photopolymerization of bisketotriazoles. N-ethyl-4-benzoyl-5-methyl-triazole undergoes photopinacolization on irradiation in isopropanol-benzene solution (186). Extension of this reaction to bisketotriazoles [100] gave polymers (Table 18) (184-186).

TABLE 18. Photopolymerization of Bisketotriazoles.

[100]	n	$\bar{M}_n$	Me_2CHOH/C_6H_6;v/v	$[\eta]_{DMF}^{23°C}$	Ref.
a	5	82,200[a]	1:3	0.32	(184)
b	6	-	1:3	0.26	(184)

[a]Measured in toluene with Hewlett-Packard 502 Membrane Osmometer; irradiation with RUL 3500 Å lamps under high vacuum.

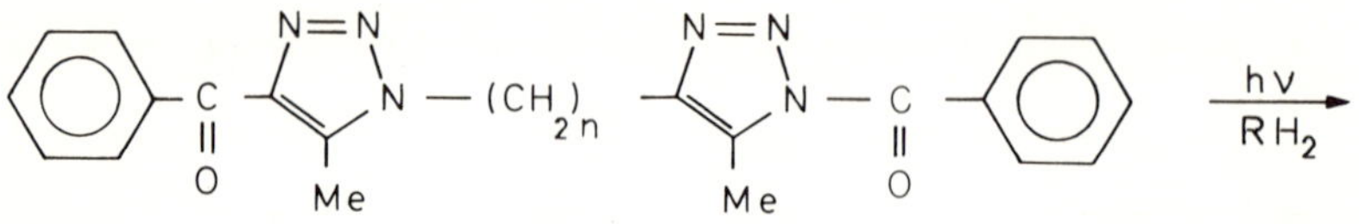

100

a : n = 5

b : n = 6

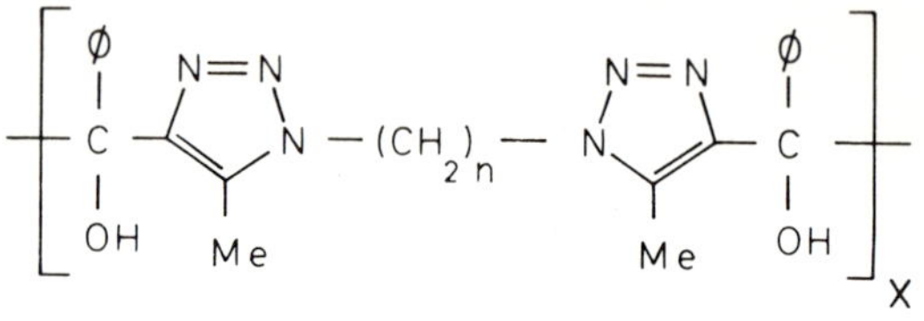

1b3. Photocopolymerization (184,186). The photolysis of diketone-dibenzhydrol mixtures involving two different monomeric units lead to the formation of high molecular weight copolypinacols (Table 19).

TABLE 19. Photocopolymerization of Diketone-dihydrol Systems in Benzene.

Diketone	Dihydrol	$\bar{M}_n$	Ref.
[98h]	[101]	174,000	(184)
[98h]	[102]	93,500	(184)
[100a]	[101]	141,000	(184)
[100b]	[102]	140,000	(184)

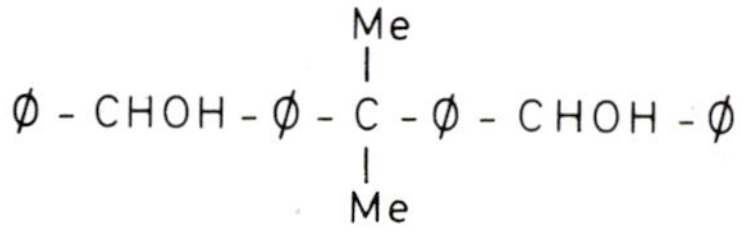

101

$$\text{Ø - CHOH - Ø - (CH_2)_{10} - Ø - CHOH - Ø}$$

102

The copolymers [101-100a] were shown to have three distinct linkages A-A [103], A-B[104] and B-B[105] and their respective percentages are reported in Table 20.

TABLE 20. Sequence Distribution in Copolypinacol[a] [101-100a].

	Copolymer
% A-A [103]	26.3
% A-B [104]	42.4
% B-B [105]	31.3

[a]Determined by nmr spectroscopy.

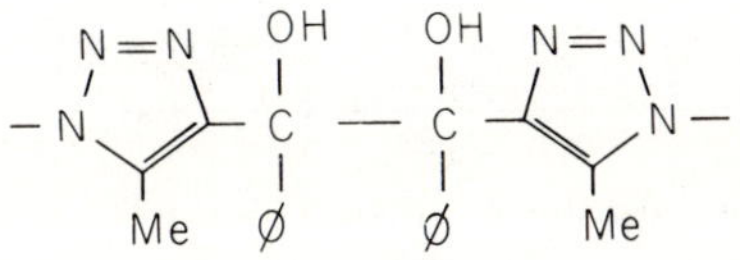

103

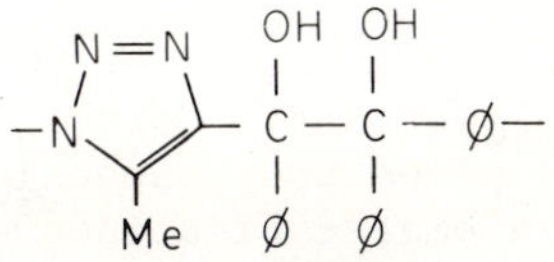

104

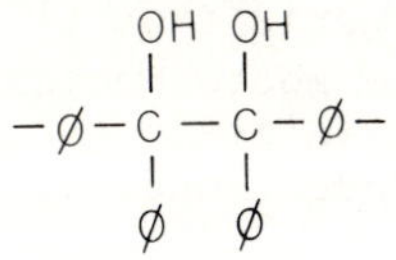

105

The following mechanism was proposed to account for the formation of random copolymers irrespective of the comonomer triplet energy (184) (Scheme III).

Scheme III

$$(Ar_1)_2CO^{*3} + (Ar_2)_2CO \rightarrow (Ar_1)_2CO + (Ar_2)_2CO^{*3} \quad \text{energy transfer } (E_{T_1} > E_{T_2})$$

$$(Ar_2)_2CO^{*3} + Me_2CHOH \quad (Ar_2)_2\dot{C}OH + Me_2\dot{C}OH \quad \text{hydrogen abstraction}$$

$$Me_2\dot{C}OH + (Ar_1)_2CO \rightarrow Me_2CO + (Ar_1)_2\dot{C}OH$$

$$Me_2\dot{C}OH + (Ar_2)_2CO \rightarrow Me_2CO + (Ar_2)_2\dot{C}OH$$

$$(Ar_2)_2\dot{C}OH + (Ar_1)_2CO \rightleftarrows (Ar_2)_2CO + (Ar_1)_2\dot{C}OH \quad \text{hydrogen exchange}$$

The first hydrogen abstraction occurs via the monomer with the lowest triplet energy. The acetone ketyl radical thus generated transfers its hydrogen indifferently to both comonomers. A hydrogen atom exchange between the aromatic ketones completes the equilibrium and leads to an equal amount of the two ketyl radicals. The requirement of the formation of random copolymers is that the comonomer of lowest triplet energy is capable of hydrogen abstraction.

1b4. Photopolymerization of bisazides. Photolysis of some aromatic azides gives the respective azocompounds (187). Irradiation of several aromatic bisazides [106] in solution resulted in the formation of dark-colored reaction products containing azo groups (188). Due to their limited solubility structure determination and molecular weight determination was impossible.

2. Kinetics of the Photopolymerization.

a. Type <u>a</u> photopolymerization (175,189,190). In the type <u>a</u> photopolymerization the photochemical reaction can occur from both the excited singlet (A^1) and excited triplet (A^3) states of the chromophore A. Under conditions of continuous and homogeneous irradiation with conventional light sources, the probability of exciting simultaneously the two chromophores of the bichromophoric molecule is negligible. The general kinetic

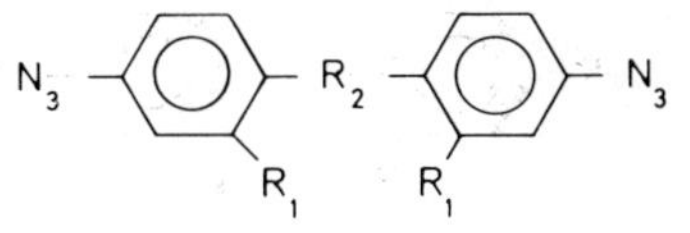

106

a : R_1 = SO_3H ; R_2 = $(CH_2)_2$

b : R_1 = H ; R_2 = $-(CH_2)-$

c : R_1 = H ; R_2 = $O-(CH_2)_2-O$

d : R_1 = H ; R_2 = $O-(CH_2)_{10}-O$

reaction scheme for the type a photopolymerization of a bichromophoric system A-A where the two chromophores react independently is described by Scheme IV where A^o represents the chromophore in its ground state. It is assumed that the added quencher Q can quench both the excited singlet and triplet states of the chromophore.

In this scheme it is further assumed that each excited state can yield a photocycloaddition product with a different configuration and that every process is irreversible. If all the incident light energy is absorbed by the bichromophoric system and if no quencher is added ([Q] = 0), the degree of conversion p as a function of the irradiation time is given by eq. 1.

$$2.303 \frac{K}{M[A^o]_o} \log \frac{1}{1-p} + p + 2.303 \frac{LM-KN}{MN[A^o]_o} \log \frac{M+N[A^o]_o}{M+N[A^o]_o(1-p)} = \frac{2\, I_o(t-t_o)}{[A^o]_o} \tag{1}$$

with $K = k_d^T(k_d^S + k_{ST})$

$L = k_d^S\, k_r^T$

$M = k_r^S k_d^T + k_r^T k_{ST}$

$N = k_r^S k_r^T$

Scheme IV

$A^o \xrightarrow{h\nu} A^o$	light absorption
$A^1 \xrightarrow{k_d^S} A^o$	unimolecular deactivation of the singlet excited state by fluorescence and radiationless decay
$A^1 \xrightarrow{k_{ST}} A^3$	intersystem crossing
$A^1 + A^o \xrightarrow{k_r^S} P$	propagation step: formation of bonds via the singlet excited state
$A^o + Q_T \xrightarrow{k_Q^S} A^o$	singlet quenching
$A^3 \xrightarrow{k_d} A^o$	unimolecular deactivation of the triplet excited state: radiationless and by emission
$A^3 + A^o \xrightarrow{k_r^T} p'$	propagation step: formation of bonds via the triplet excited state
$A^3 + Q \xrightarrow{k_q^T} A^o$	triplet quenching

$[A^o]_o$ is the molar chromophore concentration in the ground state at time t_o and p is the degree of conversion at time t. I_o is the light intensity of the incident light.

The generalized Stern-Volmer relation for the type a photopolymerization described in Scheme IV is given by eq. 2.

$$\frac{\phi_r^o}{\phi_r} = (1+k_Q^S[Q]\tau_S^o)(1+k_Q^T[Q]\tau_T^o)\left(\frac{1 + (\phi_r^{oT}/\phi_r^{oS})}{1+(\phi_r^{oT}/\phi_r^{oS}) + k_Q^T[Q]\tau_T^o}\right) \qquad (2)$$

$\tau_S^o = (k_d^S+k_{ST}+k_r^S[A^o])^{-1}$ and $\tau_T^o = (k_d^T+k_r^T[A^o])^{-1}$ are, respectively, the singlet and triplet lifetimes of the chromophore A in the absence of quencher. ϕ_r^{oS} and ϕ_r^{oT} are the quantum yields for disappearance of the chromophore respectively via the singlet and triplet reactions in the absence of quencher and are given by eqs. 3 and 4.

$$\phi_r^{oS} = \frac{2\,k_r^S[A^o]}{k_d^S + k_{ST} + k_r^S[A^o]} \tag{3}$$

$$\phi_r^{oT} = \frac{2\,k_r^T[A^o]}{k_d^T + k_r^T[A^o]} \times \frac{k_{ST}}{k_d^S + k_{ST} + k_r^S[A^o]} \tag{4}$$

The type a photopolymerization via the triplet excited state is exemplified by the photopolymerization of N,N'-nonamethylene-bisdimethylmaleimide in dichloromethane solution (190).

Upon ultraviolet irradiation in solution, N,N'-nonamethylene-bisdimethylmaleimide polymerizes into high molecular weight macromolecules with cyclobutane rings in the main chain. The reaction could be sensitized by acetophenone and was quenched by oxygen, ferrocene, and 3,3,4,4-tetramethyl-1,2-diazetine-1,2-dioxide. Stern-Volmer plots were linear for all chromophore concentrations and were dependent on the chromophore concentration. Linear quenching relations can be obtained in the following cases (Scheme IV).

(a) Two reactive states, both are quenched and

$k_Q^S\tau_S^o\,[1 + \phi_r^{oT}/\phi_r^{oS}] = k_Q^T\tau_T^o$

If $k_r^T k_Q^S = k_Q^T\,k_r^S$ and $k_r^T k_d^S = k_d^T k_r^S$, the Stern-Volmer plot is linear for all $[A^o]$.

In this case eq. 2 is reduced to eq. 5.

$$\frac{\phi_r^o}{\phi_r} = 1 + \frac{k_Q^T[Q]}{k_d^T+k_r^T[A^o]} = 1 + \frac{k_Q^S[Q]}{k_d^S+k_r^S[A^o]} \tag{5}$$

The quantum yield for disappearance of the chromophore is then given by eq. 6.

$$1/\phi_r^o = 0.5 + \frac{k_d^T}{2\,k_r^T[A^o]} \tag{6}$$

Since the experimental values of $1/\phi_r^o$ as a function of $1/[A^o]$ did not fit eq. 6 (Fig. 4), this first assumption could be excluded.

(b) Singlet excited state is reactive; both excited states are quenched.

(c) Both excited states are reactive; only the singlet excited state is quenched.

(d) Singlet excited state reacts and is quenched.

In cases b, c, and d, the singlet lifetime τ_S^o can be calculated from the slope of the Stern-Volmer plot if k_Q^S is known. However, the singlet lifetime, calculated from absorption spectra and the quantum yield of fluorescence was in the subnanosecond range and, furthermore, the reaction could be sensitized. It was also highly improbable that triplet quenching should not occur since the triplet energy of the used quencher was lower than that of the chromophore (E_T of the chromophore: 66 ± 1 kcal).

(e) Triplet state reacts; singlet state is quenched.

In this case the Stern-Volmer equation should be independent of the chromophore concentration, which was not observed in the photopolymerization.

(f) The only possibility left was that the photopolymerization should occur via the triplet excited state of the chromophore and that this state was quenched.

In this case eq. 1 is simplified to eq. 7 ($k_r^S = 0$).

$$2.303 \frac{k_d^T}{k_r^T [A^o]_o} \log \left(\frac{1}{1-p}\right) + p = \frac{2\, I_o\, \phi_{ST} (t-t_o)}{[A^o]_o} \qquad (7)$$

$\phi_{ST} = \dfrac{k_{ST}}{k_d^S + k_{ST}}$ is the quantum yield of intersystem crossing in the absence of singlet reaction

The Stern-Volmer relation (eq. 2) is rearranged to eq. 8.

$$\frac{\phi_r^o}{\phi_r} = 1 + \frac{k_Q^T [Q]}{k_d^T + k_r^T [A^o]} = 1 + k_Q^T \tau_T^o [Q] \qquad (8)$$

Assuming diffusion controlled quenching, the rate constants for deactivation and for reaction of the triplet excited state could be calculated. The rate constant for triplet deactivation k_d^T equals 4.1 $10^5 s^{-1}$ and rate constant for triplet reactivity k_r^T equals 2.2 $10^7 M^{-1} s^{-1}$.

The quantum yield for disappearance of the dimethylmaleimide chromophore via the triplet excited state at a fixed chromophore concentration was -- within the experimental error -- not dependent on the wavelength of the exciting light but was dependent on the chromophore concentration according to eq. 9.

$$\phi_r^{oT} = \frac{2\ \phi_{ST}\ k_r^T[A^o]}{k_d^T + k_r^T[A^o]} \tag{9}$$

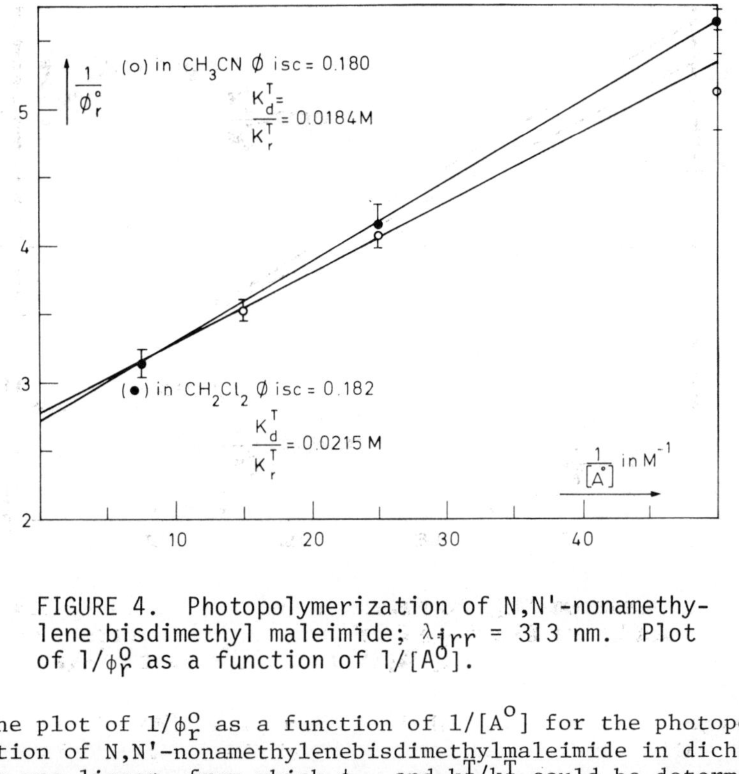

FIGURE 4. Photopolymerization of N,N'-nonamethylene bisdimethyl maleimide; λ_{irr} = 313 nm. Plot of $1/\phi_r^o$ as a function of $1/[A^o]$.

The plot of $1/\phi_r^o$ as a function of $1/[A^o]$ for the photopolymerization of N,N'-nonamethylenebisdimethylmaleimide in dichloromethane was linear, from which ϕ_{ST} and k_d^T/k_r^T could be determined. The value k_d^T/k_r^T = 0.0215M is in accordance with the value found in quenching experiments (k_d^T/k_r^T = 0.019M) with 3,3,4,4-tetramethyl-1,2-diazetine-1,2-dioxide as a quencher.

The kinetic results for the photopolymerization of N,N'-nonmethylenebisdimethylmaleimide in dichloromethane (λ = 350 nm; $[A^o]_o = 2\text{x}10^{-1}$M) are given in Figs. 5 and 6. All the incident light was absorbed by the system up to approximately 70 percent conversion. The degree of conversion p, calculated from the

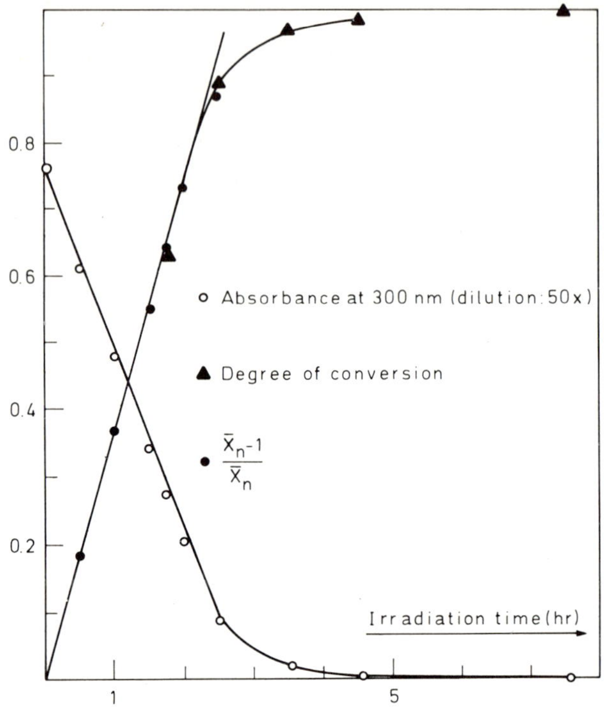

FIGURE 5. Photopolymerization of N,N'-nonmethylene bisdimethyl maleimide in dichloromethane. $[A^o]_o = 2.10^{-1}$M. Irradiation with RUL 3500=Å lamps.

change of the absorbance as a function of the irradiation time and $(\bar{x}_n-1)/\bar{x}_n$, calculated from osmotic molecular weight determinations are equal within the experimental error. This result confirms the relation $\bar{x}_n=1/(1-p)$ in this type of photopolymerization. Equation 10 can be obtained by substitution of this relation in eq. 7.

$$2.303 \frac{k_d^T}{k_r^T[A^o]_o} \log \bar{x}_n + \frac{\bar{x}_n-1}{\bar{x}_n} = \frac{2 I_o \phi_{ST}(t-t_o)}{[A^o]_o} \tag{10}$$

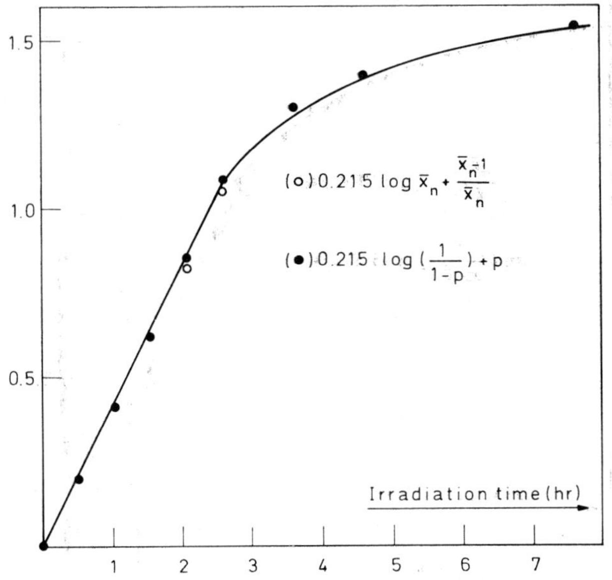

FIGURE 6. Photopolymerization of N,N'-nonmethylene bisdimethyl maleimide in dichloromethane; $[A^o]_o$ = 2.10^{-1}M. Irradiation with RUL 3500=Å lamps.

The graph of the absorbance, the degree of conversion p, and $\bar{x}_n - 1/\bar{x}_n$ as a function of the irradiation time $(t-t_o)$ was a straight line up to approximately 75 percent conversion (Fig. 5). This means that the first term in eq. 7 or eq. 10 is small as compared to p or $\bar{x}_n - 1/\bar{x}_n$. Deviations from linearity at higher conversion were due to incomplete light absorption by the system ($I_{abs} < I_o$). The total rate expression could be calculated using the values of k_d^T and k_r^T obtained from quenching experiments and the initial chromophore concentration $[A^o]_o = 2.10M^{-1}$. These data are represented graphically in Fig. 6. The intrinsic viscosity as a function of the degree of conversion is a hyperbole (Fig. 7). Similar plots have been obtained in the photopolymerization of bisketones (184) and the solid state photopolymerization of 2,5-distyrylpyrazine (201).

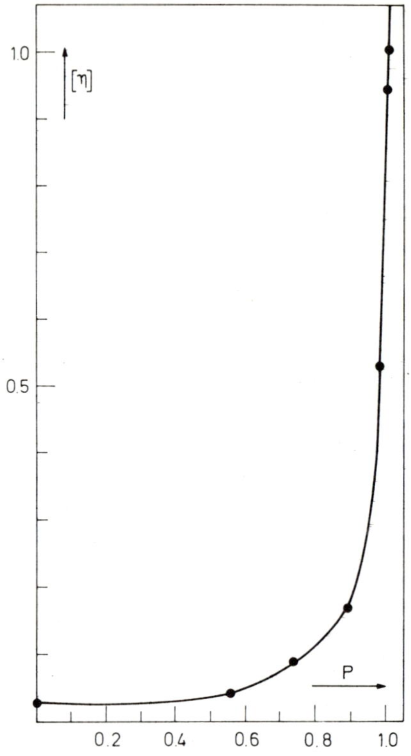

FIGURE 7. Photopolymerization of N,N'-nonamethylene bisdimethylmaleimide in dichloromethane. Intrinsic viscosity as a function of the degree of conversion.

b. Type b photopolymerization -- Photoreductive polyrecombination (191). The photoreduction kinetics of benzophenone and the photopolymerization kinetics of alkylene bisarylketones are closely interrelated. The general kinetic reaction scheme for the photoreductive polyrecombination of alkylene bisarylketones A-A where the two chromophores react independently is described by Scheme V where A^{o} represents the ketone chromophore in its ground state, AH^{o} the ketyl radical and Q the added quencher.

Scheme V

$A^{o} \xrightarrow{h\nu} A^{1}$	light absorption; excitation to singlet excited state
$A^{o} \xrightarrow{k_{ST}} A^{3}$	intersystem crossing[a)]
$A^{3} + RH_{2}{}^{b)} \xrightarrow{k_{r}} AH^{o} + RH^{o}$	hydrogen abstraction
$AH^{o} + AH^{o} \xrightarrow{k_{p}} AH - AH$	recombination: propagation step
$A^{3} \xrightarrow{k_{d}} A^{o}$	unimolecular deactivation of the triplet excited state
$A^{3} + Q \xrightarrow{k_{Q}} A^{o}$	triplet quenching

[a]The intersystem crossing quantum yield in the bichromophoric system 98h equals one.
[b]RH_2 = isopropanol; diisopropylbenzene or dihydrol 101.

The quantum yield of disappearance of the chromophore is given by eq. 11

$$\frac{1}{\phi} = 1 + \frac{k_d}{k_r[RH_2]} + \frac{k_Q[Q]}{k_r[RH_2]} \tag{11}$$

Equation 11 is valid for the benzophenone-benzhydrol or the diketone [98h] dihydrol [101] system since the plot of the reciprocal of the quantum yield versus the reciprocal of the hydrol concentration yields a straight line, with an intercept equal to one (Fig. 8, [Q] = 0). When isopropanol or diisopropyl benzene are used as reducing agents, the limiting quantum yield (intercept) deviates from unity (Fig. 8). A high limiting quantum yield ($\phi = 1.7$) in isopropanol is accounted for by a hydrogen transfer from the acetone ketyl radical to the ground state aryl ketones (192) while a low limiting quantum yield with diisopropyl benzene ($\phi_{max} < 1$) is due to an inefficient hydrogen abstraction (193,194). To take into account these discrepancies, a term ϕ_p is added which represents the probability that the metastable intermediate proceeds onto products rather than to ground state reactants (195).

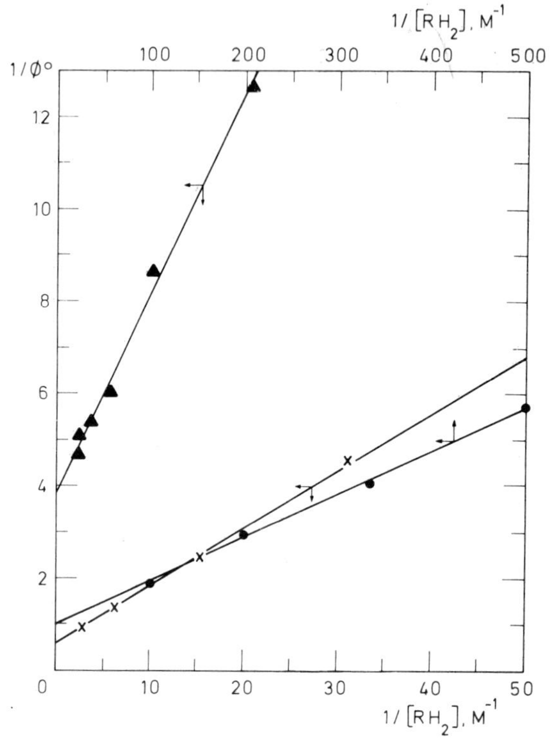

FIGURE 8. Photopolymerization of p,p'-dibenzoyl-2,2-diphenylpropane in benzene $[A^o]_o$ = 0.04 M. Dependence of $1/\phi^o$ on the reciprocal of the concentration of the reducing agent RH_2 (x) isopropanol; (o) bisbenzhydrol 101; (Δ) diisopropylbenzene.

$$\frac{1}{\phi} = \frac{1}{\phi p}\left[1 + \frac{k_d}{k_r[RH_2]} + \frac{k_Q[Q]}{k_r[RH_2]}\right] \qquad (12)$$

In the photoreduction of [98h] by isopropanol and diisopropyl benzene, the value k_d/k_r is obtained from the ratio of the slope over the intercept obtained by plotting the reciprocal of the quantum yield vs. the reciprocal of the reducing agent concentration ([Q] = 0; Fig. 8).

Equation 12 can be rearranged to eq. 13.

$$\frac{\phi^o - \phi}{\phi} = \frac{k_Q\ [Q]}{k_d + k_r[RH_2]} \tag{13}$$

where ϕ^o refers to the quantum yield in the absence of quencher.

The plot of $1/\phi$ (eq. 12); ϕ^o/ϕ (or $\phi^o - \phi/\phi$, eq. 13) for a fixed concentration of reducing agent versus the quencher (naphtalene) concentration yields a straight line from which k_Q/k_r can be calculated. Assuming diffusion controlled quenching, k_d and k_r can be calculated (Table 21).

TABLE 21. Rate Data for the Photoreduction of Diarylketone [98h] and Benzophenone with Different Hydrogen Donors in Benzene Solution.

Ketone	Bisbenzophenone (191) [98h]			Benzophenone (196)	
Substrate RH_2	Isopropanol	Diisopropyl benzene	[101]	Isopropanol	Diisopropyl benzene
k_Q/k_r	2,660	2,670	454	2,460	400
k_d/k_r, M	0.214	0.112	0.045	0.115	0.06
k_Q, $10^9 M^{-1}s^{-1}$	3.2	3.2	1.0	3.2	1.0
k_r, $10^6 M^{-1}s^{-1}$	1.2	1.2	2.2	1.3	2.5
k_d, $10^5 s^{-1}$	2.5	1.34	1.0	1.5	1.5

If all the incident light is absorbed by the bichromophoric system and if no quencher is added, the degree of conversion p as a function of the irradiation time is given by eq. 14.

$$p = \frac{2\ k_r[RH_2]I_o\ (t - t_o)}{(k_d+k_r[RH_2])[A^o]_o} = k(t - t_o) \tag{14}$$

The plot of the degree of conversion as a function of the irradiation time yields as a straight line with a slope equal to the apparent rate constant k, given by eq. 15.

$$\frac{1}{k} = \frac{[A^o]_o}{2\, I_o}\left[1 + \frac{k_d}{k_r[RH_2]}\right] \qquad (15)$$

From the slopes of the plots of conversion versus irradiation time for the photopolymerization of the [98h] isopropanol system, at constant light absorption for a fixed initial chromophore concentration and at various initial isopropanol concentrations, the apparent rate constants k can be determined (Table 24) for different isopropanol concentrations. The plot of the reciprocal of the apparent rate constant versus the reciprocal of the isopropanol concentration yields a straight line from which the ratio k_d/k_r can be obtained (Fig. 9). The value $k_d/k_r = 0.23$ obtained in Fig. 9 is in agreement with the value $k_d/k_r = 0.214$ (Table 22).

TABLE 22. Photopolymerization Kinetics of the [98h]-isopropanol System in Benzene Solution[a)]

[isopropanol] (M)	0.05	0.07	0.1	0.2	0.5
10^4 k,[b] (s^{-1})	4.44	5.86	7.65	11.66	17.83

[a][98h] = 0.05M.
[b]Light intensity $I_o = 6.2\ 10^{-5} mol\ l^{-1} s^{-1}$.

B. Photopolymerization in the Solid State

The concept of solid state photopolymerization was developed mainly by M. Hasegawa and his co-workers and was reviewed recently (197).

1. Synthesis and Properties of the Polymers.

Upon irradiation in the crystalline state trans,trans-2,5-distyrylpyrazine [107] is converted into a linear high molecular weight crystalline polymer containing cyclobutane rings in the main chain (198-210). After an extended study it was shown that bichromophoric compounds such as 1,4-bis [β-pyridyl-(2)-vinyl] benzene [108] (P2VB) (40, 199,200,211,212); p-phenylenediacrylic acid (p-PDA) and its derivatives [109] (63,215); α,α'-dicyano-p-phenylenediacrylic acid derivatives (p-CBA) [110] (216,217); α,α'-bis(4-acetoxy-3-methoxybenzylidene)-p-benzene-diacetronitrile [111] (218); polymethylene

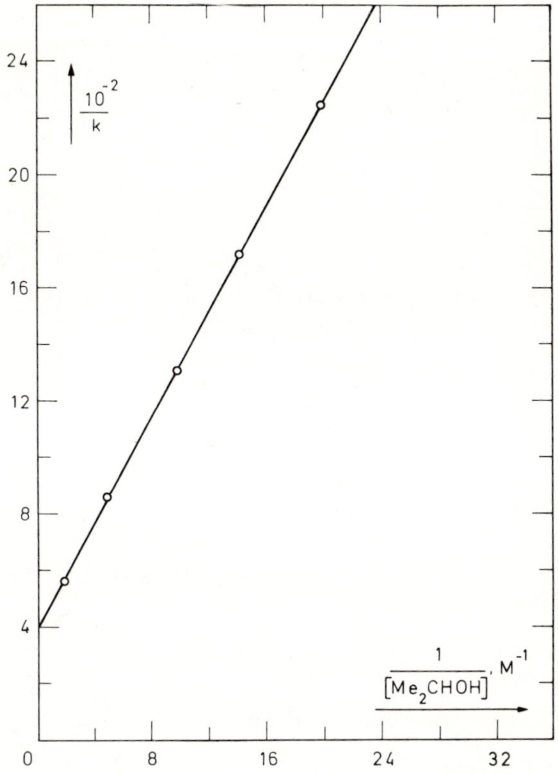

FIGURE 9. Dependence of the rate constant on the initial isopropanol concentration in the photopolymerization of p,p'-dibenzoyl-2,2-diphenylpropane in benzene.

dicinnamates [112] (219); 1,1'-trimethylenebisthymine [113] (220); and some N,N'-alkylene-bismaleimides [114] (221) formed linear polymers with cyclobutane rings in the main chain upon irradiation in the crystalline state. This type of photopolymerization was called "four-center type photopolymerization."

The untreated polymers are insoluble in most conventional organic solvents and are only soluble in strong acids such as concentrated sulphuric, dichloroacetic and trifluoroacetic acids. Some high molecular weight polymers are even totally insoluble.

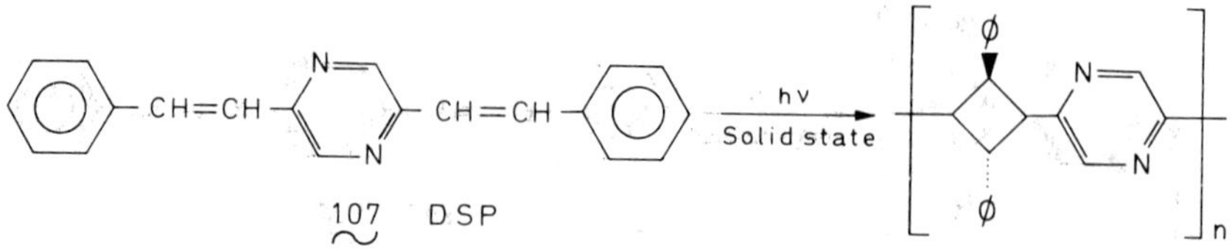

107 DSP

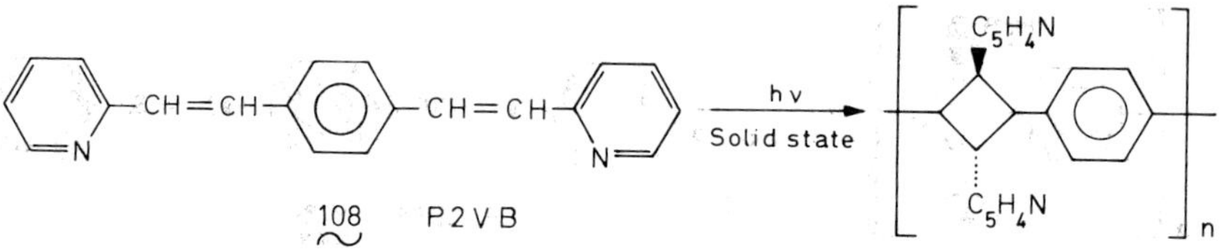

108 P2VB

with R = a : HO-;
b : CH_3O-;
c : C_2H_5O-;
d : n-C_3H_7O-;
e : i-C_3H_7O-;
f : C_6H_5O-;
g : -NH_2;

109 p-PDA

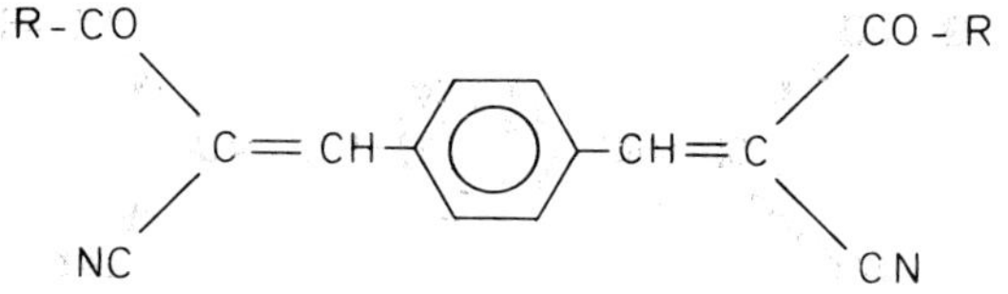

with R = a : CH_3O- ;
b : C_2H_5O-;
c : n-C_3H_7O-;
d : i-C_3H_7O-;
e : n-C_4H_9O-;
f : n-$C_8H_{17}O$-;

110 p-CBA

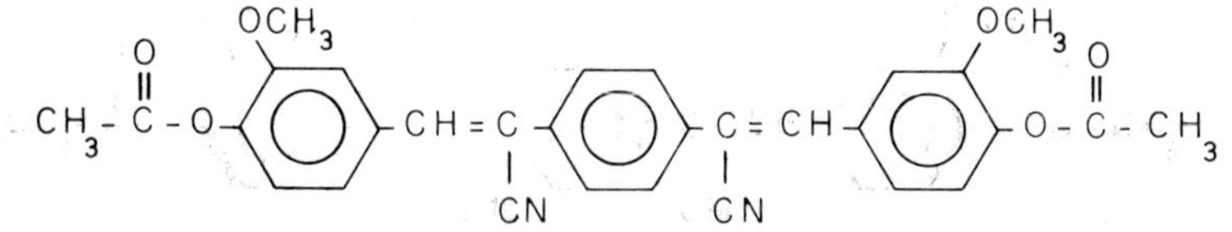

111

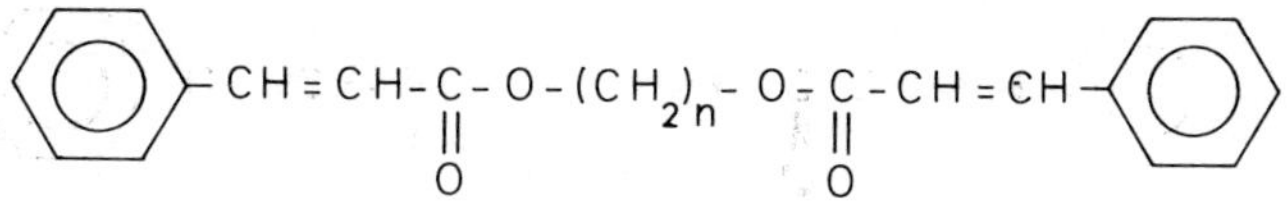

112 n = 2 ; 4

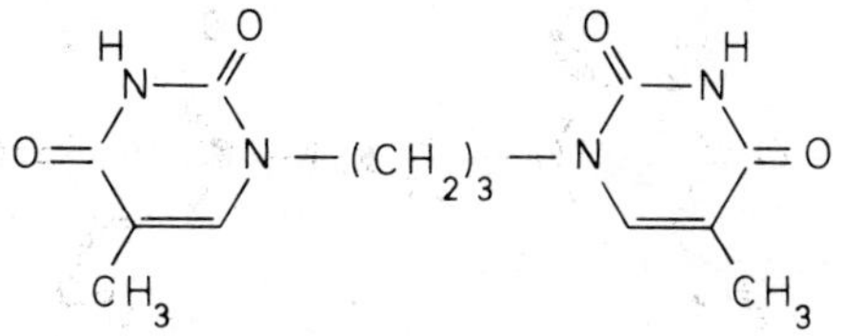

113

They all have high melting points and extremely high crystallinity. Some of them can be decomposed both thermally and photochemically into their starting products (200,210).

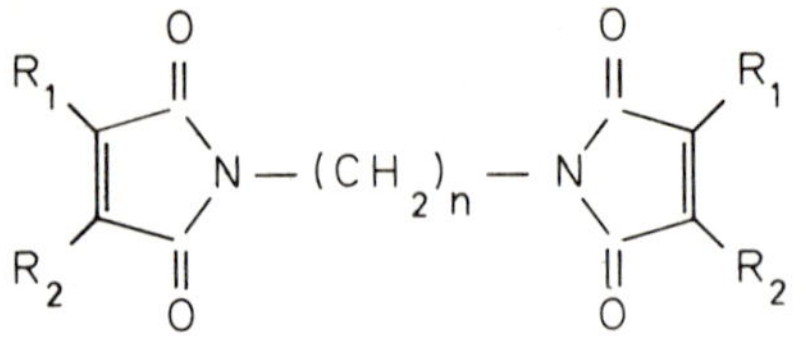

114

$R_1 = R_2 = CH_3$; n = 3,6

$R_1 = R_2 = Cl$; n = 3

$R_1 = R_2 = (CH_2)_2$; n = 3

$R_1 = R_2 = H$; n = 6

$R_1 = H$; $R_2 = CH_3$; n = 6

$R_1 = H$; $R_2 = Br$; n = 3,6

2. Photopolymerization Mechanism.

a. Polymerization kinetics. The plots of time versus conversion and conversion versus reduced viscosity for the photopolymerization of p-phenylenediacrylic acid diethylester [109c] in the crystalline state are shown in Figs. 10 and 11. The reduced viscosity increases gradually with increasing conversion and increases sharply above 80 percent conversion. Analogous curves were obtained in the solution phase photopolymerization of substituted bismaleimides (174) and bisbenzophenones (184). In Figs. 10 and 11 the broken line represents conversion to polymer in weight as separated by means of solubility differences while the solid line is the conversion calculated from the amount of monomer consumed. The reaction rate increases with light intensity whereas the reduced viscosity at a given conversion is independent of the light intensity (201). The quantum yields for disappearance of the olefinic double bond in the oligomerization and the polymerization of [107], [108] and [109b] are given in Table 23 (207). The small values (<2) indicate that this type of photopolymerization proceeds via a stepwise mechanism. This conclusion was confirmed by crystallographic measurements.

TABLE 23. Quantum Yields of Disappearance of the Olefinic Double Bond in the Oligormization and Subsequent Polymerization of [107], [108], and [109b].

Monomer	Oligormerization		Polymerization	
	λ_{irr} (nm)	ϕ	λ_{irr} (nm)	ϕ
[107]	436	1.2	365	1.6
[108]	405	0.04	-	-
[109b]	365	1.2	313	0.7

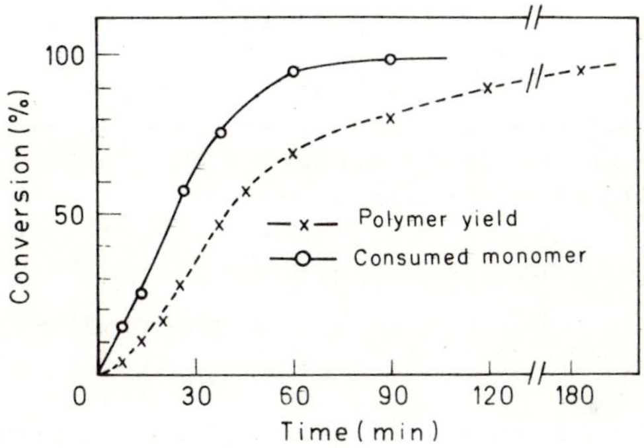

FIGURE 10. Plot of time versus conversion for polymerization of [109]c at -20°C.

b. Crystallographic studies. In the studies of the photopolymerization behavior of four-center type photopolymerization in the crystalline state, several remarkable features, which are summarized below, suggest that the molecular arrangement in the crystal of the bichromophoric compound is of the greatest importance in crystalline state photopolymerization.

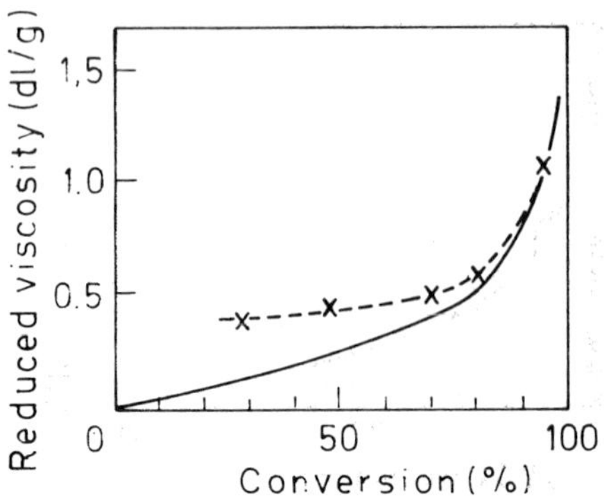

FIGURE 11. Plot of conversion versus reduced viscosity for polymerization of [109]c at -20°C.

α. Needlelike crystals of [107] obtained through sublimation are not photopolymerizable, while platelike crystals of [107] obtained in recrystallization from benzene solution photopolymerize readily in the crystalline state (199,201). The X-ray powder diffractions of platelike [107] crystals and [108] crystals are nearly identical, while they are entirely different from those of needlelike crystals of [107] and crystals of non polymerizable bifunctional compounds such as 1,4-bis[β-pyridyl-(3)-vinyl] benzene and 1,4-bis[β-pyridyl-(4)-vinyl] bnezene (199).

β. The temperature dependence of the photopolymerization of [107] (201) and [109c] (203,205) indicates the conditions are more favorable at temperatures far below the melting point of the monomer crystal so that the rigidity of the monomer crystal lattice is maintained. This is in contrast to many radiation-induced solid state polymerizations (222,223).

γ. Crystalline state photopolymerization yields a polymer with high crystallinity (199), while in solution or molten state other processes than polymerization occur (201)(209). Only the untreated oligomer crystal can grow further into a crystalline high polymer by irradiation with the appropriate wavelength of light: a matrix effect (204,207,209).

δ. During the photopolymerization the produced polymer crystallites exhibit a high degree of orientation with respect to the monomer single crystal (199,205,208).

The crystallographic data for several monomers and polymers are given in Table 24 (206,208,212,220).

The molecular arrangement of [107] viewed along the b axis is reproduced in Fig. 12 from which it is immediately apparent that the molecules are arranged favorable for polymerization along the c axis. In this stack the ethylenic double bonds have minimal separation of 3.939 Å. (A and B). The second shortest distance (5.417 Å) is found between molecules in different stacks, related by the b-glide (A and C) and is too large for the reaction to occur. Therefore, it is most probable that the double bonds related by the center of symmetry react to form a cyclobutane ring upon photoirradiation, and consequently polymer chains grow in the direction of the c axis (199,206,208).

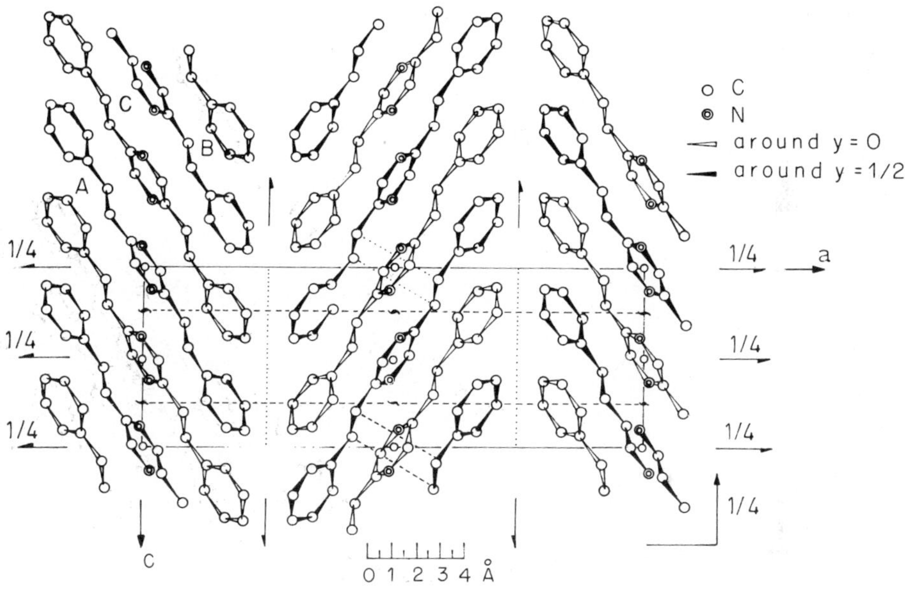

FIGURE 12. The crystal structure of 107 viewed along the b-axis.

This photopolymerization is a direct rearrangement from the monomer crystal to the polymer crystal and the resulting crystal precisely duplicates the original crystal (201,205,208). Such type of polymerization is called "lattice duplicating polymerization" (208). After polymerization of [107], the c-axis has contracted by 1.5 percent and the density has increased by 1.3 percent. The shortest intermolecular distance between two olefinic double bonds in [108] is 3.910 Å, while that in [109b] is 3.95 Å. In the case of [108], the c-axis has elongated by 2.7 percent and the density has decreased by about 1.6 percent. The changes of the a and b axial lengths of [107] and [108] upon photopolymerization are similar to each other (212). The molecular alignment in [109b] and [108] is similar to that of [107]: The reacting double bonds are related by a center of symmetry, resulting, upon photoirradiation, in a 1,3 trans type configuration with respect to the cyclobutane ring (208). The thymine rings in the crystal of 1,1'-trimethylenebisthymine [45a] are arranged in such a way that both intramolecular and intermolecular photoreaction could occur, and in each case trans-syn geometry for the product would be anticipated. Although the intramolecular distance between the midpoints of the reactive double bonds (C_5-C_6 and $C_{5'}$-$C_{6'}$) is smaller (3.501 Å) than the intermolecular distance (3.688 Å), it is reasonable to assume a closer interaction of the π-electron orbitals of the double bonds C_5-C_6 and $C_{5'}$-$C_{6'}$ in Fig. 13b (intermolecular overlap) than in Fig. 13a (intramolecular overlap) (220). Chemical studies indicate that the photoreaction of [45a] in the solid state gives a

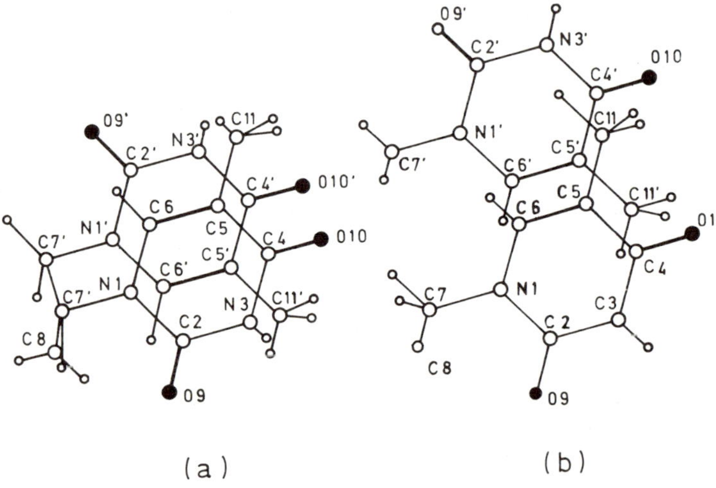

FIGURE 13. Properties of neighboring thymine rings into the basic ring in the case of (a) intramolecular overlap in 45a, (b) intermolecular overlap in 45a.

TABLE 24. Crystallographic Data.

	Crystal system	Space group	a (Å) (α(°))	b (Å) (β(°))	c (Å) (γ(°))	Z	ρ_{calc} g/cm^3
[107] Monomer (plate)	Orthorhombic	Pbca	20.638	9.599	7.655	4	1.244
Polymer	Orthorhombic	Pbca	18.36	10.88	7.52	4	1.257
Monomer (needle)	Monclinic	$P2_1/a$	13.84	18.71 (87.5)	5.84	4	1.249
[108] Monomer	Orthorhombic	Pbca	21.060	9.567	7.311	4	1.281
Polymer	Orthorhombic	Pbca	18.9	10.5	7.53	4	1.26
[109] Monomer	Triclinic	$P\bar{1}$	7.190 (95.08)	8.404 (117.06)	5.883 (78.00)	1	1.319
[45a] Monomer	Monoclinic	C2/c	7.144	16.734	12.067 (112.39)	4	1.46

polymer having cyclobutane rings with the trans-syn geometry [113]. There is a direct correlation between the molecular arrangement in the crystals of the starting bifunctional compound and the polymer. Such a correlation was already found and discussed for monofunctional compounds in relation to topochemistry (224).

C. Conclusion

The examples presented clearly substantiate the validity of the proposed concept of photopolymerization and is further proof of the synthetic use one can make of bichromophoric compounds.

ACKNOWLEDGMENT

The authors thank the N.F.W.O. for financial aid and a fellowship to one of them (N.B.).

REFERENCES

(1) Th. Förster and K. Kasper, Z. Physik. Chem. N.F., 1, 275 (1954).

(2) B. Stevens, Nature, 192, 725 (1962).

(3) Th. Förster, Ang. Chem. Int. Ed., 8, 333 (1969).

(4) J.B. Birks, in "Progress in Reaction Kinetics" (G. Porter, Ed.), Vol. 5, Pergamon, New York, 1971, p. 181.

(5) H. Leonhardt and A. Weller, Ber. Bunsenges. Phys. Chem., 67, 791 (1963).

(6) M. Mataga, T. Okada, and M. Ookari, Bull. Chem. Soc. Japan, 39, 2563 (1966).

(7) B. Stevens, Adv. Photochem., 8, 161 (1971).

(8) E. Konijnenberg, Thesis, University of Amsterdam (1963).

(9) W. Klöpffer, "Organic Molecular Photophysics", (J. B. Birks, Ed.), Wiley, New York, 1973.

(10) B. Stevens, Spectrochim. Acta, 18, 439 (1962).

(11) F. Hirayama, J. Chem. Phys., 42, 3163 (1965).

(12) W. Klöpffer, Chem. Phys. Lett., 4, 193 (1969).

(13) W. Klöpffer, Ber. Bunsenges, Phys. Chem., 74, 683 (1970).

(14) (a) W. Klöpffer and W. Liptay, Z. Naturforsch., 25a, 1091 (1970); (b) G. E. Johnson, J. Chem. Phys., 61, 3002 (1974).

(15) Landolt Börnstein, 6th ed., Springer-Verlag, Berlin, Longuart, and Bovey.

(16) E. Chandross and C. Dempster, J. Am. Chem. Soc., 92, 3586 (1970).

(17) D. T. Browne, J. Eisinger, and N. J. Leonard, J. Am. Chem. Soc., 90, 7302 (1968).

(18) M. T. Vala Jr., I. M. Hilliers, S. A. Rice, and J. Jortner, J. Chem. Phys., 44, 23 (1966).

(19) M. T. Vala, J. Houlig, and S. A. Rice, J. Chem. Phys., 43, 886 (1965).

(20) J. Eisinger and B. G. Shulman, Science, 161, 1311 (1968).

(21) K. Zachariasse and W. Kühnle, IUPAC Congress on Photochemistry Enschede, Abstracts, 1974, p. 197.

(22) J. R. Froines and P. J. Hagerman, Chem. Phys. Lett., 4, 135 (1969).

(23) K. Imakubo, J. Phys. Soc. Japan, 24, 1124 (1968).

(24) E. A. Chandross and H. T. Thomas, Chem. Phys. Lett., 9, 393 (1971).

(25) E. A. Chandross and H. T. Thomas, Chem. Phys. Lett. 9, 397 (1971).

(26) D.R.G. Brimage and R. S. Davidson, Chem. Commun., 1971, 1385.

(27) G. S. Beddard, R. S. Davidson, and A. Lewis, J. Photochem., 1, 491 (1972/1973).

(28) T. Okada, T. Fiyita, M. Kubota, S. Masaki, and N. Mataga, Chem. Phys. Lett., 14, 563 (1972).

(29) R. Ide, Y. Sakata, S. Misumi, T. Okada, and N. Mataga, Chem. Commun., 1972, 1009.

(30) H. Borkent, J. W. Verhoeven, and Th. J. De Boer, Tetrahedron Lett., 1972, 3363.

(31) J. W. Verhoeven, I. P. Dirkx, and Th. J. De Boer, Tetrahedron 25, 4037 (1969).

(32) H.A.H. Craenen, J. W. Verhoeven, and Th. J. De Boer, Rec. Trav. Chim., 91, 405 (1972).

(33) M. Itoh, T. Mimura, H. Usui, and T. Okamoto, J. Am. Chem. Soc., 95, 4388 (1973).

(34) M. Itoh, T. Mimura, and T. Okamoto, Bull. Chem. Soc. Japan, 47, 1078 (1974).

(35) E. Doller and Th. Förster, Z. Physik. Chem. N.F., 34, 132 (1962).

(36) M. Hauser and G. Heidt, Z. Physik. Chem. N.F., 69, 201 (1970).

(37) P. Avouris, J. Kordas, and M. A. El-Bayoumi, Chem. Phys. Lett., 26, 373 (1974).

(38) R. Speed and B. Selinger, Austral. J. Chem., 22, 9 (1969).

(39) I. H. Hillier, L. Glass, and S. A. Rice, J. Chem. Phys., 45, 3015 (1966).

(40) E. A. Chandross and C. J. Dempster, J. Am. Chem. Soc., 92, 703 (1970).

(41) E. A. Chandross and C. J. Dempster, J. Am. Chem. Soc., 92, 705 (1970).

(42) P. Avouris and M. A. El-Bayoumi, Chem. Phys. Lett., 20, 59 (1973).

(43) O. Schnepp and M. Levy, J. Am. Chem. Soc., 84, 172 (1962).

(44) P. A. Leermakers, G. W. Byers, A. A. Lamola, and G. S. Hammond, J. Am. Chem. Soc., 85, 2670 (1963).

(45) A. A. Lamola, P. A. Leermakers, G. S. Byers, and G. S. Hammond, J. Am. Chem. Soc., 87, 2322 (1965).

(46) D. E. Breen and R. A. Keller, J. Am. Chem. Soc., 90, 1935 (1968).

(47) See: J. B. Birks, "Photophysics of Aromatic Molecules", Wiley, New York, 1970.

(48) Th. Förster, Natuwissenschaften, 33, 166 (1946).

(49) P. M. Rentzepis, Science, 169, 234 (1970).

(50) C. Thierry, Mol. Photochem., 2, 1 (1970).

(51) An extensive review can be found in the "Biological Molecules in the Excited State" (I. Weinryb and R. F. Steiner, Eds.), Plenum, New York, 1971.

(52) J. Z. Steinberg, Ann. Rev. Biochem., 40, 83 (1971).

(53) R. H. Conrad and L. Brand, Biochemistry, 7, 5777 (1968).

(54) (a) R. M. Epand, Photochemistry and Photobiology, 18, 245 (1973). (b) W. C. Galley and L. Stryer, Proc. Nat. Acad. Sci. U.S., 60, 108 (1968).

(55) F. P. Schwarz, M. Gouterman, Z. Muljiami, and D. H. Dolphin, Bioinorganic Chemistry, 2, 1 (1972).

(56) T. Tamaki, Bull. Chem. Soc. Japan, 46, 2527 (1973).

(57) P. J. Wagner and T. Nakahina, J. Am. Chem. Soc., 95, 8474 (1975).

(58) S. A. Latt, H. T. Cheung, and E. R. Blout, J. Am. Chem. Soc., 90, 6897 (1968).

(59) R. Rauh, T. R. Evans, and P. A. Leermakers, J. Am. Chem. Soc., 90, 6897 (1968).

(60) R. D. Rauh, T. R. Evans, and P. A. Leermakers, J. Am. Chem. Soc., 91, 1868 (1969).

(61) R. A. Keller and L. J. Dolby, J. Am. Chem. Soc., 89, 2768 (1967).

(62) R. A. Keller and L. J. Dolby, J. Am. Chem. Soc., 91, 1273 (1969).

(63) R. A. Keller, J. Am. Chem. Soc., 90, 1940 (1968).

(64) J. R. De Member and N. Filipescu, J. Am. Chem. Soc., 90, 6425 (1968).

(65) A. A. Lamola, J. Am. Chem. Soc., 91, 4786 (1969).

(66) N. Filipescu, J. R. De Member, and F. L. Merin, J. Am. Chem. Soc., 91, 4169 (1969).

(67) N. W. Tyer, Jr., and R. S. Becker, J. Am. Chem. Soc., 92, 1295 (1970).

(68) H. Morrison, Tetrahedron Lett., 3653 (1964).

(69) H. Morrison, J. Am. Chem. Soc., 87, 932 (1965).

(70) S. R. Kurowsky and H. Morrison, J. Am. Chem. Soc., 94, 507 (1972).

(71) H. Morrison and R. Pfeiffer, J. Am. Chem. Soc., 90, 3428 (1968).

(72) C. S. Nakayama and P. Sigal, J. Chem. Phys., 52, 3277 (1970).

(73) C. S. Nakayama and P. Sigal, J. Chem. Phys., 53, 2993 (1970).

(74) W. Ferree Jr., J. B. Grutzner, and H. Morrison, J. Am. Chem. Soc., 93, 5502 (1971).

(75) D. O. Cowan and A. A. Baum, J. Am. Chem. Soc., 92, 6281 (1970).

(76) D. O. Cowan and A. A. Baum, J. Am. Chem. Soc., 93, 1153 (1971).

(77) P. A. Leermakers, J. P. Monteillier, and R. D. Rauh, Mol. Photochem., 1, 57 (1969).

(78) R. S. Cooke and G. S. Hammond, J. Am. Chem. Soc., 92, 2739 (1970).

(79) K. Manal, B. Levy, and A. D. McLaren, J. Am. Chem. Soc., 72, 1790 (1950).

(80) R. Brainard and H. Morrison, J. Am. Chem. Soc., 93, 2685 (1971).

(81) P. J. Wagner, P. A. Kelso, and A. E. Kempamen, Mol. Photochem., 2, 81 (1970).

(82) F. Steinmetz, D. Nicodeu, V. Muralidharan, and C. O'Donell, Mol. Photochem., 2, 87 (1970).

(83) D. Whitten and W. Punch, Mol. Photochem., 2, 77 (1970).

(84) G. Carlson, F. Quina, B. Zarnegar, and D. Whitten, submitted for publication.

(85) N. Filipescu and J. Bunting, J. Chem. Soc. B, 1498 (1970).

(86) N. Filipescu and D. Chang, J. Am. Chem. Soc., 94, 5990 (1973).

(87) O. L. Chapman, Adv. Photochem., 1, 359 (1963).

(88) G. S. Hammond and N. J. Turro, Science, 142, 1541 (1963).

(89) R. Huisgen, R. Grasbey, and J. Sauer, "The Chemistry of Alkenes" (Patai, Ed.), Interscience, New York, 1964, p. 800.

(90) P. De Mayo, Adv. Org. Chem., 2, 390 (1960).

(91) P. De Mayo and S. T. Read, Quart. Rev., 15, 408 (1961).

(92) A. Mustafa, Chem. Rev., 51, 1 (1952).

(93) J. Saltiel, "Survey of Progress in Chemistry", Vol. 2, Academic Press, New York, 1964, p. 265.

(94) K. Schaffner, Fortschr. Chem. Org. Naturstoffe, 22, 81 (1964).

(95) A. Schönberg, "Preparative Organische Chemie," Springer Verlag, Berlin, 1958, p. 2.

(96) C. Walling, "Molecular Rearrangements", Interscience, New York, 1963, p. 453.

(97) (a) W. J. Dilling, Chem. Rev., 66, 373 (1966).
(b) W. J. Dilling, Chem. Rev., 69, 845 (1969).

(98) H. Prinzbach, Pure Appl. Chem., 16, 17 (1968).

(99) F. D. Greene, J. L. Misrock, and J. R. Wolfe, Jr., J. Am. Chem. Soc., 77, 3852 (1955).

(100) F. D. Greene, Bull. Soc. Chim. France, 1356 (1960).

(101) D. E. Applequist, M. A. Lintner, and R. Searle, J. Org. Chem., 33, 254 (1968).

(102) R. Livingston and Kei Sin Wei, J. Am. Chem. Soc., 89, 3098 (1967).

(103) I. Roitt and W. Waters, J. Chem. Soc., 2695 (1952).

(104) J. H. Golden, J. Chem. Soc., 3741 (1961).

(105) S. J. Cristol, J. S. Perry, Sr., Tetrahedron Lett., 1921 (1974).

(106) J. B. Aladekomo, J. Lumines. 6, 83 (1973). An alternative mechanism for the photodimerization of anthracene has been proposed by S. Kaupp, Angew. Chem., 84, 259 (1972).

(107) F. C. De Schryver, M. De Brackeleire, S. Toppet, and M. Van Schoor, Tetrahedron Lett., 1253 (1973).

(108) N. Boens, M. De Brackeleire, J. Huybrechts and F.C. De Schryver, Zeitschrift Phys. Chem. NF; 101, 417 (1976).

(109) F. C. De Schryver, R. Arien, and J. Huybrechts, Unpublished results.

(110) E. A. Chandross and C. J. Dempster, J. Am. Chem. Soc., 92, 703 (1970).

(111) E. A. Chandross and C. J. Dempster, J. Am. Chem. Soc. 92, 704 (1970).

(112) N. J. Leonard, K. Golaneckiewicz, R. S. McCredie, S. M. Johnson, and I. C. Paul, J. Am. Chem. Soc., 91, 5855 (1969).

(113) N. J. Leonard, R. S. McCredie, M. W. Logue, and R. Cundall, J. Am. Chem. Soc., 95, 2320 (1973).

(114) N. J. Leonard and R. L. Cundall, J. Am. Chem. Soc., 96, 5904 (1974).

(115) M. W. Logue and N. J. Leonard, J. Am. Chem. Soc., 94, 2843 (1972).

(116) F. C. De Schryver, I. Bhardwaj, and J. Put, Angew. Chem. Int. Ed., 8, 213 (1969).

(117) J. Put and F. C. De Schryver, J. Am. Chem. Soc., 95, 137 (1973).

(118) J. Put, N. Boens, and F. C. De Schryver, Unpublished results.

(119) L. Leenders and F. C. De Schryver, Angew. Chem., 83, 359 (1971).

(120) L. Leenders, E. Schouteden, and F. C. De Schryver, J. Org. Chem., 38, 957 (1973).

(121) F. C. De Schryver, J. Put, L. Leenders, and H. Loos, J. Am. Chem. Soc., 96, 6994 (1974).

(122) Th. Förster, Pure Appl. Chem., 24, 443 (1970).

(123) R. Dougherty, J. Am. Chem. Soc., 93, 7187 (1972).

(124) J. Michl, Mol. Photochem., 4, 243 (1972).

(125) F. C. De Schryver and E. Schouteden, Unpublished results.

(126) J. Gervais and F. C. De Schryver, Photochem. Photobiol., 21, 71 (1975).

(127) D. Billen, F. C. De Schryver, Unpublished results.

(128) M. Van Meerbeck, S. Toppet and F. C. De Schryver, Tetrahedron Lett., 2247 (1972).

(129) J. L. Doty, S. Y. Farid, D. Specht, and J. R. Williams, XXIII Int. Symposium IUPAC Abstr., 1971, p. 156.

(130) M. Freedman, Y. Mohadge, J. Rennert, S. Solowzay, and I. Walchter, Org. Prep. Proced., 1, 267 (1969).

(131) J. Rennert, S. Soloway, I. Walchter, and B. Leong, J. Am. Chem. Soc., 94, 7242 (1972).

(132) J. R. Scheffer and R. A. Wostradowski, Chem. Commun., 1971, 144.

(133) J. R. Scheffer, R. A. Wostradowdki, and K. L. Dooley, Chem. Commun., 1971, 1271.

(134) J. R. Scheffer and B. A. Boire, Tetrahedron Lett., 4741 (1970).

(135) J. R. Scheffer and B. A. Boire, J. Am. Chem. Soc., 93, 5490 (1971).

(136) J. Scheffer and R. A. Wostradowski, J. Org. Chem., 37, 4317 (1972).

(137) J. J. Bloomfield and O. C. Oswley, Tetrahedron Lett., 1795 (1973).

(138) E. Block and H. W. Orf, J. Am. Chem. Soc., 94, 8483 (1972).

(139) P. S. Engel and M. A. Schexnayder, J. Am. Chem. Soc., 94, 4357 (1972).

(140) (a) P. De Mayo, Kagaku No Roycki Zokan, 93, 31 (1970); (b) P. De Mayo, Acc. Chem. Res., 4, 41 (1971).

(141) Y. Tamura, H. Ishibashi, Y. Kita and M. Ikeda, Chem. Commun., 1971, 1167.

(142) Y. Tamura, H. Ishibashi, Y. Kita, and M. Ikeda, Tetrahedron Lett., 1977 (1972).

(143) Y. Tamura, H. Ishibashi, Y. Kita and M. Ikeda, Chem. Commun., 1973, 101.

(144) R. Ramana and A. Satta, Tetrahedron Lett., 649 (1971).

(145) G. L. Hodgson, D. F. McSweeney, and T. Money, Chem. Commun., 1973, 236.

(146) P. Crabbe, G. Garcia, and E. Vilarde, Chem. Commun., 1973, 480.

(147) K. Golankiewicz and L. Strekowski, Bull. Acad. Pol. Sci., 19, 97 (1971).

(148) E. Chandross and A. H. Schiebel, J. Am. Chem. Soc., 95, 611 (1973).

(149) W. Lippka, W. Ferree, Jr., and H. Morrison, J. Am. Chem. Soc., 96, 2134 (1974).

(150) R. Srinivasan, J. Am. Chem. Soc., 82, 775 (1960).

(151) N. C. Yang, M. Nussim, and D. R. Coulson, Tetrahedron Lett., 1525 (1965).

(152) H. Morrison, J. Am. Chem. Soc., 87, 932 (1965).

(153) Y. Bahurel, G. Desotes, and F. Pautet, Compt. Rend., 270, 1528 (1970); Bull. Soc. Chim. France, 2222 (1971).

(154) J. K. Crandall and C. F. Mayer, J. Org. Chem., 34, 2814 (1969).

(155) J. Meinwald and A. T. Hammer, Chem. Commun., 1969, 1302.

(156) A. Van Wageningen and H. Cerfontain, Tetrahedron Lett., 3674 (1972).

(157) D. Becker, M. Nagler and D. Birnbaum, J. Am. Chem. Soc., 94, 4771 (1972).

(158) G. W. Griffin, J. Covell, R. C. Peterson, R. M. Hodson, and G. Klose, J. Am. Chem. Soc., 87, 1410 (1965).

(159) H. Kristinsson and G. S. Hammond, J. Am. Chem. Soc., 89, 5968 (1967).

(160) D. De Keukeleire, E. C. Sanford, and G. S. Hammond, J. Am. Chem. Soc., 95, 7904 (1973).

(161) G. Oster and Y. Nan Loth, Chem. Rev., 68, 125 (1968).

(162) F. C. De Schryver, "Fotopolymerisatiekinetika," Verhandelingen Koninklijk Vlaamse Academie voor Wetenschappen, Letteren en Schone Kunsten van Belgiё, 33, 120 (1971).

(163) F. C. De Schryver, Pure Appl. Chem., 34, 213 (1973).

(164) J. B. Birks, "Photophysics of Aromatic Molecules," Wiley, New York, 1970, pp. 320, 629.

(165) D. O. Cowan and W. W. Schmiegel, Angew. Chem., 83, 545 (1971).

(166) F. C. De Schryver, L. Anand, G. Smets, and J. Switten, J. Polym. Sci., Polym. Lett., 9, 777 (1971).

(167) D. Bryce-Smith, Pure Appl. Chem., 34, 193 (1973).

(168) D. Bryce-Smith, A. Gilbert and B. H. Orger, Chem. Commun., 1966, 512.

(169) D. Bryce-Smith and A. Gilbert, Chem. Commun., 1966, 643.

(170) G. O. Schenck and R. Steinmetz, Bull. Soc. Chim. Belg., 71, 781 (1962).

(171) R. Steinmetz, Fortschr. Chem. Forsch., 3, 445 (1967).

(172) R. N. Warrener and J. B. Bremner, Rev. Pure Appl. Chem., 16, 117 (1966).

(173) F. C. De Schryver, W. J. Feast, and G. Smets, J. Polym. Sci., A-1, 8, 1939 (1970).

(174) F. C. De Schryver, N. Boens, and G. Smets, J. Polym. Sci., A-1, 10, 1687 (1972).

(175) N. Boens, Ph.D. Thesis, Leuven, 1974.

(176) F. C. De Schryver and E. Schouteden, Unpublished results.

(177) D. J. Andrews and W. J. Feast, Comm. IUPAC Symposium, Leuven, 1972.

(178) H. Leblanc, Ph.D. Thesis, Leuven, 1972.

(179) Y. Musa and M. P. Stevens, J. Polym. Sci., A-1, 10, 319 (1972).

(180) N. Kardush and M. P. Stevens, J. Polym. Sci., A-1, 10, 1093 (1972).

(181) J. Higgins, A. H. Johannes, J. F. Jones, R. Schultz, D. A. McCombs, and C. S. Menon, J. Polym. Sci., A-1, 8, 1987 (1970).

(182) D. A. McCombs, C. S. Menon and J. Higgins, J. Polym. Sci., A-1, 9, 1261 (1971).

(183) F. C. De Schryver, Tran V. Thien, and G. Smets, J. Polym. Sci., Polym. Lett., 9, 425 (1971).

(184) F. C. De Schryver, Tran V. Thien, S. Toppet, and G. Smets, J. Polym. Sci., A-1, 13, 227 (1975).

(185) D. A. McCombs, C. S. Menon, and J. Higgins, J. Polym. Sci., A-1, 9, 1799 (1971).

(186) Tran V. Thien, Ph.D. Thesis, Leuven, 1974.

(187) L. Horner, A. Christmann, and A. Gross, Chem. Ber., 96, 399 (1963).

(188) H. H. Bössler and R. C. Schulz, Makromol. Chem., 158, 113 (1972).

(189) F. C. De Schryver, N. Boens, and G. Smets, Macromol., 7, 399 (1974).

(190) F. C. De Schryver, N. Boens, and G. Smets, J. Am. Chem. Soc., 96, 6463 (1974).

(191) F. C. De Schryver, Tran V. Thien, and G. Smets, J. Polym. Sci., 43, 215 (1975).

(192) E. S. Huyser and D. C. Neckers, J. Am. Chem. Soc., 85, 3641 (1963).

(193) G. S. Hammond, W. P. Baker, and W. M. Moore, J. Am. Chem. Soc., 83, 2795 (1961).

(194) G. L. Closs and L. E. Closs, J. Am. Chem. Soc., 91, 456 (1969).

(195) P. J. Wagner and R. A. Leavitt, J. Am. Chem. Soc., 92, 8506 (1970); 95, 3669 (1973).

(196) N. J. Turro, Molecular Photochemistry, W. A. Benjamin, New York, 1967, p. 142.

(197) M. Hasegawa, Y. Suzuki, H. Nakanishi, and F. Nakanishi, "Progress in Polym. Sci., Japan", Vol. 5, Wiley, New York, 1973, p. 143.

(198) M. Hasegawa and Y. Suzuki, J. Polym. Sci., Polym. Lett., 5, 813 (1967).

(199) M. Iguchi, H. Nakanishi, and M. Hasegawa, J. Polym. Sci., A-1, 6, 1055 (1968).

(200) M. Hasegawa, Y. Suzuki, F. Suzuki, H. Nakanishi, J. Polym. Sci., A-1, 7, 743 (1969).

(201) H. Nakanishi, Y. Suzuki, F. Suzuki, and M. Hasegawa, J. Polym. Sci., A-1, 7, 753 (1969).

(202) S. Fujishige, M. Hasegawa, J. Polym. Sci., A-1, 7, 2037 (1969).

(203) H. Kanetsuna, M. Hasegawa, S. Mitsuhashi, T. Kurita, K. Sasaki, K. Maeda, H. Obata, and T. Hatakeyama, J. Polym. Sci., A-2, 8, 1027 (1970).

(204) M. Hasegawa, Y. Suzuki, and T. Tamaki, Bull. Chem. Soc., Japan, 43, 3020 (1970).

(205) H. Nakanishi, N. Nakano, M. Hasegawa, J. Polym. Sci., Polym. Lett., 8, 755 (1970).

(206) Y. Sasada, H. Shimanouchi, H. Nakanishi, and M. Hasegawa, Bull. Chem. Soc. Japan, 44, 1262 (1971).

(207) T. Tamaki, Y. Suzuki, and M. Hasegawa, Bull. Chem. Soc. Japan, 45, 1988 (1972).

(208) H. Nakanishi, M. Hasegawa, and Y. Sasada, J. Polym. Sci., A-2, 10, 1537 (1972).

(209) Y. Suzuki, T. Tamaki, and M. Hasegawa, Bull. Chem. Soc. Japan, 47, 210 (1974).

(210) M. Hasegawa, H. Nakanishi, T. Yurugi, and K. Ishida, J. Polym. Sci., Polym. Lett., 12, 57 (1974).

(211) R. H. Baughman, J. Appl. Phys., 42, 4579 (1971).

(212) H. Nakanishi, K. Ueno, M. Hasegawa, and Y. Sasada, Chem. Lett., 1972, 301.

(213) M. Hasegawa, F. Suzuki, H. Nakanishi, and Y. Suzuki, J. Polym. Sci., Polym. Lett., 6, 293 (1968).

(214) F. Suzuki, Y. Suzuki, H. Nakanishi, and M. Hasegawa, J. Polym. Sci., A-1, 7, 2319 (1969).

(215) H. Nakanishi, F. Nakanishi, Y. Suzuki, and M. Hasegawa, J. Polym. Sci., Polym. Chem. Ed., 11, 2501 (1973).

(216) F. Nakanishi and M. Hasegawa, J. Polym. Sci., A-1, 8, 2151 (1970).

(217) Y. Suzuki, M. Hasegawa, and N. Kita, J. Polym. Sci., A-1, 10, 2473 (1972).

(218) M. J. Holm and F. B. Zienty, J. Polym. Sci., A-1, 10, 1311 (1972).

(219) M. Miura, T. Kitami, and K. Nagakubo, J. Polym. Sci., Polym. Lett., 6, 463 (1968).

(220) J. K. Frank and I. C. Paul, J. Am. Chem. Soc., 95, 2324 (1973).

(221) N. Boens, F. C. De Schryver, and G. Smets, J. Polym. Sci., in press.

(222) S. Okamura, K. Hayashi, and Y. Nakamura, Isotop. Radiat., 3, 416 (1959).

(223) K. Hayashi, H. Ochi, M. Nishii, Y. Miyake, and S. Okamura, J. Polym. Sci., B, 1, 427 (1963).

(224) M. D. Cohen and G.M.J. Schmidt, J. Chem. Soc., 1966 (1964).

APPENDIX

To bridge the time gap between the introduction and the appearance of this review some recent articles related to the different topics treated are logged according to their sphere of interest.

II. Excited state complex formation

K. Zachariasse and W. Kühnle, Zeit. Phys. Chem. N.F., 104, 267 (1976).

J. M. Borkent, J. W. Verhoeven and Th.J. De Boer, Chem. Phys. Lett., 42, 50 (1976).

M. Yanamoto, Y. Hatano and V. Nishyima, Chem. Lett., 4, 351 (1976).

Y. C. Wang and H. Morawetz, J. Am. Chem. Soc., 98, 364 (1976).

M. Inoh, Y. Kumano and T. Okamoto, Bull. Chem. Soc. Japan, 49, 42 (1976).

S. Tazuke, K. Sato and F. Banba, Chem. Lett., 12, 1321 (1976).

J. Ferguson, M. Morita and M. Puza, Chem. Phys. Lett., 42, 288 (1976).

T. Hayashi, T. Suzuki, N. Mataga, Y. Sakata and S. Misumi, Chem. Phys. Lett., 38, 599 (1976).

T. Hayashi, N. Mataga, Y. Sakata, S. Misumi, M. Morita, and J. Tanaka, J. Am. Chem. Soc., 98, 5910 (1976).

G. E. Jo-nson, J. Chem. Phys., 63, 4047 (1975).

III.

R. W. Ricci and J. M. Nista, J. Phys. Chem., 80, 974 (1976).

IV.

K. Golankiewicz and F. Kazmierczak, Mol. Photochem., 7, 181 (1976).

D. Bichan and M. Winnik, Tetr. Lett. 3857 (1974).

M. A. Winnik, S. N. Basu, C. K. Lee and D. S. Saunders, J. Am. Chem. Soc., 98, 2929 (1976).

M. A. Winnik, R. E. Tinamari, S. Jackowski, D. S. Saunders and S. G. Whittington, J. Am. Chem. Soc., 96, 4843 (1974).

V.

N. Boens, F. C. De Schryver and G. Smets, J. Polym. Sci., 13, 201 (1975).

F. C. De Schryver, N. Boens and G. Smets, Macromolecules, 7, 399 (1974).

F. C. De Schryver, N. Boens and G. Smets, J. Am. Chem. Soc. 96, 6463 (1974).

F. C. De Schryver and N. Boens, J. Oil Col. Chem. Assoc., 59, 171 (1976).

D. J. Andrews and W. J. Feast, J. Polym. Sci., 14, 319 (1976).

D. J. Andrews and W. J. Feast, J. Polym. Sci., 14, 331 (1976).

Subject Index

Cumulative Index, Volumes 1–10